Advances in Welding Process and Materials

Advances in Welding Process and Materials

Editors

Cosmin Codrean
Anamaria Feier
Carmen Opriș

MDPI

Basel • Beijing • Wuhan • Barcelona • Belgrade • Novi Sad • Cluj • Manchester

Editors

Cosmin Codrean
Materials and Manufacturing Engineering Department
Politehnica University Timișoara
Timișoara
Romania

Anamaria Feier
Materials and Manufacturing Engineering Department
Politehnica University Timișoara
Timișoara
Romania

Carmen Opriș
Materials and Manufacturing Engineering Department
Politehnica University Timișoara
Timișoara
Romania

Editorial Office
MDPI AG
Grosspeteranlage 5
4052 Basel, Switzerland

This is a reprint of articles from the Special Issue published online in the open access journal *Materials* (ISSN 1996-1944) (available at: www.mdpi.com/journal/materials/special_issues/BG1CJRFSY4).

For citation purposes, cite each article independently as indicated on the article page online and as indicated below:

Lastname, A.A.; Lastname, B.B. Article Title. *Journal Name* **Year**, *Volume Number*, Page Range.

ISBN 978-3-7258-2066-5 (Hbk)
ISBN 978-3-7258-2065-8 (PDF)
doi.org/10.3390/books978-3-7258-2065-8

Contents

About the Editors

Cosmin Codrean

Dr. Cosmin Codrean's research activity focuses on two directions: (i) the development of soft magnetic amorphous alloys in ribbons or bulk form; and (ii) the development of amorphous alloys for brazing stainless steels, refractory steels, and copper alloys. His scientific activities related to the first direction have been conducted mainly in the Research Grant Type A: Production and characterization of ferromagnetic nanostructured alloys from Fe-Cr (Nb, Mo) in the P-Si-B family, Contract No. 27668/2005; the IDEI program: Ferromagnetic amorphous and nanocrystalline alloys applicable for making magnetic shields, Contract No. 66/2007 as the key person; and the Research Grant PCD-TC-2017: Obtaining and characterizing bulk amorphous steels, Contract No. 16179/2017 as the project manager.

His scientific activities related to the second research direction were conducted during his Ph.D. thesis with the title "Research on the brazing process of stainless and refractory steel using amorphous brazing alloy" and the CEEX Program—Brazing amorphous novative materials for special industrial applications—NOVABRAZ, Contract No. 221/2006. Amorphous Ni- and Cu-based brazing alloys were obtained in the ribbon form, with a self-fluxing character that can be used for brazing stainless steels and Ni-Cr superalloys.

As a result of the scientific and research activity carried out, both in the doctoral thesis and in the national research programs, he published more than 120 scientific papers, of which 67 are indexed in SCOPUS and 51 are ISI-indexed. The scientific papers indexed in international databases accumulated 383 citations (h-index = 8). Also, he has been developing a monograph, amorphous alloys, and nanocrystalline alloys (VDM Verlag Dr. Mueller Aktiengesellschaft & Co.KG, 2009); a patent (Ro126021/30.08.2012); and a utility model (No. 201300017/28.03.2014) regarding the development and characterization of amorphous and nanocrystalline alloys.

Anamaria Feier

Anamaria Feier is an associate professor at the Materials and Manufacturing Engineering Department, Faculty of Mechanics, at the Politehnica University Timisoara. She is also associated with research at the ASR (Romanian Welding Society). Her skills were acquired as a result of the sedimentation of knowledge from related fields, i.e., hydrotechnical construction, steel construction, and welding (civil and industrial engineering).

She was a project manager and was responsible for six national projects. She was also a research team member for ten international projects (H2020, FP6, ERASMUS+, LLP, etc.) and five national projects (POSDRU, PNCDI, PN II, and PN III). She has published more than 90 papers that were reviewed in the last 3 years.

She is the author and co-author of more than 120 papers published in international conferences and journals in industrial engineering, welding engineering, and civil engineering; 34 of them are indexed on the Web of Science (one article in Q1, three articles in Q2, two articles in Q3, and eight articles in Q4).

Carmen Opriș

Carmen Opriș is SR III, with a good expertise in welding of metallic materials and obtaining and characterization of advanced materials (composite materials and biocompatible metallic materials with crystalline and amorphous structure). She has published numerous scientific papers in the mentioned research fields, of which 19 are indexed in SCOPUS and 24 are ISI-indexed. The articles indexed in international databases accumulated 22 citations (h-index = 3). She has participated as a key person in several national research programs (CEEX program—module 1, No. 221/2006, IDEI Program Contract: 11/30.08.2013, Partnership Program Contract: 219/2014 INOVA-FSW) and coordinated research activities (Grant GNaC2018—ARUT, No. 1358/01.02.2019) as a project manager.

Preface

Recent developments in the engineering industry require new and advanced materials with special properties. Welding and joining these new materials presents a major challenge to engineers and technicians involved in product design and manufacturing. Therefore, the developments in the research of new materials should be conducted hand in hand with the work on weldability and joining capacity aspects, as well as the development of new welding techniques. At the same time, a better understanding of the relationship between the microstructure and mechanical properties of the brazed or welded joints will provide useful information for the material development activities in both the conventional and new material areas.

Consequently, this Special Issue provides a useful platform where researchers, engineers, academics, as well as industrial professionals from Romania, Poland, Serbia, China, India, Taiwan, Indonesia, and the USA were able to present their research results and developments in the field of welding and joining of both advanced and conventional engineering materials.

The papers published in this Special Issue present the authors' contributions to the development and optimization of welding processes and technologies applicable to conventional and advanced materials for the realization of similar and dissimilar joints. Moreover, it was shown that different welding technologies can be successfully used in the field of additive manufacturing in order to obtain different components. Also, the possibility of using welding and related processes to obtain new materials and functional layers with mechanical, physical, or chemical properties superior to conventional materials was highlighted.

The results published in this volume can be of real interest both for specialists working in the research field and for those working in the production of goods.

The editors of this volume congratulate the authors of the published papers for their work and results obtained in the fields of welding engineering and material engineering and thank them for their contributions to the realization of this Special Issue.

Cosmin Codrean, Anamaria Feier, and Carmen Opriș
Editors

materials

Article

Fabrication of Multi-Walled Structure through Parametric Study of Bead Geometries of GMAW-Based WAAM Process of SS309L

Jay Vora [1], Rudram Pandey [1], Pratik Dodiya [1], Vivek Patel [1], Sakshum Khanna [2], Vatsal Vaghasia [1] and Rakesh Chaudhari [1,*]

1 Department of Mechanical Engineering, School of Technology, Pandit Deendayal Energy University, Raisan, Gandhinagar 382007, India; jay.vora@sot.pdpu.ac.in (J.V.)
2 Journal of Visualized Experiments, Delhi 110016, India
* Correspondence: rakesh.chaudhari@sot.pdpu.ac.in

Abstract: In the present study, an attempt is made to investigate and optimize the bead geometries of bead width (BW) and bead height (BH) of SS-309L using an SS316L substrate by employing a gas metal arc welding (GMAW)-based wire-arc additive manufacturing (WAAM) process. The Box–Behnken design approach was used to conduct the trials of single-layer depositions with input variables of travel speed (TS), voltage (V), and gas mixture ratio (GMR). The developed multi-variable regression models were tested for feasibility using ANOVA and residual plots. The data obtained indicated that V had the most significant impact on BW, followed by TS and GMR. For BH, TS had the most significant impact, followed by GMR and V. The results of single-response optimization using a passing vehicle search (PVS) algorithm showed a maximum BH of 9.48 mm and a minimum BW of 5.90 mm. To tackle the contradictory situation, a multi-objective PVS algorithm was employed, which produced non-dominated solutions. A multi-layered structure was successfully fabricated at the optimal parametric settings of TS at 20 mm/s, of voltage at 22 V, and of GMR at 3. For multi-layer structures, fusion among the layers was observed to be good, and they were found to be free from the disbonding of layers. This revealed the suitability of the PVS algorithm for generating suitable optimal WAAM variables. We consider the current work highly beneficial for users fabricating multi-layer structures.

Keywords: Gas Metal Arc Welding (GMAW); SS309LL; Wire-Arc Additive Manufacturing (WAAM); passing vehicle search (PVS) optimization; bead geometries; multi-walled structure

Citation: Vora, J.; Pandey, R.; Dodiya, P.; Patel, V.; Khanna, S.; Vaghasia, V.; Chaudhari, R. Fabrication of Multi-Walled Structure through Parametric Study of Bead Geometries of GMAW-Based WAAM Process of SS309L. *Materials* **2023**, *16*, 5147. https://doi.org/10.3390/ma16145147

Academic Editors: Cosmin Codrean, Carmen Opriș and Anamaria Feier

Received: 21 June 2023
Revised: 17 July 2023
Accepted: 19 July 2023
Published: 21 July 2023

1. Introduction

Additive manufacturing (AM) is now preferred as a widely accepted technique over traditional manufacturing methods because it produces near-net-shaped components quickly and efficiently using various materials [1,2]. Presenting opportunities while minimizing expenses and enhancing efficiency has become increasingly appealing to manufacturers seeking to optimize their processes [3]. As a result, AM has gained significant traction in recent years, with many companies adopting it on a large scale to gain a competitive edge in their respective industries [4]. The aerospace, automotive, and biomedical areas are becoming more interested in metal additive manufacturing [5]. Wire-arc additive manufacturing (WAAM) is a technique that can save time, costs, and material when making components compared to other AM methods [6,7]. Three key categories used to classify AM for heat sources are electron beams, laser beams, and electric arcs. Electron and laser beams require metal powder as feedstock material, which restricts their production capability [8,9]. Due to this reason, these two techniques have certain limitations for larger-scale productions [10]. However, using an electric arc as a heat source is mainly suitable for producing intricate and complex large-scale components at lower cost and reducing material waste due to the more significant deposition rate [11,12]. The electric arc method uses metal wire as feedstock material, which has reduced cost compared to metal powder as feedstock material for the same proportion [13]. The WAAM technique is founded on automated welding methods, for example, gas metal arc welding (GMAW), gas tungsten arc welding, and plasma arc welding. The fabrication of thin, multi-walled structures with reduced cost and easier material deposition at a higher rate in the GMAW-based WAAM method makes it more favorable over other techniques [14–16]. However, GMAW-based welding is a complex process that depends on several factors, such as current intensity, shielding gas type, voltage (V), gas flow rate, gas mixture ratio (GMR), travel speed (TS), wire feed speed (WFS), contact tip-to-work distance, and torch angle. This means that the process parameters that govern the quality of specimens must be carefully optimized, as the required parameters differ for different grades of materials. Controlling these variables for a suitable multi-layered structure is essential [17]. Two important characteristics of bead morphologies are bead width (BW) and bead height (BH). Optimal parametric settings of WAAM parameters improve bead morphologies for the fabrication of multi-walled structures [18]. In recent studies, passing vehicle search (PVS) algorithms have been successfully executed for multiple production systems [19–22].

Kumar et al. [23] studied the optimization of GMAW-based WAAM for multi-layer bead deposition on steels. They aimed to optimize the process parameters for improved dimensional accuracy and mechanical properties of the deposited beads. They utilized an RSM design for single-layer depositions using the input variables of TS, V, current, and gas flow rate. Their findings revealed that TS was a vital factor, with 52.29% influence on BW response and 43% on BH response. The desirability function was employed to achieve optimal parametric settings, and a multi-layered structure was then fabricated. Le et al. [24] created parts from 308L stainless steel using the WAAM technique. The authors used a mix of experimental design and optimization with the help of ANOVA. Their results showed that the mechanical properties of the part made from the optimized process parameters were excellent, showing the importance of optimization. Chaudhari et al. [25] studied the effect of WAAM process parameters on a bead's geometry for single-layer deposition, such as BW or BH, with the variance of parameters such as TS, WFS, and V. They used a Box–Behnken design for singe-layer deposits and found that WFS was the most significant factor for both BW and BH, followed by V and TS. Their study revealed that a multi-layered structure was successfully fabricated at optimized parametric settings at a TS at 141 mm/min, a WFS at 5.50 m/min, and a voltage of 19 V. Natryan et al. [26] used the Taguchi technique to study the effects of TS, welding current, and filler diameter on the quality of welded joints. By applying an orthogonal array design and statistical analysis, they found the optimal mix of parameters that enhanced weld quality by minimizing defects and improving bead geometry and weld penetration. Vora et al. [27] fine-tuned the bead shape for GMAW-based WAAM. They applied a Box–Behnken design (BBD) to perform bead-on-plate tests. They used analysis of variance (ANOVA) to examine the regression equations and employed a teaching–learning-based optimization (TLBO) method to determine the best input parameters. A minimum BW of 4.73 mm and a maximum BH of 7.81 mm for single-layer depositions were obtained. A multi-layered structure was fabricated at simultaneously optimized parameters, and the structure was found to be free from the disbonding of layers. Kumar et al. [28] used a genetic algorithm to obtain the best process parameters for WAAM with near-net-shaped deposition. They discovered that the GA efficiently obtained process parameters that led to near-net-shaped deposition with fewer layers. Liberini et al. [29] aimed to select optimal process parameters for wire arc additive manufacturing using a multi-objective optimization approach to find the optimal values for bead width and height, porosity, and deposition rate. Their approach effectively found the most optimal values for the targeted properties, such as BW and BH, porosity, and deposition rate. Another study conducted by Wang et al. [30] employed a multi-wire indirect-arc-directed energy deposition method. Their obtained results showed significant impacts of WFS, current, and angle between the wires on the employed process. The results of the microstructures and mechanical properties showed favorable results for the used methodology as compared to the conventional one. Mishra et al. [31] used the optimization approach for simultaneous topology and deposition direction in WAAM using a mathematical model and a combination of GA- and gradient-based optimization techniques. This resulted in an improvement in part quality and a reduction in manufacturing time compared to conventional methods.

Stainless steel (SS) has increased its popularity in WAAM processing. SS is highly valued for its capacity to resist corrosion, rust, and staining [32]. Owing to its excellent strength, durability, and aesthetic appeal, it is primarily preferred for various applications [24]. It is also used in the medical and aerospace industries due to its biocompatibility and excellent ratio of strength-to-weight [33]. SS309L is a heat-resistant alloy commonly used in the chemical and petrochemical industries. SS309L has increased carbon presence compared to other steels, which gives it improved high-temperature strength [34]. The alloy is also resistant to suffixation and carburization, which makes it suitable for use in environments where these processes occur [35–37].

Based on past studied work, minimal work has been presented on the experimental investigation and optimization of process parameters for bead morphologies using the GMAW-based WAAM process for SS309L. Thus, the current study attempts an experimental investigation and optimization of the bead geometries of BW and BH of SS-309L using an SS316L substrate by employing a GMAW-based WAAM process. The optimized set of parameters is used to fabricate a multi-walled structure. TS, V, and GMR are elected as WAAM variables based on preliminary experimental trials, machine limits, and recently studied work. At the same time, BH and BW are taken as responses of bead morphology. A BBD is used to generate an experimental matrix for single-layer depositions, and the obtained results are analyzed through ANOVA, residual plots, and main effect plots. WAAM variables are optimized through a PVS algorithm using the empirical relations developed through the BBD. The multi-walled structure is then successfully fabricated at optimal parametric settings. We consider the current work with optimized parametric settings to be highly beneficial for users fabricating multi-layer structures.

2. Materials and Methods

2.1. Experimental Setup and Plan

In the present study, a metal wire of SS309L with a 1.2 mm diameter was used, and the bead on the plate was deposited on stainless-steel 316L-grade substrate plates using GMAW-based WAAM. Table 1 shows the chemical compositions of the filler wire and substrate, respectively.

Table 1. Chemical compositions of substrate and metallic wire (SS316L).

Element	Cr	Ni	Mo	Mn	Si	C	P	S	N	Fe
Metallic wire (SS309L)	22–25	12–15	-	2	1	0.2	0.045	0.03	-	Balance
Substrate plate (SS316L)	17.09	10.61	2.38	1.17	0.59	0.013	0.011	0.011	0.09	Balance

The GMAW process uses a temporary heat source to heat, melt, and solidify two parent metals and a filler material in a limited fusion zone to make a joint between the parent metals. An autonomous wire feeder constantly feeds the wire electrode via the tip of the torch, where the heat from the welding arc melts it. The distance between the end of the melted electrode and the molten weld pools and the transfer of molten metal to the weld pools controls the heat. The GMAW welding parameters determine the quality and cost of the welded joint. An ideal arc is formed if all the welding parameters are optimal and in accordance. Figure 1 displays the experimental setup used in the present study for the GMAW-based WAAM process.

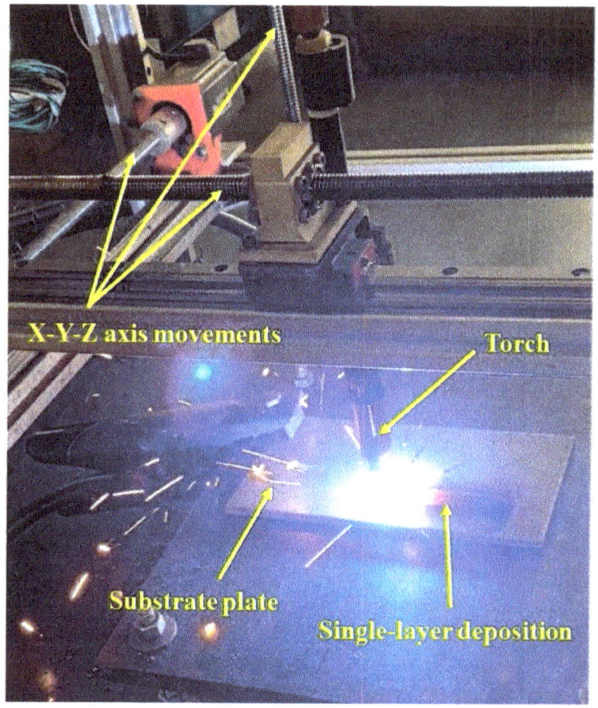

Figure 1. Experimental setup for GMAW-based WAAM process.

The setup used in the study had the components of a wire feeder, a GMAW torch, shielding gas, a mixing chamber, and a controller. The build volume for the machine used was 220 × 220 × 500 mm. The torch was enabled to move in the x, y, and z axes to deposit the material on substrate plates. The controller was provided with input through G-code programming via a computer interface in the experimental setup. COLTON iFLEX 350 was used as a power source to heat and melt the metal wire.

A BBD was used to generate an experimental matrix for single-layer depositions on substrate plates using metal wire of SS309L. By using orthogonal arrays of a BBD, multiple factors can be tested with minimal experimental runs [38,39]. This approach helps systematically identify the essential factors that affect product or process quality [40]. A BBD also gives the relationships between a response variable and multiple input variables, such as polynomial regression models, to identify optimal input conditions that maximize the response variable [41,42]. TS, V, and GMR were elected as WAAM variables based on preliminary experimental trials, machine limits, and recently studied work. At the same time, BH and BW were taken as responses of bead morphology. GMR represented the proportions of CO_2 gas, and the remainder was argon. Preliminary trials were carried out to identify the range of the selected variables following the BBD. Throughout the single-layer WAAM depositions, a weld bead length of 150 mm, an arc length of 3 mm, and a gas flow rate of 15 L/min were maintained. Table 2 displays the machining conditions used in the present study.

The selected input parameters were varied at 3 levels, and 15 runs were carried out following the BBD matrix. The bead geometries of BH and BW were investigated for all the experimental trials. Figure 2 displays the single-layer depositions of fifteen trials by following the BBD matrix, as shown in Table 3 (run order). All the bead-on-plate samples were visually checked and found to be free of any lack of fusion, porosity, or any such defects. This established the workable parameter range of the selected study.

Table 2. Experimental conditions of GMAW-based WAAM process.

Input Factors	Values/Levels
Travel speed, TS (mm/s)	16; 20; 24
Voltage, V (V)	22; 24; 26
Gas mixture ratio, GMR	1; 5; 9
Gas flow rate (L/min)	15
Weld bead length (mm)	150
Arc length, (mm)	3

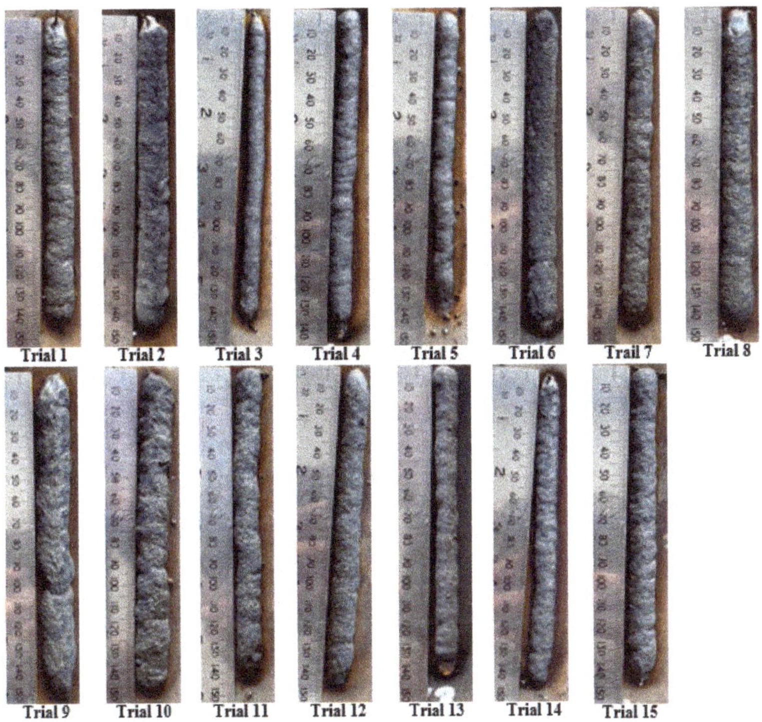

Figure 2. Single-layered deposition following the BBD.

Table 3. Results of bead geometries following the BBD.

Std. Order	Run Order	V (V)	TS (mm/min)	GMR	BH (mm)	BW (mm)
14	1	24	20	5	5.61	7.850
11	2	24	16	9	7.46	8.820
3	3	22	24	5	4.71	6.050
10	4	24	24	1	4.31	7.259
4	5	26	24	5	4.71	9.554
9	6	24	16	1	9.20	9.010
15	7	24	20	5	5.78	7.735
6	8	26	20	1	5.51	9.910
2	9	26	16	5	7.47	10.270
1	10	22	16	5	7.52	7.250
13	11	24	20	5	5.90	7.690
8	12	26	20	9	5.51	10.050
12	13	24	24	9	4.73	8.704
7	14	22	20	9	5.51	6.987
5	15	22	20	1	7.13	6.783

Optical microscopy was used to measure the bead morphologies of each single-layer deposition. Figure 3 represents the methodology used for the determination of bead geometries. Three different cross-sections of the bead depositions were cut to measure the bead geometries at various locations of the deposition. For better accuracy and more reliable results, their average values were considered in the present study.

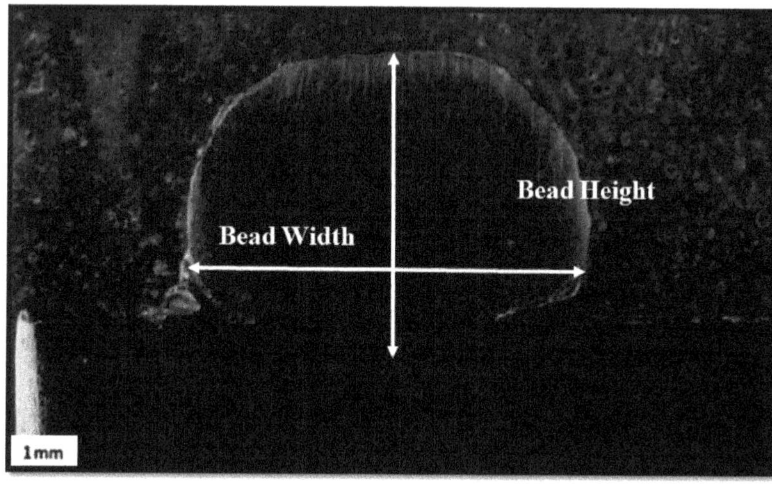

Figure 3. Procedure for determining bead geometries (BW and BH).

2.2. *Optimization Using PVS Algorithm*

Savsani and Savsani [43] studied the passing vehicle search (PVS) algorithm, particularly for design problems related to engineering. The algorithm imitates the passing of a vehicle on a two-lane highway, emphasizing the principle of safe overtaking opportunities. The mechanism involves several interdependent and complex parameters, including the availability of gaps in oncoming traffic, the speed and acceleration of individual vehicles, road conditions, overall traffic, weather, and driver skill. While the PVS algorithm has demonstrated its usefulness, its application in real-world scenarios with various complexities and uncertainties still requires further investigation and refinement. The algorithm presented in this paper accounts for three types of vehicles (Oncoming Vehicle—OV; Front Vehicle—FV; and Back Vehicle—BV) on a two-lane highway. When a Back Vehicle desires to overtake a Front Vehicle, it must have a faster speed than the latter. Overtaking cannot occur if the Back Vehicle has a lower speed than the Front Vehicle. The algorithm also considers the speed and position of the Oncoming Vehicle and its respective distances and velocities when determining the feasibility of an overtaking maneuver.

On a two-lane road, there are three different vehicles (OV, FV, and BV) with different velocities (V_1, V_2, and V_3), and their respective distances can be determined at any given time. The distance between BV and FV can be denoted as 'x', and that between FV and OV can be marked as 'y'. Their velocities impose a primary constraint, where either FV's velocity is slower than BV's ($V_1 > V_3$) or BV's velocity is slower than FV's. If FV exceeds the speed of BV, overtaking becomes impossible, and BV maintains its desired velocity. It is possible to pass only if FV's velocity is less than that of BV. An additional requirement for passing is that the distance between FV and BV during the maneuver is shorter than the distance covered by OV within the same timeframe. Thus, different conditions emerge for the selected vehicles. Employing a human-activity-based technique, the PVS algorithm models the passing behavior of vehicles and offers a meta-heuristic optimization approach capable of finding optimal or near-optimal solutions for given objective functions.

3. Results and Discussion

The BBD matrix of the RSM approach was utilized to obtain the results of BW and BH, as shown in Table 3. The most suitable condition for fabricating a multi-walled structure is to achieve maximum BH and minimum BW. The study also generated multi-variable non-linear regressions between the machining factors and the responses.

3.1. Empirical Relations for BH and BW

Multi-variable non-linear regressions were generated using Minitab v17 with the BBD of RSM to establish the relationships between the WAAM variables and bead geometries (BH and BW). These equations provided a starting point for evaluating response values beyond the experimental matrix of the BBD. Equations (1) and (2) are the regression equations for BH and BW, respectively, which were derived using the stepwise method of statistical approach with Minitab software.

$$\begin{gathered} BH = 36.81 - 0.356{\cdot}V - 1.613{\cdot}TS - 1.977{\cdot}GMR + 0.0258{\cdot}TS{\cdot}TS \\ +0.0505\ {\cdot}V{\cdot}GMR + 0.0337{\cdot}TS{\cdot}GMR \end{gathered} \quad (1)$$

$$\begin{gathered} BH = 37.2 - 2.25{\cdot}V - 0.919{\cdot}TS - 0.724{\cdot}GMR + 0.0634{\cdot}V{\cdot}V + 0.01683 \\ {\cdot}TS{\cdot}TS + 0.02629{\cdot}GMR{\cdot}GMR + 0.0255{\cdot}GMR{\cdot}GMR \end{gathered} \quad (2)$$

3.2. ANOVA for BW and BH

A statistical method, analysis of variance (ANOVA), was used to determine the significance of the factors affecting the response variables. It measured the variability between different levels of the factors to determine which factors significantly impacted the response variables. The adequacy and reliability of the resulting regression equations were tested through an ANOVA analysis. Minitab v17 software was utilized to assess the significance of the model terms at a confidence level of 95%. Terms with probability values less than 0.05 were considered to significantly impact the response variables, while non-significant terms were treated as irrelevant [44].

Table 4 showcases the ANOVA results for BW and BH, and Table 5 shows the results of the model summaries for BW and BH. The statistical analysis of the output factors of BW and BH revealed significant contributions from the regression, linear, square, and two-way interaction models.

Table 4. ANOVA for BW and BH.

Source	DF	SS	MS	F	P
Bead Width					
Regression	7	24.0000	3.4286	125.53	0.000
Linear	3	22.3142	7.4381	272.33	0.000
V	1	20.2057	20.2057	739.39	0.000
TS	1	1.7889	1.7889	65.50	0.000
GMR	1	0.3169	0.3196	11.70	0.011
Square	3	1.0175	0.3392	12.42	0.003
V × V	1	0.2372	0.2372	8.68	0.021
TS × TS	1	0.2676	0.2676	9.80	0.017
GMR × GMR	1	0.6535	0.6535	23.92	0.002
Two-way Interaction	1	0.6683	0.6683	24.47	0.002
TS × GMR	1	0.6683	0.6683	24.47	0.002
Error	7	0.1912	0.0273		
Lack of Fit	5	0.1776	0.0355	5.22	0.169
Pure Error	2	0.0136	0.0068		
Total	14	24.1912			

Table 4. *Cont.*

Source	DF	SS	MS	F	P
Bead Height					
Regression	6	34.6626	5.7771	30.27	0.000
Linear	3	31.3517	10.4506	54.75	0.000
V	1	0.4689	0.4689	2.46	0.156
TS	1	29.4147	29.4174	154.13	0.000
GMR	1	1.4654	1.4654	7.68	0.024
Square	1	0.8587	0.8587	4.50	0.067
TS × TS	1	0.8587	0.8587	4.50	0.067
Two-way Interaction	2	2.4521	1.2261	6.42	0.022
V × GMR	1	0.8806	0.8806	4.61	0.064
TS × GMR	1	1.5715	1.5715	8.23	0.021
Error	8	1.5269	0.1909		
Lack of Fit	6	1.4711	0.2452	8.79	0.106
Pure Error	2	0.0558	0.0279		
Total	14	26.7947			

Table 5. Model summaries for BW and BH.

Model Summary for BW			
S	**R-sq.**	**R-sq. (adj.)**	**R-sq. (pred.)**
0.165266	99.21%	98.42%	94.36%
Model Summary for BH			
S	**R-sq.**	**R-sq. (adj.)**	**R-sq. (pred.)**
0.436877	95.78%	92.62%	89.73%

The three WAAM variables of V, TS, and GMR were all significant factors in the case of BW, while for the BH response, TS and GMR were substantial variables. TS had the highest contribution to BH, while V was observed to have the highest effect on BW. The small impact of the error term on all the responses indicated high accuracy in predicting values with minimal errors, and lack of fit was statistically non-significant, confirming the accuracy of the ANOVA results [45]. The model was deemed appropriate for predicting the output value as a result. The generated regression equations were deemed reliable and dependable for predicting the values of BW and BH, as evidenced by the significance of the model terms and a close-to-one R^2 value, indicating effective prediction. The model's effectiveness was evaluated by analyzing its R^2 values. The R^2 values obtained for BW and BH were 0.9921 and 0.9578. Such high R^2 values suggest that the model's predictions were accurate and closely matched the actual data. The model was a good fit for the data, as the R^2 values were close to one.

3.3. Residual Plots for Output Measures

To ensure the validity of the ANOVA model, specific assumptions must be met, and residual plots must be utilized to confirm the analysis outcomes [45]. Residual plots were used for the validation of the statistical results given by the ANOVA. The residual plots contained four individual analyses. Successful validation of these four plots confirmed the adequacy and reliability of ANOVA results and developed regression equations. Additionally, these validations meant that the regression equations could be used for the future prediction of outcomes for any parametric levels within the range of input parameters. The residual plots for BH, consisting of four plots, are presented in Figure 4.

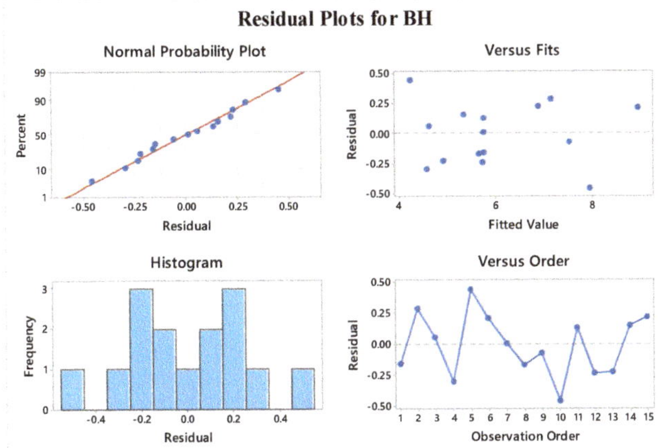

Figure 4. Residual plots for bead height.

The normality plot exhibited a linear trend that supported the ANOVA model's suitability, indicating that the residuals followed a normal distribution. The versus fit plot also showed that the fits were randomly distributed around the source. In contrast, the histogram plot displayed a bell-shaped curve, which indicated the ANOVA data well. Furthermore, the absence of any particular pattern in the versus order plot further confirmed the ANOVA statistics, leading to better predictions of future outcomes. Figure 5 shows similar findings in the case of BW response. Thus, this shows that the generated regression equations were found to be adequate and reliable for future predictions of outcomes.

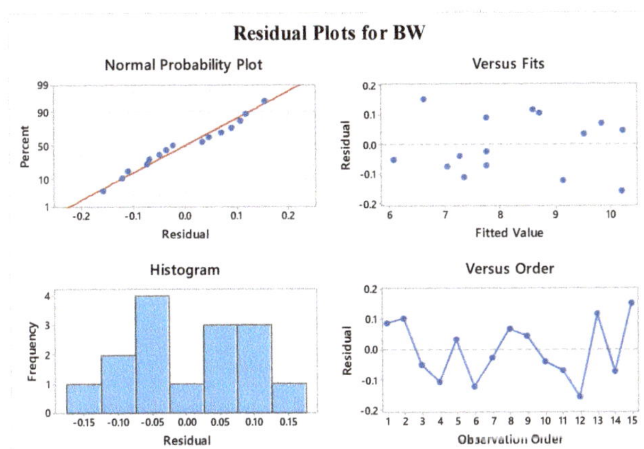

Figure 5. Residual plots for bead width.

3.4. Main Effect Plots for Bead Width and Bead Height

The main effect plot, as shown in Figure 6, demonstrated the trends followed by V, TS, and GMR variance for bead height response, and it can be observed that a slight decrement in the BH of the deposited material was observed with an increase in V.

This phenomenon could be attributed to the increased heat generated through electrical resistance heating and the absorption of electrons onto the wire tip, providing the necessary energy for the melting and superheating of the wire electrode material [46]. The TS showed a trend wherein its increase led to decreased BH. This was because the higher movement speed of the torch allowed less time for deposition, leading to lesser BH [27,47].

The graph of BH vs. GMR shows that, with an increase in GMR value, the BH increased, and it also became obsolete when talking in relation to BH, as it just safeguarded the pool against air contamination. As per the main effect plots, the WAAM variables of V at 22 V, TS at 16 mm/s, and GMR at 1 were desirable to obtain the maximum BH value.

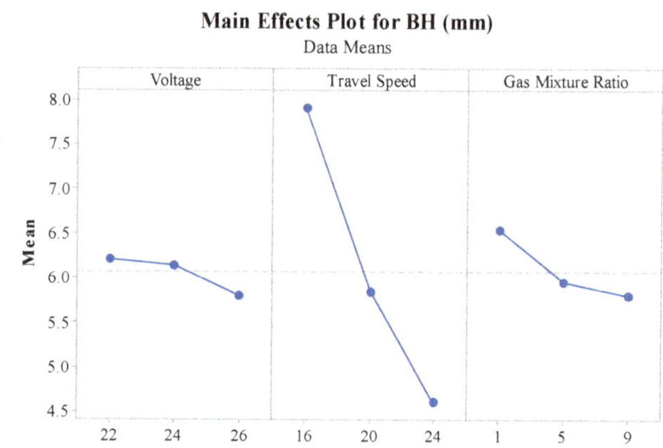

Figure 6. Effect of variations in WAAM variables on bead height.

The main effect plot, as shown in Figure 7, demonstrates the trends followed by V, TS, and GMR variance for bead height response. The plot of BW vs. V shows that the BW of deposition increased with an increase in voltage from 22 V to 26 V. This was due to higher voltage making wide arcs, leading to significant drops of molten metal [48]. The TS trend shows that the BW decreased with an increase in TS. This happened because, with higher TS, less metal could be dropped on the same position, leading to less deposition and, thus, smaller BW [25]. Finally, with higher values of GMR, the BW caused an insignificant change as, at first, it decreased and then increased, negating the changes or uncertainties caused. The gas mixture ratio, on the other hand, had minimal impact on the BW response. Its primary function was to protect the weld pool from atmospheric contamination, and it did not significantly affect the weld BW. As per the main effect plots, the WAAM variables of V at 22 V, TS at 24 mm/s, and GMR at 5 were desirable to obtain the minimum BW value.

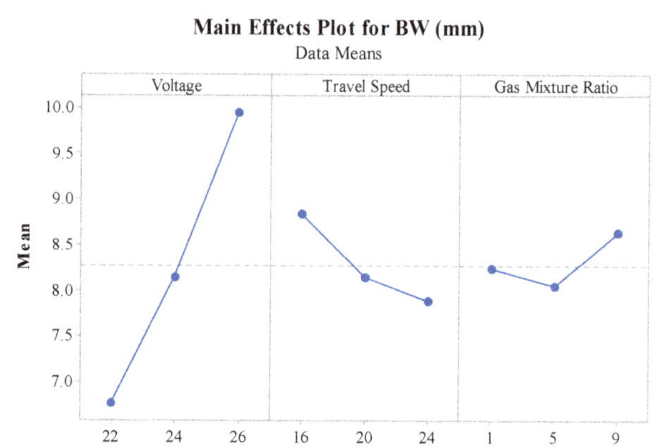

Figure 7. Effect of variations in WAAM variables on bead width.

3.5. Optimization of BH and BW Responses Using PVS Algorithm

The main effect plots of the response variables demonstrated the contradictory levels of WAAM variables for the optimal levels of BH and BW responses. This meant that the PVS algorithm was employed to optimize the responses. During the execution of the PVS technique, the upper and lower bounds of the design variables were considered as follows: $22 \leq V \leq 26$; $1 \leq GMR \leq 9$; and $16 \leq TS \leq 24$. Table 6 shows the results of the single-response optimization.

Table 6. HTS results for individual response geometries.

Conditions	Input Factors			Predicted Responses through PVS		Experimental Results	
	V	TS	GMR	BH	BW	BH	BW
Maximization of BH	22	16	1	9.48	7.70	9.69	7.59
Minimization of BW	22	24	2	5.01	5.90	5.17	6.02

Apart from voltage, the other two WAAM variables were observed at different levels. Additionally, the BW response value increased for the maximization of the BH response, which was not desirable, and vice versa. To validate the findings of the PVS algorithm, actual experiments were carried out. The smallest error between the results achieved from the PVS algorithm and validation trials resulted in good agreement between the bead geometry and the WAAM variables. This demonstrated the suitability of the PVS algorithm for the developed regression models.

Although the responses of BW and BH exhibited conflicting tendencies concerning the levels of WAAM design variables, it was crucial to determine the optimal design variable combination that could enhance both responses. This was necessary to achieve the desired quality in both BW and BH simultaneously. Owing to such reasons, multi-objective optimization was necessary. Table 7 shows the results of such an optimization, or Pareto points. Pareto fronts were employed to determine non-dominated solutions that satisfied the requirements of various industrial applications.

Table 7. Pareto optimal points.

Sr. No.	V	TS	GMR	BH	BW
1	22	16	1	9.45	7.70
2	22	24	2	5.01	5.90
3	22	16	2	9.12	7.46
4	22	16	3	8.79	7.28
5	22	16	4	8.47	7.15
6	22	16	5	8.14	7.07
7	22	19	3	6.97	6.52
8	22	18	3	7.53	6.74
9	22	17	3	8.13	6.99
10	22	17	5	7.55	6.83
11	22	24	1	5.07	5.94
12	22	20	3	6.46	6.33
13	22	19	3	6.97	6.52
14	22	20	4	6.27	6.30
15	22	19	4	6.74	6.46
16	22	23	2	5.35	5.98
17	22	19	2	7.19	6.63
18	22	18	4	7.27	6.66
19	22	21	4	5.85	6.18
20	22	22	2	5.73	6.09
21	22	17	4	7.84	6.89
22	22	21	2	6.17	6.24

Table 7. *Cont.*

Sr. No.	V	TS	GMR	BH	BW
23	22	17	4	7.84	6.89
24	22	23	1	5.44	6.04
25	22	20	2	6.65	6.41
26	22	19	2	7.19	6.63
27	22	21	3	6.01	6.18
28	22	20	2	6.65	6.41
29	22	18	4	7.27	6.66
30	22	17	3	8.13	6.99
31	22	18	2	7.78	6.87
32	22	20	3	6.46	6.33
33	22	22	3	5.61	6.06
34	22	21	2	6.17	6.24
35	22	19	4	6.74	6.46
36	22	18	3	7.53	6.74
37	22	16	5	8.14	7.07
38	22	23	3	5.25	5.98
39	22	22	2	5.73	6.09
40	22	20	4	6.27	6.30
41	22	22	3	5.61	6.06
42	22	21	3	6.01	6.18
43	22	22	1	5.85	6.18
44	22	18	2	7.78	6.87

Figure 8 displays the Pareto graph of the BH vs. BW responses. Depending on the desired bead geometries needed for multi-layer thin-walled structure fabrication, a user can choose an optimal value from the available options.

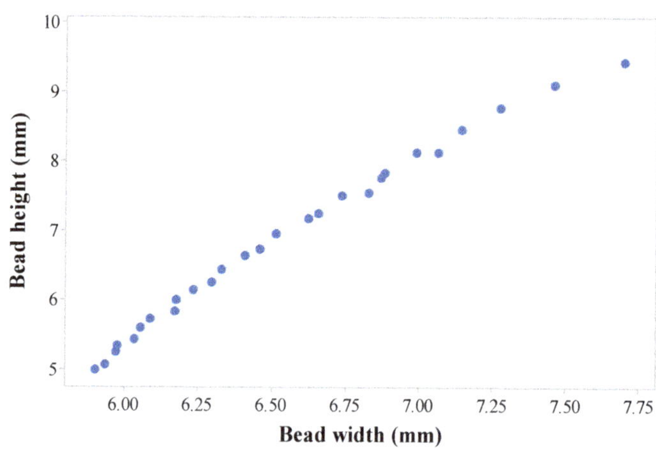

Figure 8. Pareto graph for BH vs. BW.

3.6. Fabrication of Multi-Walled Structure

An objective function was selected to fabricate a multi-layered structure by assigning equal importance to the BH and BW responses. The PVS algorithm was used for obtaining the levels of the WAAM variables, and it showed at the voltage of 22 V, the TS of 20 mm/s, and the DOP of 3 response values of BH at 6.46 mm and BW at 6.33 mm for single-layer deposition. A multi-layered structure was fabricated at these parametric settings. The multi-layer structure was fabricated at the optimal parameter settings of WAAM variables, as shown in Figure 9. For better geometry accuracy, the multi-layer structure was fabricated through layer-on-layer deposition following 180-degree turns of filler wire.

Figure 9. Multi-layer structure at optimal parametric settings.

A cooling time of 60 s was applied between successive layers to reduce residual stresses. Fusion among the layers was observed to be good, and it was found to be free from the disbonding of layers. Some extra lumps of metal core were noticed on the extreme sides of the structure. However, these were effectively eliminated in post-processing. This revealed the suitability of the PVS algorithm for generating the suitable optimal WAAM variables. Therefore, the present work effectively demonstrated the requirement of having optimal parametric settings and the necessity of parametric optimization for fabricating thin, multi-walled structures using a GMWA-based WAAM process for SS-309L using an SS316L substrate. We believe that the present work may be useful for researchers and industrial applications to find optimal sets of parameters.

4. Conclusions

The present study used a GMAW-based-WAAM process for SS309L wire. V, TS, and GMR were identified as machining parameters, and BH and BW were output factors. Experiments were conducted following a Box–Behnken design. Optimization of bead geometries was obtained through the application of a PVS algorithm. Based on the key findings and results, the following conclusions can be drawn:

- Multi-variable non-linear regressions were generated among the WAAM variables and output responses.
- ANOVA was employed to validate the appropriateness and reliability of the obtained regression equations. The ANOVA revealed that the quadratic model, including linear, squared, and interaction model terms, was statistically significant for both the bead height and width responses. The lack of fit results signified the model's suitability and acceptability for both responses. The validation results from the ANOVA of an R-squared value close to one showed that the model was adequate and acceptable.
- TS had the largest impact on BH response, followed by GMR, while TS followed V and GMR showed a substantial impact on BW response.

- The influences of the WAAM variables (TS, V, and GMR) were studied on the BH and BW responses. It showed conflicting situations for attaining the desired levels of bead geometries.
- Single-response optimization using the PVS technique obtained a maximum BH and a minimum BW of 9.48 mm and 5.90 mm, respectively. Pareto fronts were employed to determine non-dominated solutions that satisfied the requirements of various industrial applications.
- The multi-layered structure was successfully fabricated at optimal parametric settings of V at 22 V, TS at 20 mm/s, and DOP at 3. For the multi-layer structure, fusion among the layers was observed to be good, and it was found to be free from the disbonding of layers. This revealed the suitability of PVS for generating suitable, optimal WAAM variables.
- The present work effectively demonstrated the requirement of having optimal parametric settings and the necessity of parametric optimization for the fabrication of thin, multi-walled structures using a GMWA-based WAAM process for SS-309L using an SS316L substrate. We believe that the present work may be useful to researchers' industrial applications to find optimal sets of parameters. As a result, in future work, fabricated multi-layered structures will be examined using microstructure investigations and mechanical properties such as tensile, impact, and microhardness testing.

Author Contributions: Conceptualization, R.C., R.P., P.D., J.V., V.V., V.P. and S.K.; methodology, R.C., R.P., P.D., J.V., V.V., S.K. and V.P.; software, S.K. and V.P.; validation, R.C. and J.V.; formal analysis, J.V. and V.P.; investigation, J.V., V.V., S.K. and R.C.; resources, R.C., R.P., P.D., J.V., V.V. and S.K.; data curation, J.V. and V.P.; writing—original draft preparation, R.C., R.P., P.D., J.V., V.V., S.K. and V.P. writing—review and editing, R.C.; visualization, R.C. and J.V.; supervision, J.V. and R.C. All authors have read and agreed to the published version of the manuscript.

Funding: The reported study was funded by the Institution of Engineers (India): R.6/2/DR/2022-23/DR2023005.

Institutional Review Board Statement: Not applicable.

Informed Consent Statement: Not applicable.

Data Availability Statement: Data presented in this study are available in this article.

Conflicts of Interest: The authors declare no conflict of interest.

References

1. Korkmaz, M.E.; Waqar, S.; Garcia-Collado, A.; Gupta, M.K.; Krolczyk, G.M. A technical overview of metallic parts in hybrid additive manufacturing industry. *J. Mater. Res. Technol.* **2022**, *18*, 384–395. [CrossRef]
2. Ladani, L.; Sadeghilaridjani, M. Review of powder bed fusion additive manufacturing for metals. *Metals* **2021**, *11*, 1391. [CrossRef]
3. Chaturvedi, M.; Scutelnicu, E.; Rusu, C.C.; Mistodie, L.R.; Mihailescu, D.; Subbiah, A.V. Wire arc additive manufacturing: Review on recent findings and challenges in industrial applications and materials characterization. *Metals* **2021**, *11*, 939. [CrossRef]
4. Moeinfar, K.; Khodabakhshi, F.; Kashani-bozorg, S.; Mohammadi, M.; Gerlich, A. A review on metallurgical aspects of laser additive manufacturing (LAM): Stainless steels, nickel superalloys, and titanium alloys. *J. Mater. Res. Technol.* **2021**, *16*, 1029–1068. [CrossRef]
5. Mohd Yusuf, S.; Cutler, S.; Gao, N. The impact of metal additive manufacturing on the aerospace industry. *Metals* **2019**, *9*, 1286. [CrossRef]
6. Vora, J.; Parmar, H.; Chaudhari, R.; Khanna, S.; Doshi, M.; Patel, V. Experimental investigations on mechanical properties of multi-layered structure fabricated by GMAW-based WAAM of SS316L. *J. Mater. Res. Technol.* **2022**, *20*, 2748–2757. [CrossRef]
7. Wang, J.; Pan, Z.; Wang, Y.; Wang, L.; Su, L.; Cuiuri, D.; Zhao, Y.; Li, H. Evolution of crystallographic orientation, precipitation, phase transformation and mechanical properties realized by enhancing deposition current for dual-wire arc additive manufactured Ni-rich NiTi alloy. *Addit. Manuf.* **2020**, *34*, 101240. [CrossRef]
8. Knapp, G.; Mukherjee, T.; Zuback, J.; Wei, H.; Palmer, T.; De, A.; DebRoy, T. Building blocks for a digital twin of additive manufacturing. *Acta Mater.* **2017**, *135*, 390–399. [CrossRef]
9. Popov, V.V.; Fleisher, A. Hybrid additive manufacturing of steels and alloys. *Manuf. Rev.* **2020**, *7*, 6. [CrossRef]
10. Wang, L.; Xue, J.; Wang, Q. Correlation between arc mode, microstructure, and mechanical properties during wire arc additive manufacturing of 316L stainless steel. *Mater. Sci. Eng. A* **2019**, *751*, 183–190. [CrossRef]

11. Ke, W.; Oliveira, J.; Cong, B.; Ao, S.; Qi, Z.; Peng, B.; Zeng, Z. Multi-layer deposition mechanism in ultra high-frequency pulsed wire arc additive manufacturing (WAAM) of NiTi shape memory alloys. *Addit. Manuf.* **2022**, *50*, 102513. [CrossRef]
12. Chaudhari, R.; Parikh, N.; Khanna, S.; Vora, J.; Patel, V. Effect of multi-walled structure on microstructure and mechanical properties of 1.25 Cr-1.0 Mo steel fabricated by GMAW-based WAAM using metal-cored wire. *J. Mater. Res. Technol.* **2022**, *21*, 3386–3396. [CrossRef]
13. Ding, D.; Pan, Z.; Cuiuri, D.; Li, H. Wire-feed additive manufacturing of metal components: Technologies, developments and future interests. *Int. J. Adv. Manuf. Technol.* **2015**, *81*, 465–481. [CrossRef]
14. Xiong, J.; Li, Y.; Li, R.; Yin, Z. Influences of process parameters on surface roughness of multi-layer single-pass thin-walled parts in GMAW-based additive manufacturing. *J. Mater. Process. Technol.* **2018**, *252*, 128–136. [CrossRef]
15. Yang, K.; Wang, F.; Duan, D.; Zhang, T.; Luo, C.; Cressault, Y.; Yu, Z.; Yang, L.; Li, H. Experimental Investigation of Integrated Circular Triple-Wire Pulse GMAW of Q960E High-Strength Steel for Construction Machinery. *Materials* **2021**, *14*, 375. [CrossRef]
16. Chen, Y.; Sun, S.; Zhang, T.; Zhou, X.; Li, S. Effects of post-weld heat treatment on the microstructure and mechanical properties of laser-welded NiTi/304SS joint with Ni filler. *Mater. Sci. Eng. A* **2020**, *771*, 138545. [CrossRef]
17. Choudhury, S.S.; Marya, S.K.; Amirthalingam, M. Improving arc stability during wire arc additive manufacturing of thin-walled titanium components. *J. Manuf. Process.* **2021**, *66*, 53–69. [CrossRef]
18. Srivastava, S.; Garg, R. Process parameter optimization of gas metal arc welding on IS: 2062 mild steel using response surface methodology. *J. Manuf. Process.* **2017**, *25*, 296–305. [CrossRef]
19. Yin, Z.-P.; Ding, H.; Xiong, Y.-L. Visibility theory and algorithms with application to manufacturing processes. *Int. J. Prod. Res.* **2000**, *38*, 2891–2909. [CrossRef]
20. Kumar, S.; Tejani, G.G.; Pholdee, N.; Bureerat, S. Multi-objective passing vehicle search algorithm for structure optimization. *Expert Syst. Appl.* **2021**, *169*, 114511. [CrossRef]
21. Parsana, S.; Radadia, N.; Sheth, M.; Sheth, N.; Savsani, V.; Prasad, N.E.; Ramprabhu, T. Machining parameter optimization for EDM machining of Mg-RE-Zn-Zr alloy using multi-objective passing vehicle search algorithm. *Arch. Civ. Mech. Eng.* **2018**, *18*, 799–817. [CrossRef]
22. Kumar, S.; Tejani, G.G.; Pholdee, N.; Bureerat, S.; Mehta, P. Hybrid heat transfer search and passing vehicle search optimizer for multi-objective structural optimization. *Knowl.-Based Syst.* **2021**, *212*, 106556. [CrossRef]
23. Kumar, V.; Mandal, A.; Das, A.K.; Kumar, S. Parametric study and characterization of wire arc additive manufactured steel structures. *Int. J. Adv. Manuf. Technol.* **2021**, *115*, 1723–1733. [CrossRef]
24. Mai, D.S.; Doan, T.K.; Paris, H. Wire and arc additive manufacturing of 308L stainless steel components: Optimization of processing parameters and material properties. *Eng. Sci. Technol. Int. J.* **2021**, *24*, 1015–1026.
25. Chaudhari, R.; Parmar, H.; Vora, J.; Patel, V.K. Parametric study and investigations of bead geometries of GMAW-based wire–arc additive manufacturing of 316L stainless steels. *Metals* **2022**, *12*, 1232. [CrossRef]
26. Natrayan, L.; Anand, R.; Kumar, S.S. Optimization of process parameters in TIG welding of AISI 4140 stainless steel using Taguchi technique. *Mater. Today Proc.* **2021**, *37*, 1550–1553. [CrossRef]
27. Vora, J.; Parikh, N.; Chaudhari, R.; Patel, V.K.; Paramar, H.; Pimenov, D.Y.; Giasin, K. Optimization of Bead Morphology for GMAW-Based Wire-Arc Additive Manufacturing of 2.25 Cr-1.0 Mo Steel Using Metal-Cored Wires. *Appl. Sci.* **2022**, *12*, 5060. [CrossRef]
28. Kumar, A.; Maji, K. Selection of process parameters for near-net shape deposition in wire arc additive manufacturing by genetic algorithm. *J. Mater. Eng. Perform.* **2020**, *29*, 3334–3352. [CrossRef]
29. Liberini, M.; Astarita, A.; Campatelli, G.; Scippa, A.; Montevecchi, F.; Venturini, G.; Durante, M.; Boccarusso, L.; Minutolo, F.M.C.; Squillace, A. Selection of optimal process parameters for wire arc additive manufacturing. *Procedia Cirp* **2017**, *62*, 470–474. [CrossRef]
30. Wang, L.; Wu, T.; Wang, D.; Liang, Z.; Yang, X.; Peng, Z.; Liu, Y.; Liang, Y.; Zeng, Z.; Oliveira, J. A novel heterogeneous multi-wire indirect arc directed energy deposition for in-situ synthesis Al-Zn-Mg-Cu alloy: Process, microstructure and mechanical properties. *Addit. Manuf.* **2023**, *72*, 103639. [CrossRef]
31. Mishra, V.; Ayas, C.; Langelaar, M.; Van Keulen, F. Simultaneous topology and deposition direction optimization for Wire and Arc Additive Manufacturing. *Manuf. Lett.* **2022**, *31*, 45–51. [CrossRef]
32. Jin, W.; Zhang, C.; Jin, S.; Tian, Y.; Wellmann, D.; Liu, W. Wire arc additive manufacturing of stainless steels: A review. *Appl. Sci.* **2020**, *10*, 1563. [CrossRef]
33. Zhang, D.; Liu, A.; Yin, B.; Wen, P. Additive manufacturing of duplex stainless steels-a critical review. *J. Manuf. Process.* **2022**, *73*, 496–517. [CrossRef]
34. Sohrabi, M.J.; Naghizadeh, M.; Mirzadeh, H. Deformation-induced martensite in austenitic stainless steels: A review. *Arch. Civ. Mech. Eng.* **2020**, *20*, 124. [CrossRef]
35. Khidhir, G.I.; Baban, S.A. Efficiency of dissimilar friction welded 1045 medium carbon steel and 316L austenitic stainless steel joints. *J. Mater. Res. Technol.* **2019**, *8*, 1926–1932. [CrossRef]
36. Vieira, D.; Siqueira, R.; Carvalho, S.; Lima, M. Effect of the initial substrate temperature on heat transfer and related phenomena in austenitic stainless steel parts fabricated by additive manufacturing using direct energy deposition. *J. Mater. Res. Technol.* **2022**, *8*, 5267–5279.

37. Saboori, A.; Aversa, A.; Marchese, G.; Biamino, S.; Lombardi, M.; Fino, P. Microstructure and mechanical properties of AISI 316L produced by directed energy deposition-based additive manufacturing: A review. *Appl. Sci.* **2020**, *10*, 3310. [CrossRef]
38. Huang, M.; Chen, L.; Zhang, M.; Zhan, S. Multi-Objective Function Optimization of Cemented Neutralization Slag Backfill Strength Based on RSM-BBD. *Materials* **2022**, *15*, 1585. [CrossRef]
39. Singh, A.; Nath, T.; Dommeti, S.G.; Sekar, S. Bulk Fabrication of SS410 Material Using Cold Metal Transfer-Based Wire Arc Additive Manufacturing Process at Optimized Parameters: Microstructural and Property Evaluation. *Machines* **2022**, *10*, 1136. [CrossRef]
40. Wu, W.; Xue, J.; Xu, W.; Lin, H.; Tang, H.; Yao, P. Parameters optimization of auxiliary gas process for double-wire SS316L stainless steel arc additive manufacturing. *Metals* **2021**, *11*, 190. [CrossRef]
41. Wu, J.; Gao, P.P.; Gao, X. Optimization of Response Surface Methodology for Pulsed Laser Welding of 316L Stainless Steel to Polylactic Acid. *Metals* **2023**, *13*, 214. [CrossRef]
42. Alam, M.A.; Ya, H.H.; Yusuf, M.; Sivraj, R.; Mamat, O.B.; Sapuan, S.M.; Masood, F.; Parveez, B.; Sattar, M. Modeling, optimization and performance evaluation of tic/graphite reinforced al 7075 hybrid composites using response surface methodology. *Materials* **2021**, *14*, 4703. [CrossRef] [PubMed]
43. Savsani, P.; Savsani, V. Passing vehicle search (PVS): A novel metaheuristic algorithm. *Appl. Math. Model.* **2016**, *40*, 3951–3978. [CrossRef]
44. Tian, H.; Lu, Z.; Chen, S. Predictive Modeling of Thermally Assisted Machining and Simulation Based on RSM after WAAM. *Metals* **2022**, *12*, 691. [CrossRef]
45. Al-Amin, M.; Abdul-Rani, A.M.; Ahmed, R.; Shahid, M.U.; Zohura, F.T.; Rani, M.D.B.A. Multi-objective optimization of process variables for MWCNT-added electro-discharge machining of 316L steel. *Int. J. Adv. Manuf. Technol.* **2021**, *115*, 179–198. [CrossRef]
46. Wu, B.; Pan, Z.; Ding, D.; Cuiuri, D.; Li, H.; Xu, J.; Norrish, J. A review of the wire arc additive manufacturing of metals: Properties, defects and quality improvement. *J. Manuf. Process.* **2018**, *35*, 127–139. [CrossRef]
47. Zhou, Y.; Lin, X.; Kang, N.; Huang, W.; Wang, J.; Wang, Z. Influence of travel speed on microstructure and mechanical properties of wire+ arc additively manufactured 2219 aluminum alloy. *J. Mater. Sci. Technol.* **2020**, *37*, 143–153. [CrossRef]
48. Pan, Z.; Ding, D.; Wu, B.; Cuiuri, D.; Li, H.; Norrish, J. Arc welding processes for additive manufacturing: A review. In *Transactions on Intelligent Welding Manufacturing*; Chen, S., Zang, Y., Feng, Z., Eds.; Springer: Singapore, 2018; pp. 3–24.

materials

MDPI

Article

Effect of Mechanical Stirring on High-Speed GMAW Hump Bead

Jun Xiao [1,2], Xiaolei Wang [1,2], Shengnan Gai [1,2,*], Shujun Chen [1,2] and Wenhao Huang [3]

[1] Engineering Research Center of Advanced Manufacturing Technology for Automotive Components, Ministry of Education, Beijing University of Technology, Beijing 100124, China; jun.xiao@bjut.edu.cn (J.X.)
[2] Welding Equipment R&D Center, Beijing University of Technology, Beijing 100124, China
[3] College of Mechanical and Electrical Engineering, Wenzhou University, Wenzhou 325035, China; huangwenhaojay@163.com
* Correspondence: supergai@bjut.edu.cn

Abstract: High-speed GMAW tends to be accompanied by periodic humping defects, thereby reducing the weld bead quality. A new method was proposed to actively control the weld pool flow for eliminating humping defects. A high-melting point solid pin was designed and inserted into the weld pool to stir the liquid metal during the welding process. The characteristics of the backward molten metal flow were extracted and compared by a high-speed camera. Combined with particle tracing technology, the momentum of the backward metal flow was calculated and analyzed, and the mechanism of hump suppression in high speed GMAW was further revealed. The stirring pin interacted with the liquid molten pool, resulting in a vortex zone behind the stirring pin, which significantly reduced the momentum of the backward molten metal flow, and thus it inhibited the formation of humping beads.

Keywords: high-speed GMAW; mechanical stirring; humping bead; weld pool flow

Citation: Xiao, J.; Wang, X.; Gai, S.; Chen, S.; Huang, W. Effect of Mechanical Stirring on High-Speed GMAW Hump Bead. *Materials* **2023**, *16*, 4493. https://doi.org/10.3390/ma16124493

Academic Editors: Cosmin Codrean, Carmen Opriș and Anamaria Feier

Received: 19 April 2023
Revised: 12 June 2023
Accepted: 14 June 2023
Published: 20 June 2023

1. Introduction

As one of the most important arc welding processes in the manufacturing industry, gas metal arc welding (GMAW) has the advantages of high automation, adaptability, and low production cost. Hump welding is one of the commonly formed defects in high-speed GMAW, and it has become a bottleneck to improve production efficiency [1,2]. Therefore, understanding the formation process and mechanism of the hump can help us to take action to prevent its occurrence. At present, it is widely recognized that the reverse metal flow in the molten pool at a high speed is the main cause of hump formation. Nguyen et al. [3] confirmed that the powerful momentum of the reverse flow of molten metal in the molten pool was the main factor causing the hump by collecting sequential images during the hump formation. With the maturity of hump research, many scholars simulated hump welding by numerical calculations [4–8], calculated the velocity distribution of liquid metal in the molten pool, observed the flow state of molten pool more intuitively, and verified the effect of reverse metal flow on hump formation. If the characteristics of the reverse metal flow are directly extracted by experimental method, the relationship between the liquid metal flow and welding parameters and forming can be quantitatively analyzed.

Zähr et al. [9] studied the influence of protective gas composition on humping in TIG welding and found that the addition of helium increased the heat transfer of the workpiece, thus increasing the welding speed without the hump. Based on the current research on the formation of the hump bead, many improved arc welding processes or equipment have been reported to prevent the hump bead. For example, double- or three-wire GMAWs [10–13] are considered because the arc pressure and the impact force of the droplet were evenly distributed, and the relative value was small in the double- or three-wire co-melting cell, the interconnected molten pool act as a coordinated buffer, which can delay the reflux tendency of liquid metal and inhibit hump defect. Double electrode GMAW [14], due to bypass diversion, resulting in part of the main current being flown to the bypass arc, which was applied to heat and melt the bypass wire, which improves the welding rate and avoids the hump. TIG-MIG hybrid welding [15,16] and Laser GMAW hybrid welding [17,18] can improve the fluid flow and the heat transfer in the molten pool, thereby reducing the momentum of backward molten metal flow and inhibiting the hump. However, the double electrodes or wires and hybrid heat source welding not only increase equipment investment, but they also make equipment more complex and difficult to operate.

The Shandong University team applied an external transverse magnetic field (TMF) to the molten pool, using electromagnetic forces generated by electromagnetic induction to suppress the flow of reverse liquid metal and successfully prevented the hump beads [19,20]. Then, combining the visual sensing system with the tracing particles, a two-dimensional fluid flow field [21–23] was measured on the surface of the molten pool. The relationship between the reverse liquid metal and the hump was quantitatively analyzed. The additional electromagnetic force produced by the interaction of the external magnetic field and the welding current can inhibit the flow of reverse metal in the weld pool. Studies have shown that the maximum velocity of the tracing particles dropped below the threshold level when a magnetic field was applied. However, the way in which magnetic fields are added is still relatively cumbersome and requires some cost.

Wang et al. [24] explored the effect of compensating gas on hump defects in TW-GIA and revealed its effect on molten pool flow and arc stability. The results have shown that the compensating gas with an injection angle of 90° has the ability to suppress the hump bead and reduces the oxygen content of the weld.

In this study, a new mechanical stirring process was proposed for the first time. Tungsten rods are designed to be desired shapes. The designed tungsten rod was inserted directly into the molten pool and rotated by the stepping motor to stir with the welding. Its equipment is simple, and its effect on the molten pool has great advantages in practical application. At the same time, the velocity of reverse molten metal flow was obtained by using tracer technique. By calculating the velocity of reverse metal without stirring pin and mechanical stirring, respectively, the influence of mechanical stirring on the flow of metal liquid in a molten pool was quantitatively analyzed, and the new mechanism of hump suppression was further revealed.

2. Experimental System

2.1. Mechanical Stirring System

In order to study the hump bead, a high-speed GMAW stirring system was established, as shown in Figure 1, which consists of stirring equipment, MIG power source, a data acquisition system, and a computer control system. The high-speed camera is controlled by the computer to complete data acquisition, the step motor was joined to the tungsten pin through the joining mechanism, and the tungsten pin was rotated to realize the process of stirring with welding.

In this study, bead-on-plate flat welding experiments were carried out on a low carbon steel (Q235) base metal with a dimension of 200 mm × 60 mm × 3 mm. Argon (99.999% purity) was also used as a protective gas for the arc and welds with a flow rate of 20 L/min. The wire used is an ER50-6 carbon steel wire with a diameter of 1.2 mm, and the composition of the wire and the mechanical properties of the base metal are shown in Tables 1 and 2. It is noteworthy that the rotation of the stirring pin affects the arc plasma, making it unstable, thereby affecting the weld formation. Therefore, after many experiments, the distance between the stir pin and the center of the wire was fixed to be 4 mm. Due to the high hardness of the tungsten pin and its poor machinability, this study only focuses on the simple shape of the cone-shaped stirring pin. The cone-shaped stirring pin can continuously introduce force into the liquid molten pool and can change the flow state in the molten pool, which will hopefully reduce the flow velocity of the reverse liquid metal and achieve the effect of hump suppression. Table 3 lists other process parameters used in the experiment.

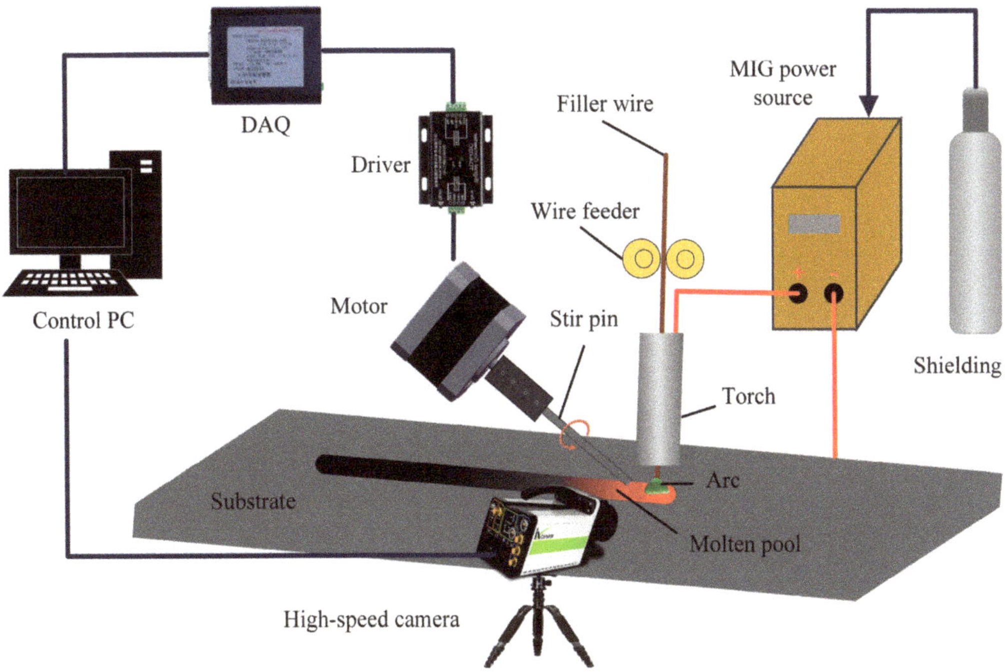

Figure 1. Schematic diagram of the experimental system.

Table 1. ER50-6 welding wire chemical composition (wt %).

	Fe	C	Mn	Si	P	S	Cr	Ni	Cu	Mo	Ti
ER50-6	Bal.	0.08	1.34	0.14	0.015	0.013	0.021	0.011	0.14	0.38	0.06

Table 2. Q235 mechanical properties.

	Tensile Strength (MPa)	Yield Point (MPa)	Elongation (%)
Q235	375–500	≥235	≥25

Table 3. Welding parameters.

Parameters	Values	Units
Arc current I	220	A
Arc voltage U	21	V
Welding speed v	2.25	m/min
Stirring speed n	0–1000	r/min
Stir pin diameter d	3	mm
Stirring pin angle θ	45	°

2.2. Observe the Convective Patterns on the Weld Pool Surfaces

The influence of fluid flow velocity on the hump was quantitatively studied by observing the motion of tracing particles. A high-speed camera (iX-Cameras, i-Speed 720) was connected to the computer, and photo shooting was performed at a speed of 5000 fps. A narrow band pass filter of 630 ± 20 nm was installed in front of the lens to reduce intense arc interference. Zirconia (ZrO_2) has a much lower density (5.85 g/cm^3) than iron solution and does not react with iron. In addition, the melting point of zirconia (2700 °C) is higher than that of iron and becomes an ideal tracer particle. In order to accurately reflect the flow velocity of surface tracing particles, 0.3 mm zirconia particles were used as tracers. Before testing, a blind hole was drilled with a diameter of 1.2 mm and a depth of 1mm every 15 mm along the centerline of the top of the workpiece, and then the particles were placed into the blind hole. The upper part of the blind hole was sealed with a small piece of wire to prevent the tracer particle from flying out under the force of the protective gas and the arc. Figure 2 is a schematic illustration of the experimental setup to observe the convective patterns on the surface of the molten pool. In order to fully observe the motion of tracing particles, the image of the molten pool was taken from two angles: welding direction and vertical welding direction, respectively. By extracting the coordinates of the tracing particles in the image, the trajectory of the tracer particle was obtained, and its flow velocity was calculated.

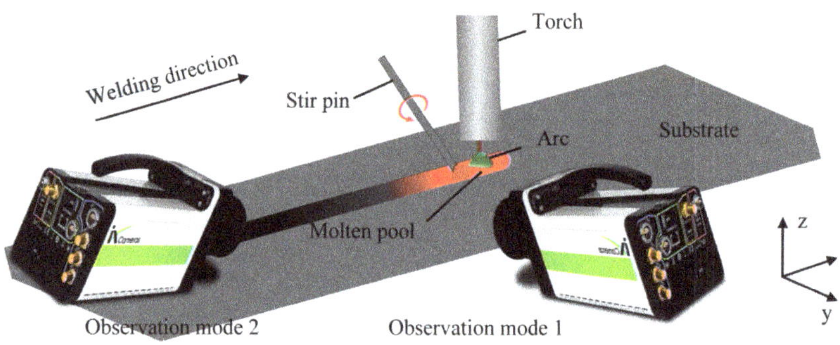

Figure 2. Schematic illustration of the experimental setup to observe the convective patterns on weld pool surfaces.

3. Results and Discussions

3.1. Observation of Weld Pool Behavior

This study adopts a new method of mechanical stirring. By changing the rotation speed of the stirring pin, an additional force source was applied to the molten pool to change the flow behavior and temperature field of the liquid metal in the molten pool. The flow pattern of the molten pool was compared with that of the molten pool with or without mechanical stirring. Figure 3 shows an image of a conventional high-speed GMAW molten pool, showing obvious hump defects. The results show that the formation of the hump bead passed through three stages: initial stage, growth stage, and solidification stage. In the initial stage, under the combined effect of arc force and droplet impact, the front surface of the molten pool was depressed, creating a high momentum reverse metal flow toward the end of the molten pool. The dynamic changes in the arc force and the impact force of the droplet cause the liquid metal to flow towards the molten pool tail wave undulation. In the growth stage, the hump continued to grow as the metal was continuously filled and the molten parent material was filled. During the solidification stage, as the heat source moved forwardly, the molten pool became elongated, and the insufficient heat was available at the end of the molten pool to solidify first, preventing the liquid metal from continuing to flow behind the pool to form a crest. Finally, the liquid metal at the crest all solidifies to form a hump. Figure 4 shows the image of a high-speed GMAW molten pool under mechanical stirring. It can be seen from the images taken by a high-speed camera that the molten pool tends to be flat, and the hump disappears after mechanical stirring. Contrast was performed in relation to the stirred or unstirred images to show the reduction in the size of the molten pool after stirring. The results show that mechanical stirring can inhibit the hump. At the same time, stirring can promote molten pool heat dissipation, reduce the weld pool temperature, and reduce the weld pool size.

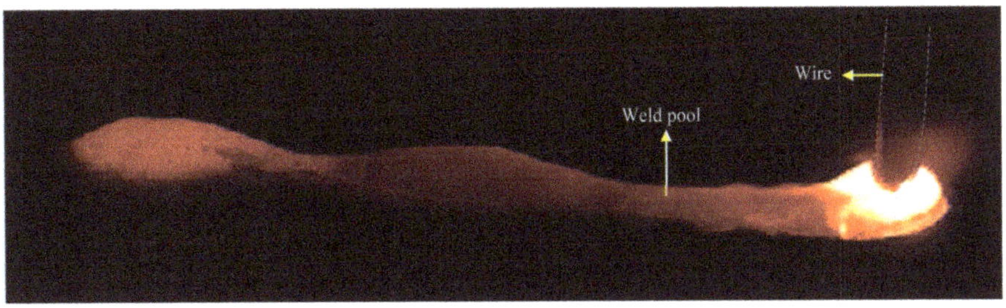

Figure 3. Traditional high-speed GMAW molten pool image.

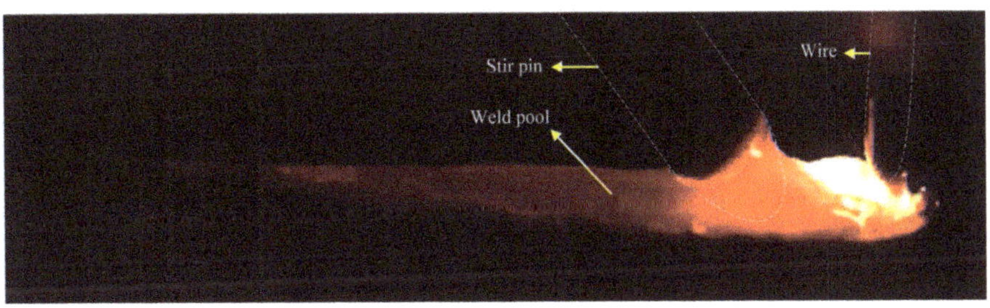

Figure 4. Image of high-speed GMAW molten pool under mechanical stirring.

Figure 5 shows the effect of mechanical stirring on bead forming at a welding speed of 2.25 m/min. It can be seen that a typical hump weld bead is formed without stirring pin, as shown in Figure 5a. When the pin keeps static (i.e., 0 r/min), the weld bead becomes uniform and continuous, and the hump is suppressed due to the blocking effect of the pin, which inhibits the backward flow of liquid metal, as shown in Figure 5b. Influenced by the machining accuracy of the pin connector, resulting in eccentric rotation of the pin, the stirring process is unstable, and the weld bead forming is not ideal when the molten pool is stirred with different pin rotation speed, but it still has an inhibitory effect on the hump, as shown in Figure 5c,d.

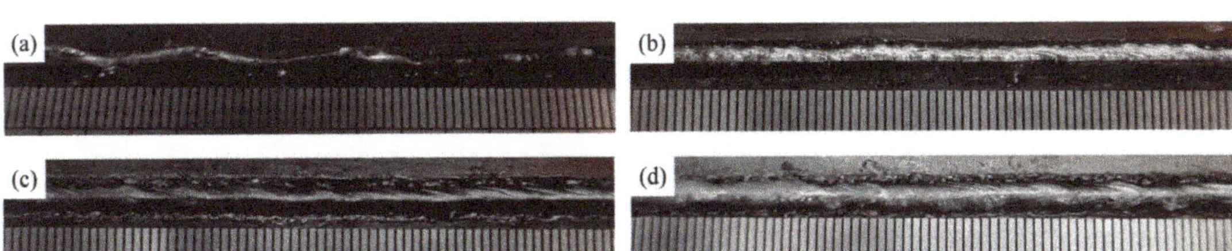

Figure 5. Effect of mechanical stirring on weld forming. (**a**), without stir pin (**b**), 0 r/min (**c**), 500 r/min (**d**), 1000 r/min.

3.2. *Influence of Mechanical Stirring on Molten Pool Flow*

At present, many scholars believe that high momentum reverse liquid metal is the main mechanism of hump formation. Therefore, all methods capable of inhibiting the flow velocity of reverse liquid metals can inhibit the occurrence of the hump. In order to clarify the effect of mechanical stirring on reverse liquid metal, in this section, the flow pattern of liquid metal on the surface of the molten pool is reflected by tracking the flow trajectory of tracing particles. Figure 6 shows the flow pattern on the surface of the molten pool without a stirring pin. The zirconia particles flowed backward by the arc force and the droplet impact force and were eventually blocked by the slag behind the molten pool. Figure 7 shows the flow pattern of the surface of the molten pool under mechanical stirring. The molten pool is divided into three parts: the arc influence zone, corresponding to region 1, the stirring pin influence zone, corresponding to region 2, and the liquid molten pool, corresponding to region 3 behind the stirring pin. By tracing zirconia particle motion, it can be seen that the zirconia particles rotate and move under the effect of mechanical stirring, and a vortex zone is formed behind the stirring pin. Therefore, the rotation of the stirring pin changes the flow state in the molten pool, decreasing the reverse liquid metal flow velocity and eventually suppressing the hump.

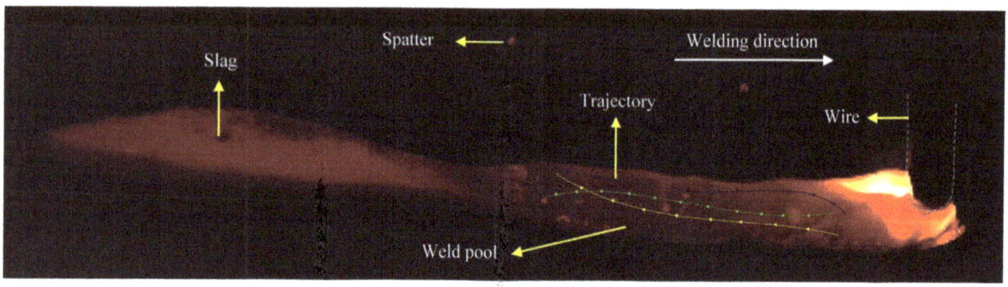

Figure 6. Flow pattern of the molten pool surface without stirring.

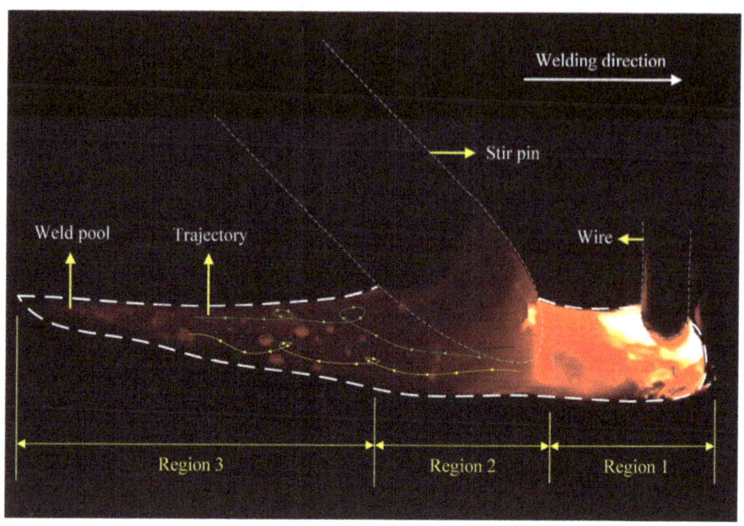

Figure 7. Flow pattern of the surface of the molten pool under mechanical stirring.

The viewing angle of the surface flow pattern of the molten pool was observed in the direction of vertical welding. Only part of the liquid molten pool can be seen. It does not accurately reflect the real flow of the molten pool. Therefore, this study combines with the viewing angle along the welding direction to observe the flow state of the entire liquid molten pool surface. By tracking the movement of zirconia particles on the surface of the molten pool, the complete flow pattern of the surface of the molten pool can be described. Figure 8 shows that the flow pattern of the surface of the molten pool was observed from the angle of view along the welding direction without stirring. Based on the addition of the filter, the arc light was filtered out, sadly, covering part of the front of the molten pool and not being able to be observed. The image of the molten pool visualized from the rear shows that the zirconia particles flowed backward rapidly without stirring, while the high momentum reverse metal in the molten pool flowed backward at a speed higher than the molten pool surface. Figure 9 shows the flow pattern of the viewing angle molten pool surface in the welding direction under mechanical stirring. The strong arc light was blocked by the stirring pin, and the image of the molten pool behind the stirring pin was clearly visible. It can be seen from the image of the molten pool that the reverse liquid metal with high momentum is divided into two by the stirring pin. Simultaneously, by observing the movement of zirconia particles, the zirconia particles were rotated after stirring pin. The above phenomenon shows that a whirlpool was produced on the back side of the stirring pin, which prevents the flow of zirconia particles on the surface of the molten pool. Therefore, it is proved, once again, that under mechanical stirring, the vortex zone was formed behind the stirring pin, which can inhibit the flow of reverse liquid metal.

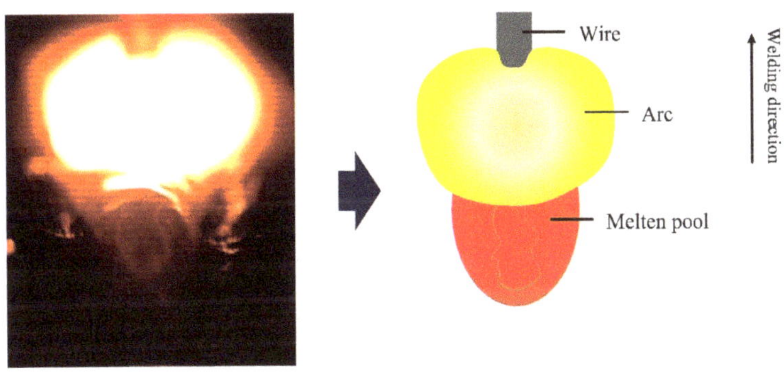

Figure 8. Flow pattern of welding molten pool surface in welding direction without stirring.

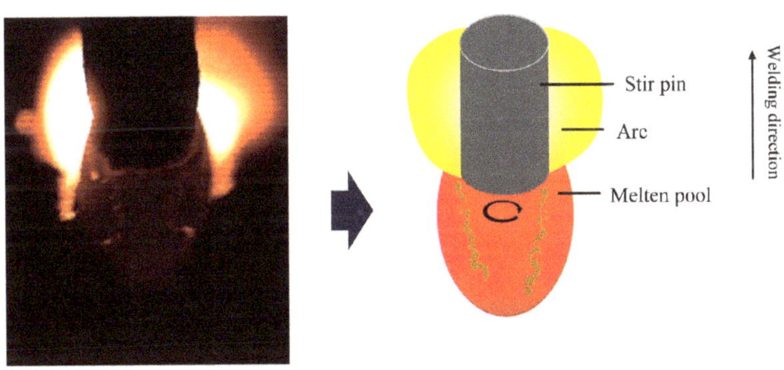

Figure 9. Flow pattern of welding molten pool surface in the direction of welding under mechanical stirring.

In order to further reveal the new mechanism of hump suppression by mechanical stirring of high-speed GMAW, the momentum of metal reverse flow on the surface of molten pool was calculated and analyzed by extracting the pixel coordinates of tracing particles from high-speed camera images. The distance was customized between the weld center and the tracer particle as D_{wp}. Zirconia particles cannot be observed within the D_{wp} = 0–4 mm due to strong arc interference. Therefore, we count the flow velocity of zirconia particles after D_{wp} = 4 mm. Figure 10 shows the effect of mechanical stirring on the flow velocity of tracing particles, where velocity is the absolute value (magnitude). The results show that the flow velocity of zirconia particles gradually increased to 950 mm/s within D_{wp} = 0–5 mm by the arc force and the droplet impact force under traditional high speed GMAW. Then, the speed gradually decreased. Under mechanical stirring, when the stirring pin did not turn (0 r/min), the flow velocity of zirconia particles within the D_{wp} = 0–5 mm exceeded the velocity peak of traditional high-speed GMAW and increased to 1046 mm/s. Within the D_{wp} = 5–7 mm, the flow velocity of zirconia particles decreased gradually. After D_{wp} = 7 mm, the vortex within the vortex zone again reduced the flow velocity of zirconia particles. In contrast to the case where the stirring pin does not rotate, in the impact area of the stir pin (D_{wp} = 4–7 mm), the rotation of the stirring pin significantly reduced the flow rate of zirconia particles, indicating that stir pin rotation can reduce the flow velocity of the reverse liquid metal. Comparing the stirring pin rotation, it could be seen that the flow velocity of zirconia particles increased to 800 mm/s and 810 mm/s when the stirring pin speed was 500 r/min and 1000 r/min, respectively, and the velocity change pattern of zirconia particles was similar for both. There was a rotation speed threshold for the stirring action of the stirring pin in the molten pool, and the stirring intensity stopped increasing after this value.

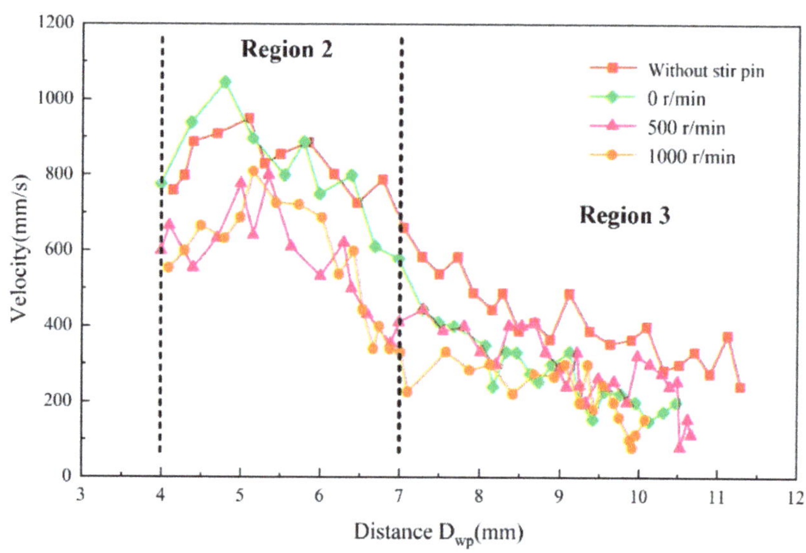

Figure 10. Effect of mechanical stirring on the flow velocity of tracing particles.

By calculating the average velocity of region 3 (D_{wp} = 7–12 mm), as shown in Figure 11, the flow velocity of zirconia particles under mechanical stirring was significantly lower compared to that without stirring pin, indicating that the average velocity of zirconia particles in region 3 decreased with the increase in stirring pin speed. Therefore, the vortex zone produced by mechanical stirring can reduce the flow velocity of the reverse liquid metal, and as the stirring pin speed increases, the more effective the vortex zone is in reducing the flow velocity.

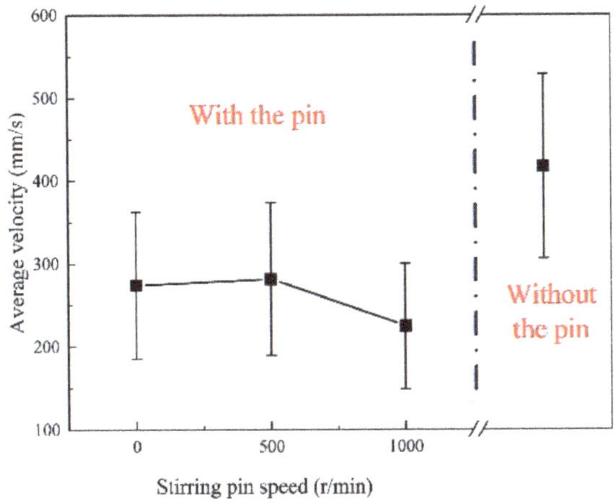

Figure 11. Comparison of average velocity with or without stirring pin.

3.3. Suppression Mechanism of Humping Bead

The partial stirring pin inserted into the molten pool at 45 degrees was reduced to a cylinder and studied as a "cylindrical flow" problem, as shown in Figure 12. The surface flow mode of directly mechanically stirred molten pool can be divided into the following four stages: in the first stage, under the combined action of the arc force and the impact force of the droplet, the surface of the molten pool in front of the pool is depressed, forming high momentum reverse metal flow moving towards the tail of the molten pool. In the second stage, as the welding proceeded, the reverse liquid metal was shunted due to the blocking of the stirring pin, and part of the kinetic energy is lost. When the reverse liquid metal surrounds the front half of the cylinder, part of the pressure was converted into kinetic energy, increasing the velocity of the tracer particle. When the reverse liquid metal surrounds the back half of the cylinder, part of the kinetic energy was converted into pressure energy, part of the kinetic energy was converted into the potential energy of the tracing particles, and the velocity of the tracer particle was reduced. In the third stage, when the reverse liquid metal flows to the back of the cylinder, the boundary layer was separated from the surface of the cylinder, causing the vortex zone to form behind the cylinder. Tracer particle velocity manifests as a decrease. In the fourth phase, the flow velocity of tracing particles continued to decrease and eventually solidify. Under the combined action of the blocking of the stir pin and the eddy flow zone formed after the stir pin, the flow velocity of the reverse liquid metal was reduced, and finally the hump was suppressed.

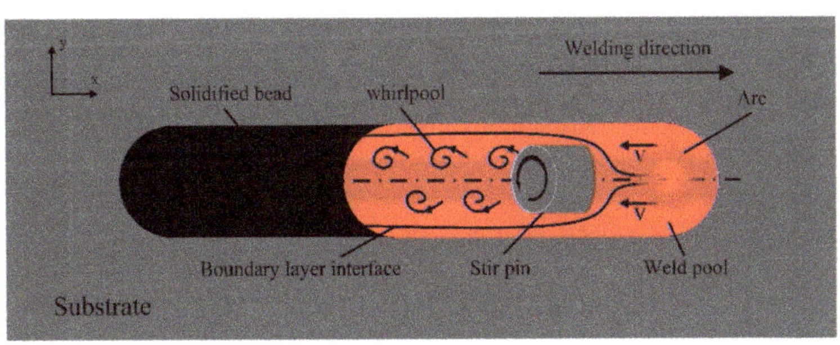

Figure 12. Schematic diagram of cylindrical flow under mechanical stirring.

4. Conclusions

This paper mainly studied the suppression effect of mechanical stirring on hump defects at different stirring pin speeds, and conclusions are drawn as follows.

1. A new method that is to use a stirring tungsten pin inserted directly into the molten pool is proposed for achieving better bead formation in the high-speed GMAW process. The results show that the mechanical stirring of the weld pool could successfully suppress the occurrence of the hump defect.
2. The flow velocity of reverse metal flow of the weld pool was analyzed by measuring the motion of the tracing particles. Under the direct mechanical stirring, the flow pattern in the molten pool was changed, and the flow velocity of the backward metal flow was significantly weakened. That is the key mechanism of suppressing the hump defect.
3. The existence of vortex zone behind the stirring pin was proved by tracking the trajectory of tracing particles and the rotating motion of tracing particles in molten pool under mechanical stirring. The vortex zone contributes to weakening the backward metal flow.

Author Contributions: Investigation, W.H. and S.G.; Writing—original draft, X.W.; Writing—review & editing, J.X. and S.C.; Project administration, S.G.; Funding acquisition, S.C. All authors have read and agreed to the published version of the manuscript.

Funding: This work was supported by the National Natural Science Foundation of China (Grant No. 51975014, 52275304, 51905012) and sponsored by the Beijing Nova Program.

Institutional Review Board Statement: Not applicable.

Informed Consent Statement: Informed consent was obtained from all subjects involved in the study.

Data Availability Statement: Data not available due to privacy or ethical restrictions.

Conflicts of Interest: The authors declare no conflict of interest.

References

1. Nguyen, T.C.; Weckman, D.C.; Johnson, D.A.; Kerr, H.W. High speed fusion weld bead defects. *Sci. Technol. Weld. Join.* **2006**, *11*, 618–633. [CrossRef]
2. Scalet Rossini, L.F.; Valenzuela Reyes, R.A.; Spinelli, J.E. Double-wire tandem GMAW welding process of HSLA50 steel. *J. Manuf. Process.* **2019**, *45*, 227–233. [CrossRef]
3. Nguyen, T.C.; Weckman, D.C.; Johnson, D.A.; Kerr, H.W. The humping phenomenon during high speed gas metal arc welding. *Sci. Technol. Weld. Join.* **2005**, *10*, 447–459. [CrossRef]
4. Cho, M.H.; Farson, D.F. Understanding Bead Hump Formation in Gas Metal Arc Welding Using a Numerical Simulation. *Met. Mater. Trans. B* **2007**, *38*, 305–319. [CrossRef]
5. Chen, J.; Wu, C.-S. Numerical Analysis of Forming Mechanism of Hump Bead in High Speed GMAW. *Weld. World* **2010**, *54*, R286–R291. [CrossRef]
6. Xu, G.; Cao, Q.; Hu, Q.; Zhang, W.; Liu, P.; Du, B. Modelling of bead hump formation in high speed gas metal arc welding. *Sci. Technol. Weld. Join.* **2016**, *21*, 700–710. [CrossRef]
7. Wu, D.; Hua, X.; Ye, D.; Li, F. Understanding of humping formation and suppression mechanisms using the numerical simulation. *Int. J. Heat Mass Transf.* **2017**, *104*, 634–643. [CrossRef]
8. Ni, M.; Qin, X.; Hu, Z.; Ji, F.; Yang, S.; Wang, S. Forming characteristics and control method of weld bead for GMAW on curved sur-face. *Int. J. Adv. Manuf. Technol.* **2022**, *3*, 119.
9. Zähr, J.; Füssel, U.; Hertel, M.; Lohse, M.; Sende, M.; Schnick, M. Numerical and Experimental Studies of the Influence of Process Gases in Tig Welding. *Weld. World* **2012**, *56*, 85–92. [CrossRef]
10. Ueyama, T.; Ohnawa, T.; Tanaka, M.; Nakata, K. Effects of torch configuration and welding current on weld bead formation in high speed tandem pulsed gas metal arc welding of steel sheets. *Sci. Technol. Weld. Join.* **2005**, *10*, 750–759. [CrossRef]
11. Liu, L.; Yu, S.; Song, G.; Hu, C. Analysis of arc stability and bead forming with high-speed TW-GIA welding. *J. Manuf. Process.* **2019**, *46*, 67–76. [CrossRef]
12. Wu, K.; Ding, N.; Yin, T.; Zeng, M.; Liang, Z. Effects of single and double pulses on microstructure and mechanical properties of weld joints during high-power double-wire GMAW. *J. Manuf. Process.* **2018**, *35*, 728–734. [CrossRef]
13. Wang, Z.; Zhang, T.; Dong, X.; Liu, L. Suppression of humping bead in high-speed triple-wire gas indirect arc welding. *Int. J. Adv. Manuf. Technol.* **2022**, *122*, 2593–2605. [CrossRef]

14. Lu, Y.; Chen, S.; Shi, Y.; Li, X.; Chen, J.; Kvidahl, L.; Zhang, Y.M. Double-electrode arc welding process: Principle, variants, control and developments. *J. Manuf. Process.* **2014**, *16*, 93–108. [CrossRef]
15. Kanemaru, S.; Sasaki, T.; Sato, T.; Mishima, H.; Tashiro, S.; Tanaka, M. Study for TIG–MIG hybrid welding process. *Weld. World* **2014**, *58*, 11–18. [CrossRef]
16. Chen, J.; Wu, C.; Chen, M. Improvement of welding heat source models for TIG-MIG hybrid welding process. *J. Manuf. Process.* **2014**, *16*, 485–493. [CrossRef]
17. Choi, H.W.; Farson, D.F.; Cho, M.H. Using a hybrid laser plus GMAW process for controlling the bead humping defect. *Weld. J.* **2006**, *85*, 174s–179s.
18. Li, F.; Tao, W.; Peng, G.; Qu, J.; Li, L. Behavior and stability of droplet transfer under laser-MIG hybrid welding with synchronized pulse modulations. *J. Manuf. Process.* **2020**, *54*, 70–79. [CrossRef]
19. Wu, C.; Yang, F.; Gao, J. Effect of external magnetic field on weld pool flow conditions in high-speed gas metal arc welding. *Proc. Inst. Mech. Eng. Part B J. Eng. Manuf.* **2016**, *230*, 188–193. [CrossRef]
20. Wang, L.; Wu, C.S.; Gao, J.Q. Suppression of humping bead in high speed GMAW with external magnetic field. *Sci. Technol. Weld. Join.* **2016**, *21*, 131–139. [CrossRef]
21. Wang, L.; Chen, J.; Wu, C.; Gao, J. Backward flowing molten metal in weld pool8 and its influence on humping bead in high-speed GMAW. *J. Mater. Process. Technol.* **2016**, *237*, 342–350. [CrossRef]
22. Wang, L.; Wu, C.; Chen, J.; Gao, J. Influence of the external magnetic field on fluid flow, temperature profile and humping bead in high speed gas metal arc welding. *Int. J. Heat Mass Transf.* **2018**, *116*, 1282–1291. [CrossRef]
23. Wang, L.; Wu, C.; Chen, J.; Gao, J. Experimental measurement of fluid flow in high-speed GMAW assisted by transverse magnetic field. *J. Manuf. Process.* **2020**, *56*, 1193–1200. [CrossRef]
24. Wang, Z.; Zhu, Y.; Lang, Q. Suppression of humping bead defects in high-speed triple-wire gas indirect arc welding through com-pensation gas. *Vacuum* **2023**, *213*, 112108. [CrossRef]

materials

MDPI

Article

Study of the TIG Welding Process of Thin-Walled Components Made of 17-4 PH Steel in the Aspect of Weld Distortion Distribution

Marek Mróz [1,*], Bartłomiej Kucel [2], Patryk Rąb [1] and Sylwia Olszewska [1]

1 Department of Foundry and Welding, Faculty of Mechanical Engineering and Aeronautics, Rzeszow University of Technology, Al. Powstańców Warszawy 12, 35-959 Rzeszow, Poland; p.rab@prz.edu.pl (P.R.); s.olszewska@prz.edu.pl (S.O.)
2 MB Aerospace Rzeszow, ul. Przemysłowa 9b, 35-105 Rzeszow, Poland; bartlomiej.kucel@mbaerospace.com
* Correspondence: mfmroz@prz.edu.pl

Abstract: This article presents the results of a study on the distribution of weld distortion in thin-walled components made of 17-4 PH steel, resulting from TIG (Tungsten Inert Gas) welding. Both manual and automatic welding processes were examined. Physical simulation of the automated welding process was conducted on a custom-built welding fixture. Analysis of weld distortion in thin-walled components made of 17-4 PH steel was based on the results of measurements of transverse shrinkage and displacement angle values. These measurements were taken on thin-walled parts before and after the welding process using a coordinate measuring machine (CMM). To determine the effect of manual and automated welding processes on the microstructure of the welded joint area, metallographic tests and hardness measurements were performed. The microstructure was analyzed using a scanning electron microscope (SEM). An analysis of the chemical composition of selected welded joint zones was also conducted. These tests were performed using an optical emission spectrometer (OES). According to the results, the use of automated welding and special fixtures for manufacturing thin-walled aircraft engine components made of 17-4 PH steel reduces the propensity of these components for distortion due to the effects of the thermal cycle of the welding process. This conclusion is supported by the results of the observation of the microstructure and analysis of the chemical composition of the various zones of the welded joint area.

Keywords: TIG method; 17-4 PH steel; thin-walled components; weld distortion

Citation: Mróz, M.; Kucel, B.; Rąb, P.; Olszewska, S. Study of the TIG Welding Process of Thin-Walled Components Made of 17-4 PH Steel in the Aspect of Weld Distortion Distribution. *Materials* **2023**, *16*, 4854. https://doi.org/10.3390/ma16134854

Academic Editors: Cosmin Codrean, Carmen Opriș and Anamaria Feier

Received: 5 June 2023
Revised: 3 July 2023
Accepted: 4 July 2023
Published: 6 July 2023

1. Introduction

The aerospace industry, especially in the design of aircraft engines, frequently employs welded structures made of sheet metal. These designs are distinguished by their lower overall weight, which improves the engine's thrust-to-weight ratio [1,2]. During the manufacturing process of welded structures, the material is exposed to the effects of the welding thermal cycle. This exposure can result in irreversible changes in shape and dimensions due to welding stresses and strains [3–5]. The intensity of these changes also depends on the effect of temperature on the values of such material parameters as longitudinal modulus of elasticity, transverse modulus of elasticity, Poisson's ratio, yield strength, and coefficient of linear thermal expansion. As temperature increases, the values of the longitudinal modulus of elasticity and yield strength decrease. Research on materials that exhibit high strength at minimum yield strengths is reported in papers [6–8].

In the welding process, rapid heating and rapid cooling occur in the thermal cycle of the welded joint area. This causes a mechanical interaction between the different areas of the welded joint. This process is further intensified by the occurring phase transformations (solid state transformations in the heat affected zone). In specific areas of the welded joint, changes in the dimensions and shape of structural elements may occur, resulting from the expansion or contraction of the welded material and the weld. These strains can be longitudinal, transverse, or angular. Longitudinal strains are caused by the shrinkage of the weld and the adjacent material in the direction of the length of the welded joint [9]. Angular strains are present in all types of joints and welds. They are induced by the non-uniform transverse shrinkage of the weld in its thickness and shrinkage of the metal adjacent to the weld. The measure of an angular strain is the β angle, which is the angle of refraction of the joint plane. The method of determining this angle using numerical methods and neural networks is described in paper [10], among others.

Numerical simulation is one of the modern tools for predicting welding stresses and distortion. The technical literature reports studies on the use of various programs, such as SYSWELD [11–15], ANSYS [16–18], ABAQUS [19,20], or AUTODESK CFD [21], for modelling welding stress and distortion distributions.

A significant factor affecting the formation of weld distortion is the type of heat source used in the welding thermal cycle. Proper selection of process parameters, especially the current parameters of the heat source, can help minimize weld strains. This is demonstrated in papers on GTAW welding [22,23] and GMAW welding [11,24], among other works. Similar conclusions were also reached by the authors of paper [25] on the FSW method, and papers [5,26,27] dealing with laser welding.

One of the ways to minimize weld distortion is to establish a proper welding procedure that accounts for the complexity of the welded structure, as well as the sequence and manner of making the welds. Paper [11] presents the results of a numerical simulation of different variants of back-step welds with respect to minimizing weld distortion.

The challenges of achieving adequate dimensional tolerance and shape tolerance of a welded structure are particularly evident in the case of thin-walled components, which are extensively utilized in the construction of aircraft engines (aerospace industry). As the aviation industry developed, new materials and design solutions were adopted. This led to changes in the manufacturing process to ensure sufficiently high performance properties of aviation parts [28].

Turbofan and turbojet engines can be divided into two sections, which differ in operating temperature: a cold section and a hot section [29,30]. Besides the required propulsion performance, the engines are characterized by compact design, low weight, reliable operation, and durability. The high demand for these designs translates into the complexity of engine components and the manufacturing processes for their production, as well as the ongoing search for new solutions and materials. The primary focus is on high strength and resistance to corrosive environments at elevated temperatures [31,32].

One of the materials employed in modern aircraft engines is low-carbon martensitic steel 17-4 PH (X5CrNiCuNb16-4 according to EN 10088-2 or UNS S17400 according to ASTM A564). This 17-4 PH steel, a registered trademark of AK Steel, is distinguished by high strength, excellent resistance to high-temperature corrosion, relatively high hardness, substantial resistance to thermal fatigue, and good durability of welded joints. Because of its properties, the steel has also been employed in industrial machinery, power, shipbuilding, geothermal, and petrochemical applications. In the aviation industry, it is used for elements of aircraft equipment and in the construction of aircraft engines for elements of the so-called engine cold zone. These include fan housings and compressor housings [33–38].

The principal alloying elements of 17-4 PH steel are chromium and copper. The steel is considered to have high corrosion resistance due to the presence of more than 12% chromium, while the 4% copper content improves corrosion resistance and increases heat resistance and creep resistance [39]. The authors of paper [40] emphasize the high corrosion resistance of 17-4 PH steel, arguing that, in terms of corrosion resistance in various environments, this steel is comparable to austenitic stainless steel 304 and group 400 stainless steel alloys. According to the authors of papers [40–43], this material is easily weldable by conventional arc welding methods such as TIG, laser welding, and hybrid methods. The authors of paper [44,45] state that the risk of hot cracks in the weld is low due to the solidification of the weld with the formation of δ-ferrite.

The authors of papers [25,46,47] have demonstrated that, by applying appropriate heat treatment, the hardness of 17-4 PH steel can be significantly increased. The treatment primarily involves supersaturation at high temperatures and aging at lower temperatures. The hardness of the 17-4 PH steel after treatment amounted to 49 HRC.

The research in this article focuses on the analysis of the effect of the type of welding process flow (manual or automated welding) on the propensity for distortion that arises as a result of the impact of the thermal cycle of welding thin-walled components made of 17-4 PH steel. This study introduces an original fixture solution for automatic TIG welding of these components. To fully illustrate the impact of the thermal cycle of manual and automatic TIG welding of thin-walled 17-4 PH steel elements, microstructure tests and hardness measurements were conducted.

2. Materials and Methods

The test material used was 0.9 mm thick 17-4 PH steel sheet. The chemical composition of the steel, as determined using a Bruker Q4 TASMAN (Kalkar, Germany) emission spectrometer, is shown in Table 1 (average of 5 analyses). The chemical composition of steel 17-4 PH (1.4021) according to EN 10088-2 is as follows (%wt.): max 0.07 %C, max 0.07 %Si, max 1.50 %Mn, max 0.04 %P, max 0.015 %S, 15.00–17.00 %Cr, 3.00–5.00 %Cu, max 0.06 %Mo, 3.00–5.00 %Ni, 5x%C %Nb, remainder Fe.

Table 1. Chemical composition of 17-4 PH steel.

Chemical Compositions (%wt.)							
C	Si	Mn	P	S	Cr	Mo	Ni
0.052	0.376	0.511	0.084	0.0081	16.18	0.075	3.857
Cu	Al	Co	Nb	Ti	V	W	Fe
4.092	0.0068	0.048	0.329	0.0075	0.059	0.029	74.24

Strips 50 mm wide and 150 mm long were cut from the sheet metal, rolled up, and joined by a butt weld to produce a pipe section with a diameter of 150 mm. The test joint comprised two pipe components connected by a circumferential weld (one bead). Prior to making the test joint, the components were cleaned and degreased.

The test joints were made using the TIG method in two variants: manual and automated. Table 2 summarizes the basic parameters of TIG welding.

Table 2. Process parameters of TIG welding of thin-walled components of 17-4 PH steel.

Parameter	Automated	Manual
Amperage I (A)	34	37
Voltage U (V)	~8.3	~8.1
Welding speed v (mm/min)	170	65
Linear energy (kJ/cm)	1.00	2.8

The process of automated welding of the test joint was implemented on the bench shown in Figure 1, which includes a special welding fixture with a welding gun as its main component. This gun, equipped with copper cooling pads, also ensured the protection of the test joint root by flushing it with argon. The gun with the tacked tubular elements was mounted on a swivel. A TIG welding torch with automated wire feeding (additional material) was used to create the joint. The torch utilizes a 2% cerium tungsten electrode (WC 20) with a diameter of 3.2 mm and a sharpening angle of 30°.

Figure 1. Bench for automated TIG welding of thin-walled components made of 17-4 PH steel.

Before starting the welding process, the tacked test joints were measured using a Mitutoyo Crysta 122010 CMM (Tokyo, Japan) with a Renishaw REVO-2 measuring head (Gloucestershire, UK). Figure 2 depicts a diagram of the distribution of measurement points in lines A, B and C. These measurements served as a reference for estimating the values of transverse shrinkage, angular strain, and profile deformation.

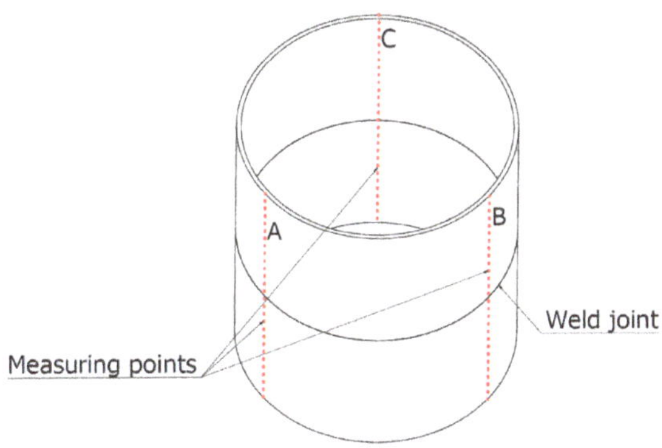

Figure 2. Diagram of the distribution of measurement points.

The methodology for determining angular strain values is shown in Figure 3. The strain angle is the angle of intersection of the trend line of the transverse profile based on points in lines A, B and C. The weld areas and sample edges were ignored in the calculations. Transverse shrinkage for both welding variants was also measured.

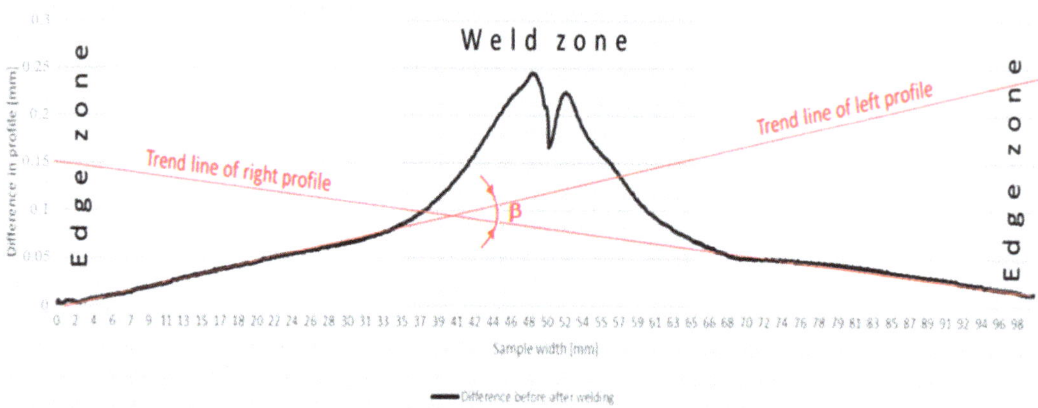

Figure 3. Methodology for calculating angular strain values.

Visual examinations were performed based on PN-EN ISO 5817 to identify potential inconsistencies in the test welded joints.

Metallographic tests were conducted to evaluate the changes in the microstructure of the joined components, with a particular focus on the impact of the welding thermal cycle on the structure and size of the heat affected zone (HAZ). Microstructure studies of test welded joints were performed on metallographic specimens. The metallographic specimens were prepared by cutting from the joints on a Struers Labotom-3 metallographic cutter (Copenhagen, Denmark), then ground and polished with abrasive paper, polishing cloths, and diamond slurries on a Metimex SM-PM 250AV1 laboratory grinder-polisher (Pyskowice, Poland). The specimens were etched with Kalling's reagent for 5 s for microstructure revelation, followed by observation under a Tescan VEGA 3 scanning microscope (Brno, Czech Republic).

Hardness measurements were also made on the test joints, in accordance with the scheme shown in Figure 4.

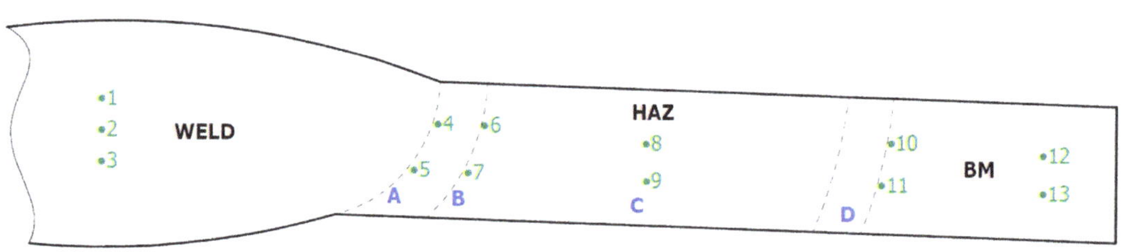

Figure 4. Scheme of hardness measurements of test welded joints manually and automatically.

Hardness measurements were made using the Vickers method, with the HV5 load. These measurements were carried out using a Zwick-Roell ZHV10 hardness tester (Ulm, Germany).

3. Results and Discussion

3.1. Measurements of Transverse Strain and Angular Strain

Plots of transverse profiles, which are an illustration of the transverse strain at the measurement points in lines A, B and C on the test joint made manually and the difference between the transverse profiles before and after welding, are shown in Figures 5–7.

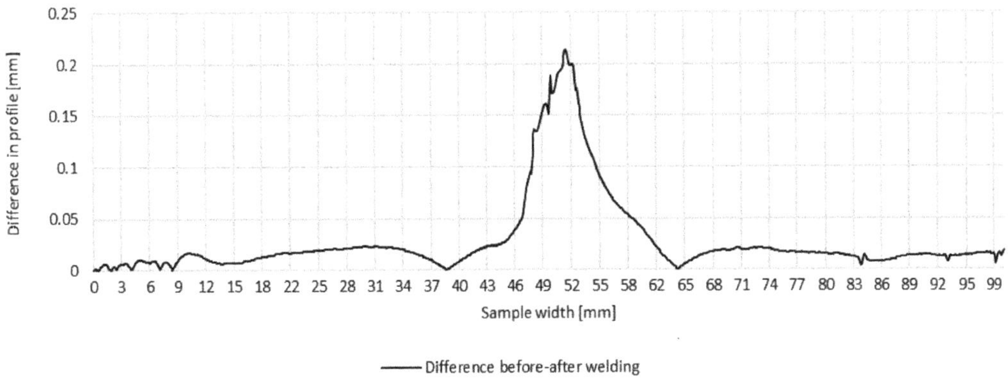

Figure 5. Difference in transverse profiles on line A before and after manual welding.

Figure 6. Difference in transverse profiles on line B before and after manual welding.

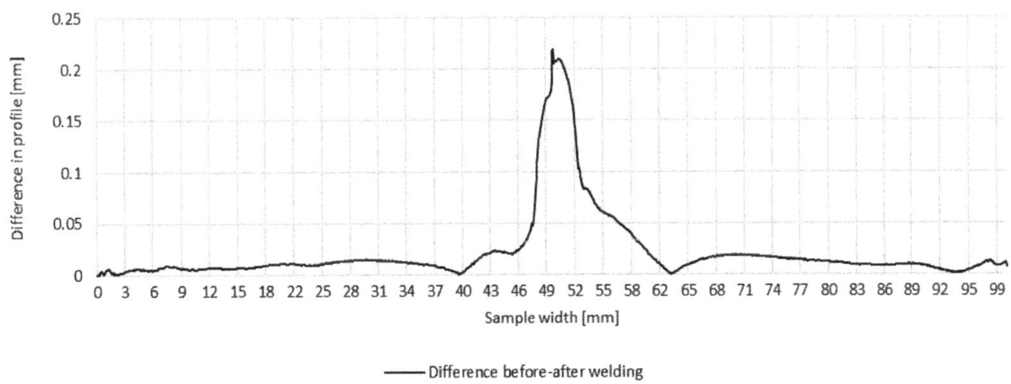

Figure 7. Difference in transverse profiles on line C before and after manual welding.

Plots of transverse profiles at the measurement points in lines A, B and C on the test joint made automatically and the difference between the transverse profiles before and after welding are shown in Figures 8–10.

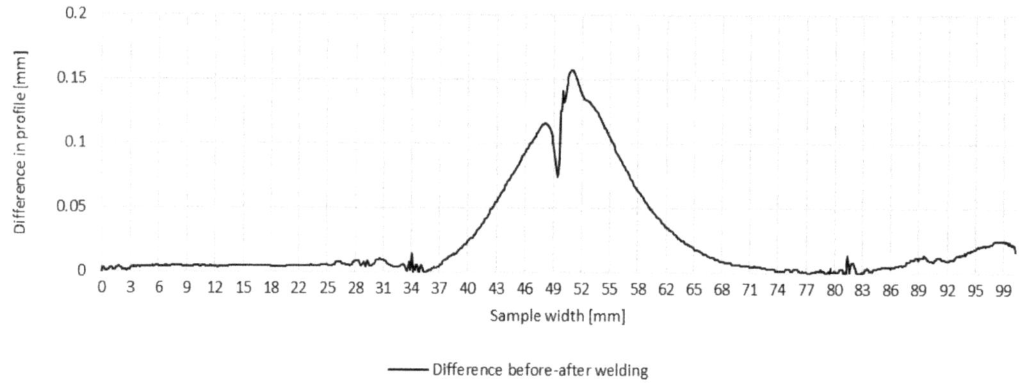

Figure 8. Difference in transverse profiles on line A before and after automated welding.

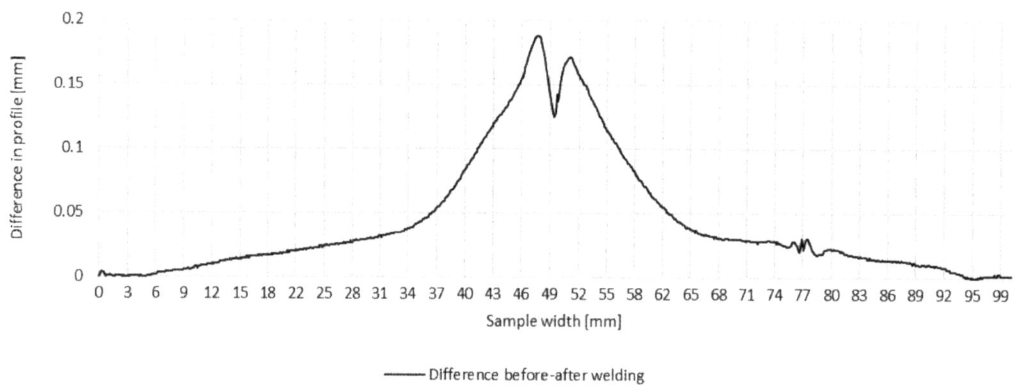

Figure 9. Difference in transverse profiles on line B before and after automated welding.

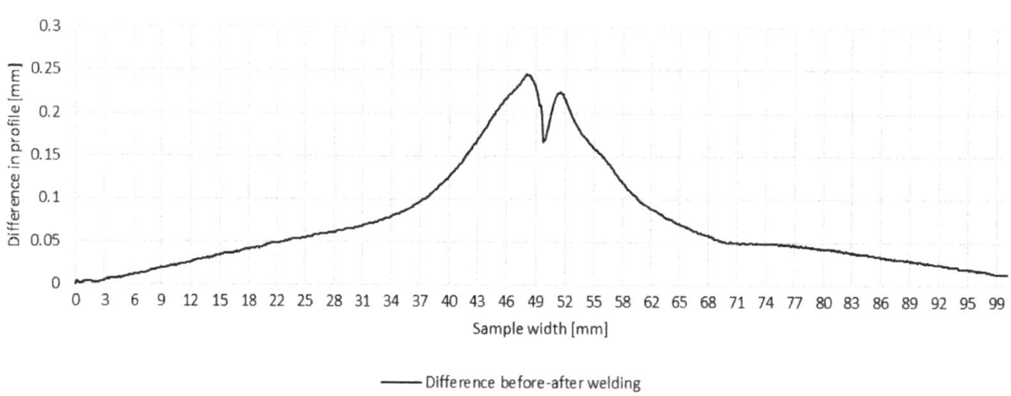

Figure 10. Difference in transverse profiles on line C before and after automated welding.

To better demonstrate the impact of the welding method (manually or automatically), Figures 11 and 12 present the transverse profiles determined on lines A, B and C on manually and automatically welded test joints. These waveforms indicate clear differences in the distribution of transverse shrinkage values of test joints welded manually and automatically.

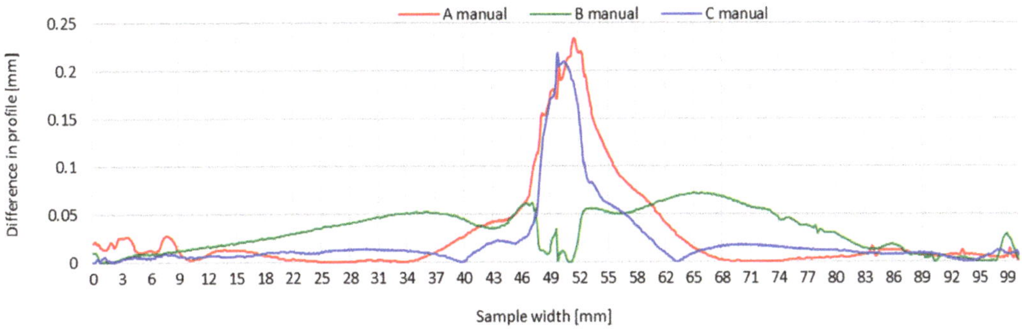

Figure 11. Transverse profiles on lines A, B and C after making a test joint of thin-walled components of 17-4 PH steel—manual welding.

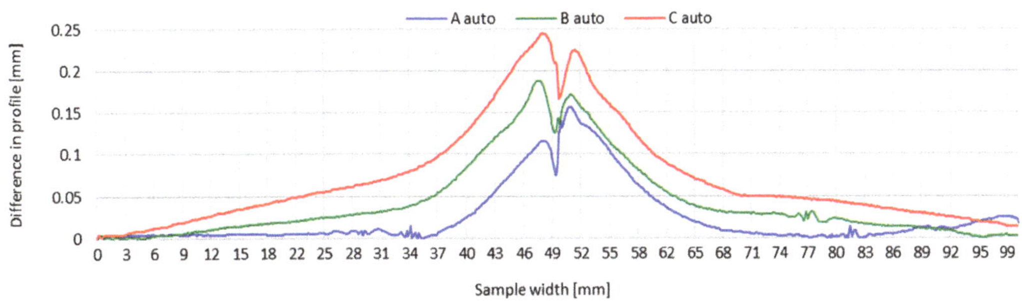

Figure 12. Transverse profiles on lines A, B and C after making a test joint of thin-walled components of 17-4 PH steel—automated welding.

In the case of the manually made test joint, the connected sheets across their width are noticeably more corrugated. The wave-like character of the distribution of transverse shrinkage is particularly evident at the beginning and end of the transverse profile on lines A and C. In the case of line C, this profile has a rising character from the edges of the joined components in the direction of the weld, reaching peaks of distortion on both sides of the weld. This testifies to the greater distortion of test joints caused by the thermal cycle of the manual welding process, compared to automated welding.

The transverse profiles of the automatically made test joint show a similar pattern on both sides of the weld, with an increasing tendency of the transverse shrinkage value in the direction from the edges of the components to be joined to the weld. For this joint, there is no distinct corrugation of the sheet metal.

The obtained results indicate that the thermal cycle of automated welding has a significantly decreased effect on the distortion of the welded joint, compared to a joint made by hand. The outcome of the automation of the welding process was a correction of the process parameters of the TIG method, resulting in a reduction in the linear energy of the welding process (Table 2). This was reflected in the reduced amount of heat introduced into the welded joint area, resulting in a lower impact of the welding thermal cycle.

The results of measuring the angular strain values of test joints made manually and automatically are given in Table 3.

The average value of the angular strain β of the test joint made manually is 0.12 °C, while that for the test joint made by automated welding is 0.11 °C. Once again, the automatically welded joint has a lower propensity for angular strain β, compared to the joint made by hand.

Table 3. Angular strain values of test joints made manually and automatically.

Welding Method	Measurement Point	Angular Strain β (Degrees)
Manual	A	0.02
	B	0.13
	C	0.21
Automated	A	0.04
	B	0.26
	C	0.04

3.2. Visual Testing

The purpose of visual testing (VT) was to evaluate the welded joints for the presence of weld discontinuities. An example view of the weld face of the test joints is presented in Figure 13. In both cases, observing the weld face and root did not reveal any significant welding inconsistencies. However, it was noted that the face of the weld made by the automated method has better smoothness and evenness and the weld is continuous. In the case of a weld made by hand, the way the weld is made with a back-step technique (sectional welding) and small deviations in the straightness of the weld are visible.

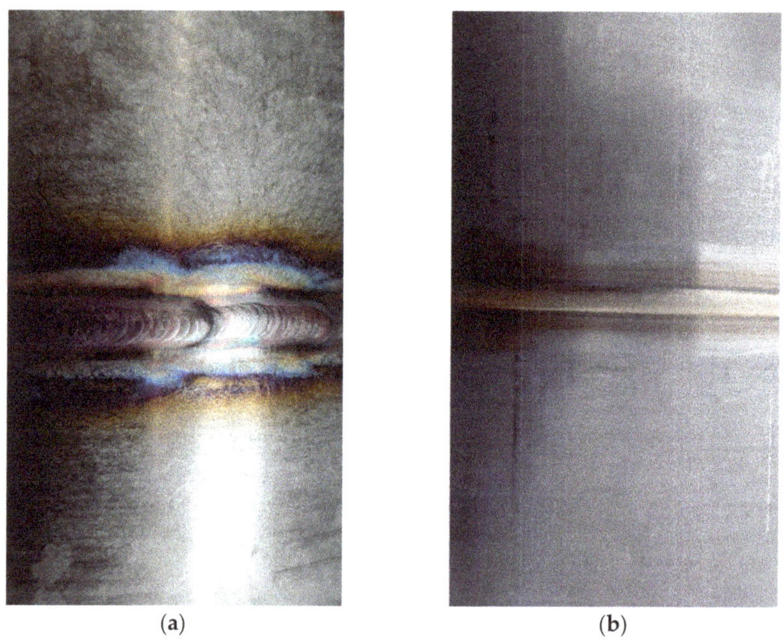

Figure 13. Example view of the face of the test joint made manually (**a**) and by automated welding method (**b**).

3.3. Macro- and Microstructure Testing

The macrostructure of the welded joint area made manually is shown in Figure 14.

The microstructure of the weld, the heat affected zone, and the parent material were analyzed in this joint. The area of the heat affected zone was analyzed on the basis of the extent of the impact of the welding thermal cycle. Consequently, four areas (A–D) were revealed in the heat affected zone.

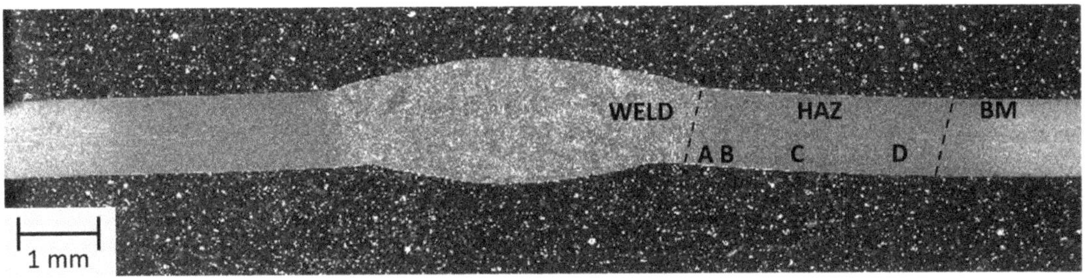

Figure 14. Macrostructure of the welded joint area of thin-walled 17-4 PH steel components made manually by TIG method.

The sample microstructure of the weld, the A–D areas of the heat affected zone and the parent material of the test joint made by hand are shown in Figures 15–20.

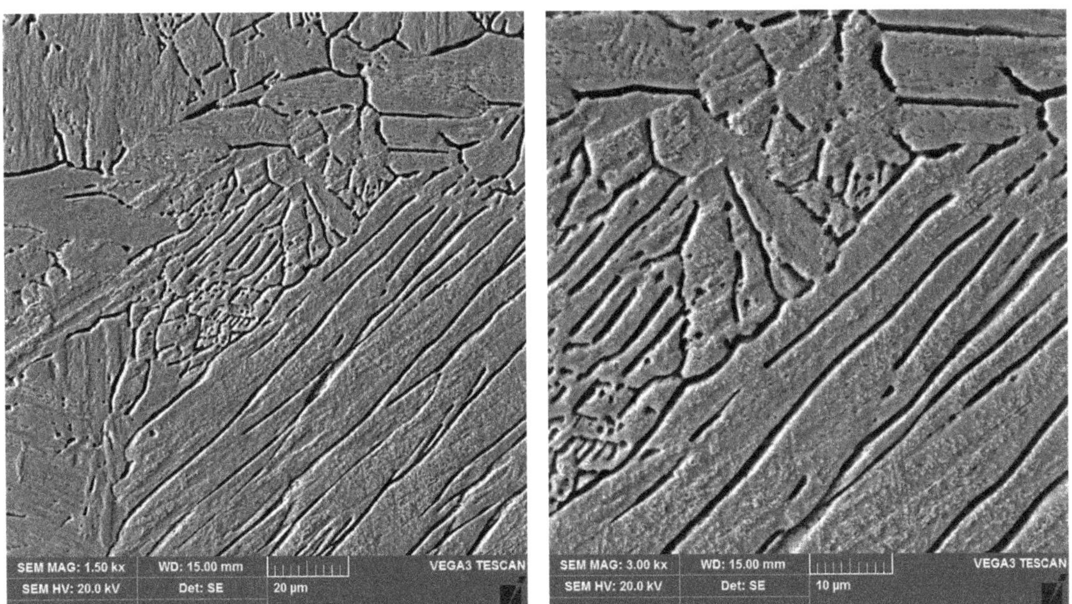

(**a**) (**b**)

Figure 15. Microstructure of the weld area of the test joint made manually: (**a**) magnification 1500×, (**b**) 3000×.

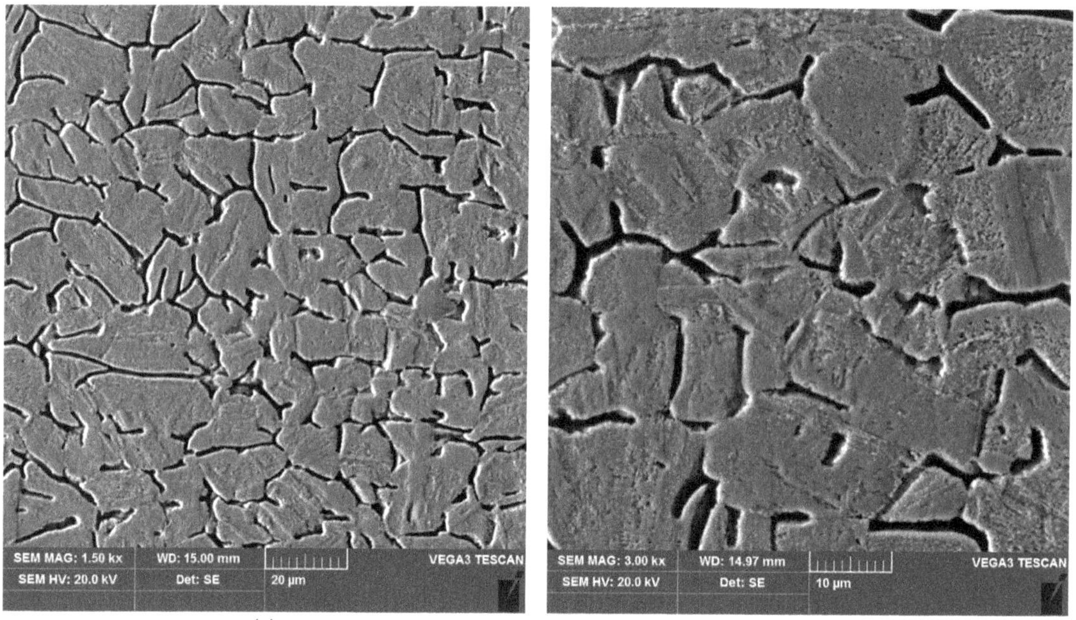

(a) (b)

Figure 16. Microstructure of the heat affected zone in area A of the test joint made manually: (**a**) magnification 1500×, (**b**) 3000×.

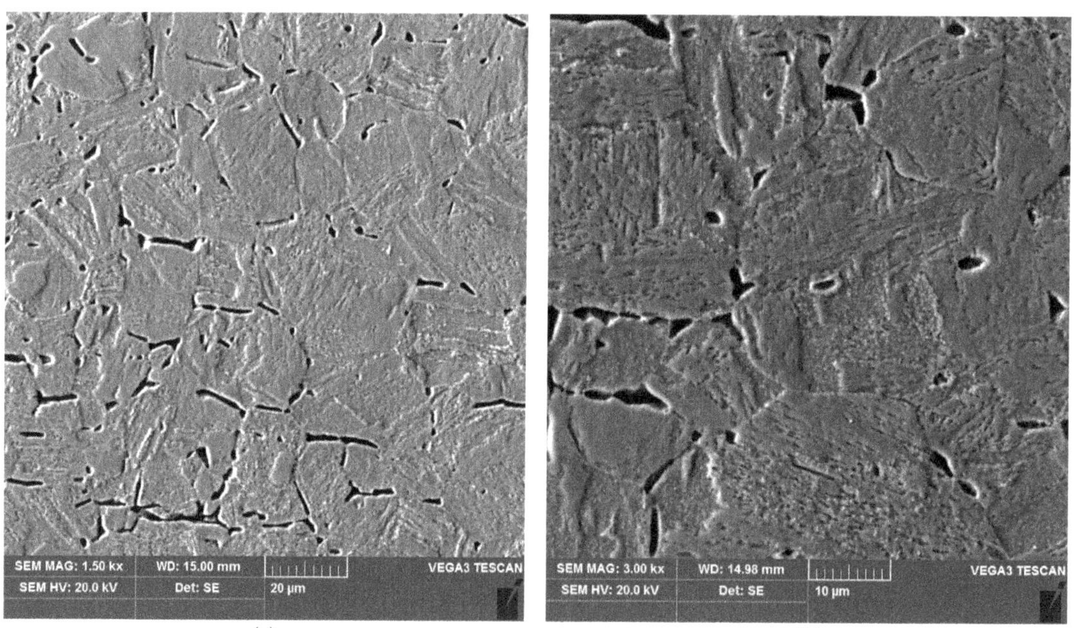

(a) (b)

Figure 17. Microstructure of the heat affected zone in area B of the test joint made manually: (**a**) magnification 1500×, (**b**) 3000×.

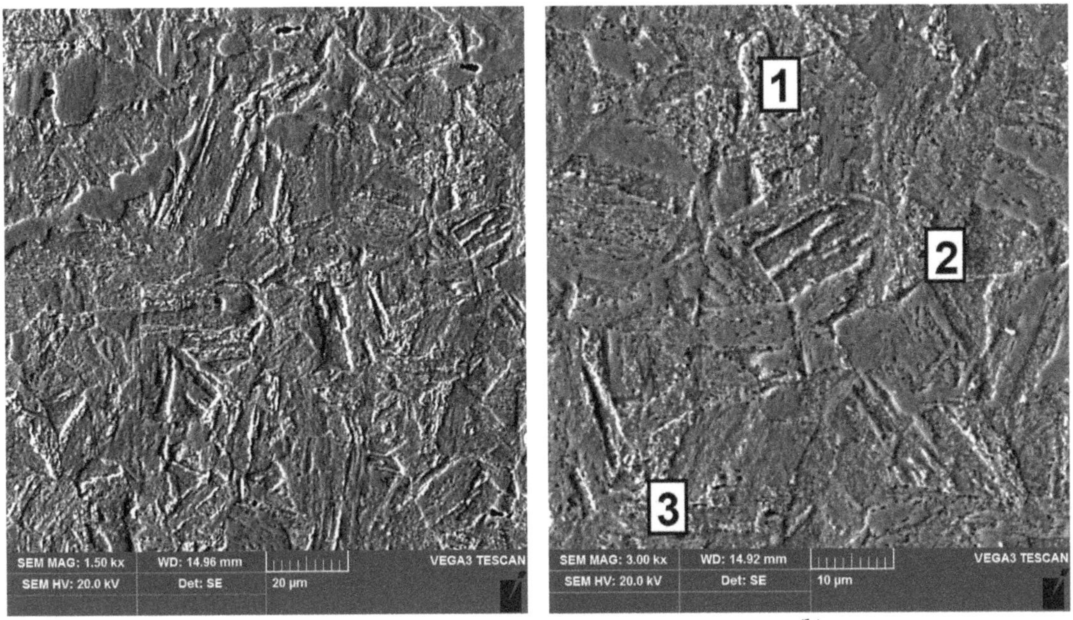

(a) (b)

Figure 18. Microstructure of the heat affected zone in area C of the test joint made manually: (**a**) magnification 1500×, (**b**) 3000×. 1–3 are the places where the analysis of the chemical composition was performed.

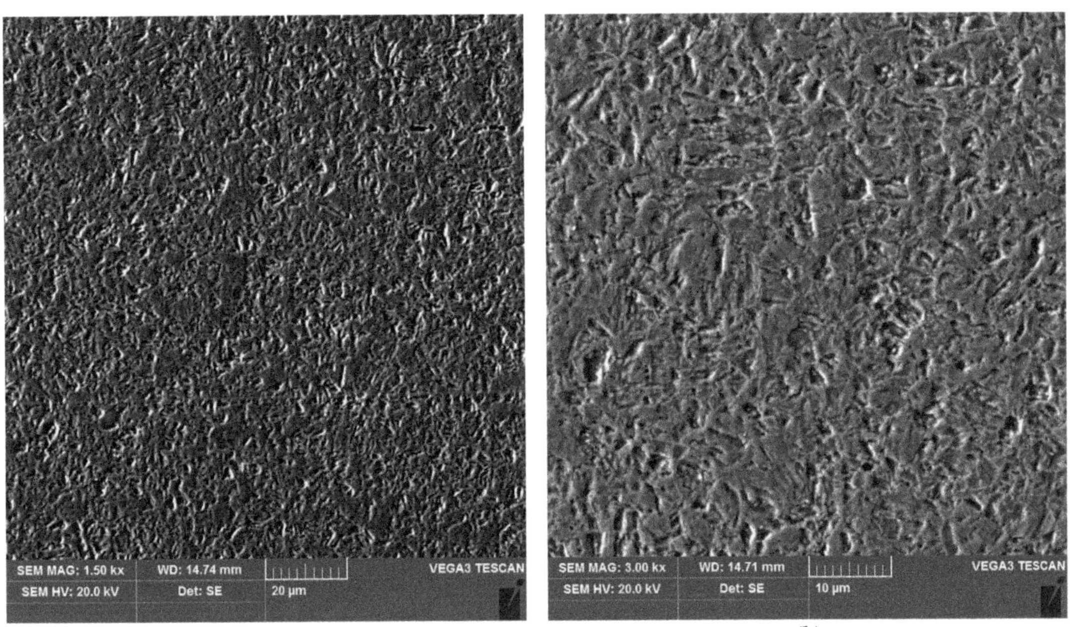

(a) (b)

Figure 19. Microstructure of the heat affected zone in area D of the test joint made manually: (**a**) magnification 1500×, (**b**) 3000×.

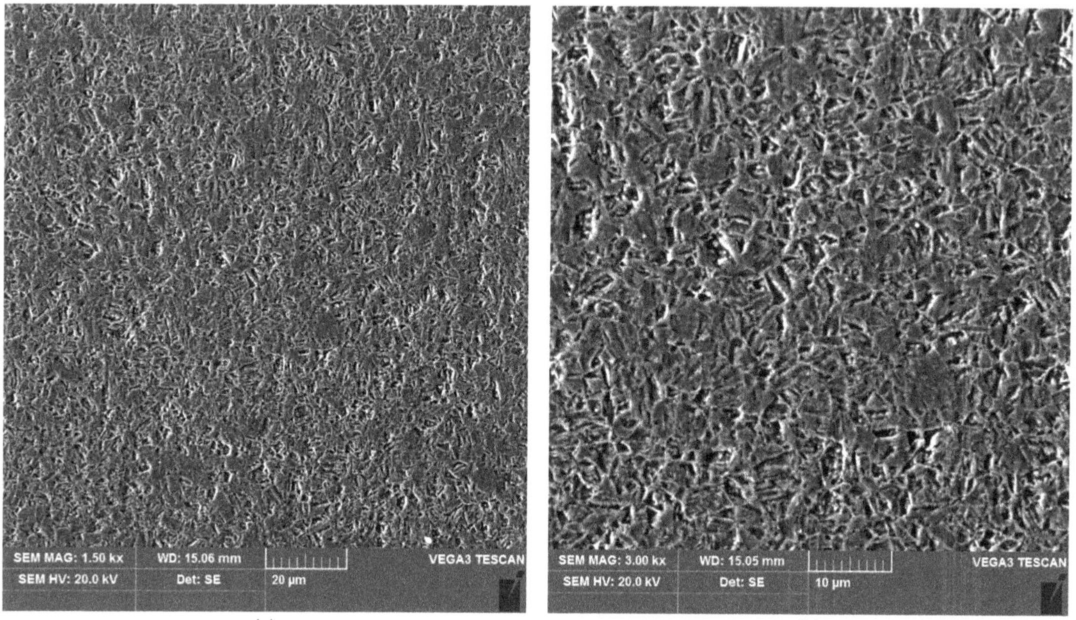

(a) (b)

Figure 20. Microstructure of the parent material area of the test joint made manually: (**a**) magnification 1500×, (**b**) 3000×.

The macrostructure of the welded joint area made by the automated welding method is shown in Figure 21. Analysis of the macrostructure of the area of this joint suggests a similar structure, compared to the joint made manually. The heat affected zone of the joint also displayed areas of varying microstructure, resulting from the extent of the welding thermal cycle. Thus, as in the case of the manually welded joint, four areas (A, B, C and D) were identified in the heat affected zone of the test joint made by the automated welding method.

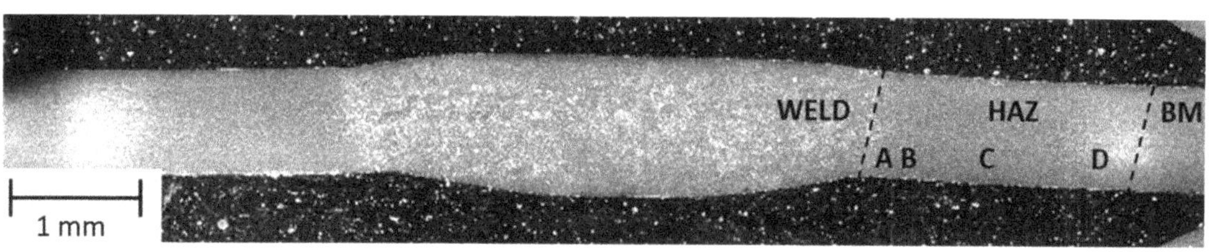

Figure 21. Macrostructure of the welded joint area of thin-walled 17-4 PH steel components made automatically by TIG method.

The sample microstructure of the weld of areas A–D of the heat affected zone of the test joint made by the automated welding method is shown in Figures 22–27.

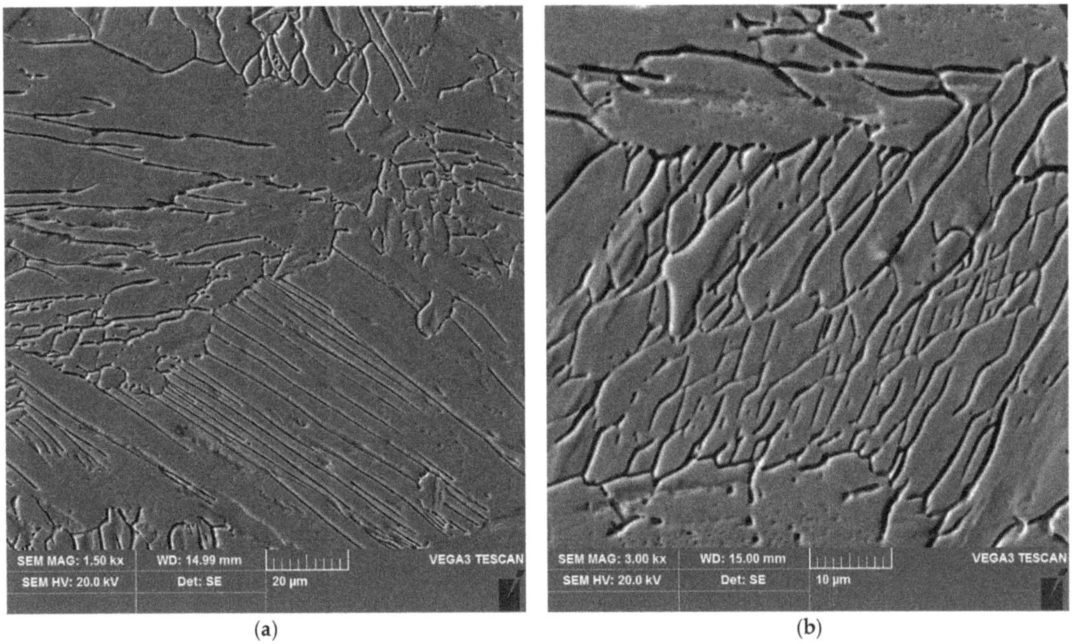

Figure 22. Microstructure of the weld area of the test joint made by the automated welding method: (**a**) magnification 1500×, (**b**) 3000×.

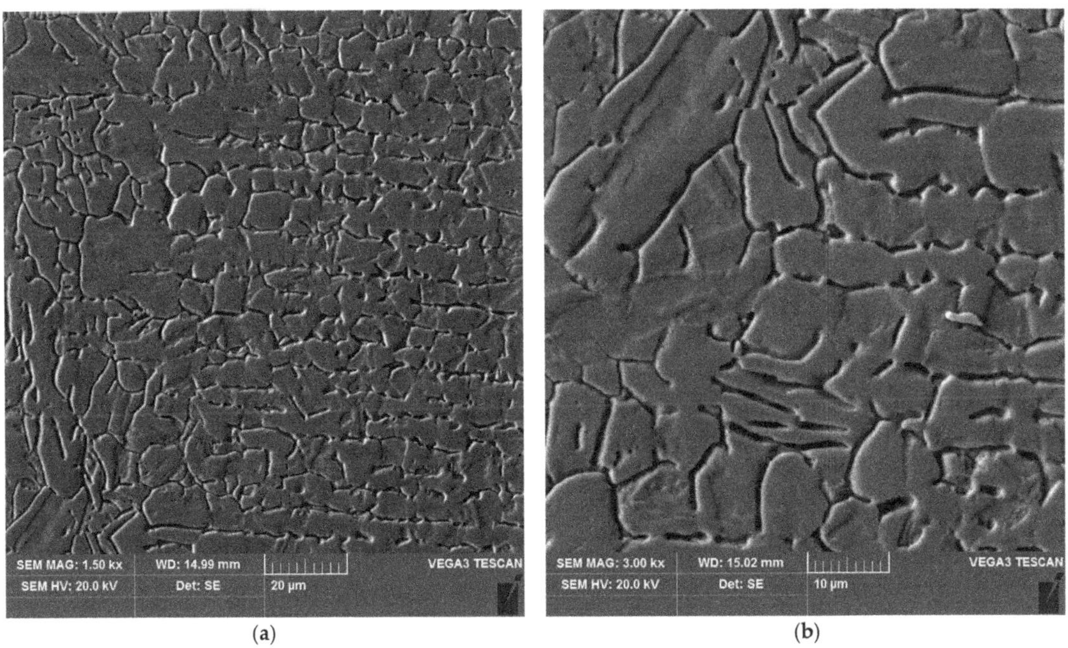

Figure 23. Microstructure of the heat affected zone in area A of the test joint made by the automated welding method: (**a**) magnification 1500×, (**b**) 3000×.

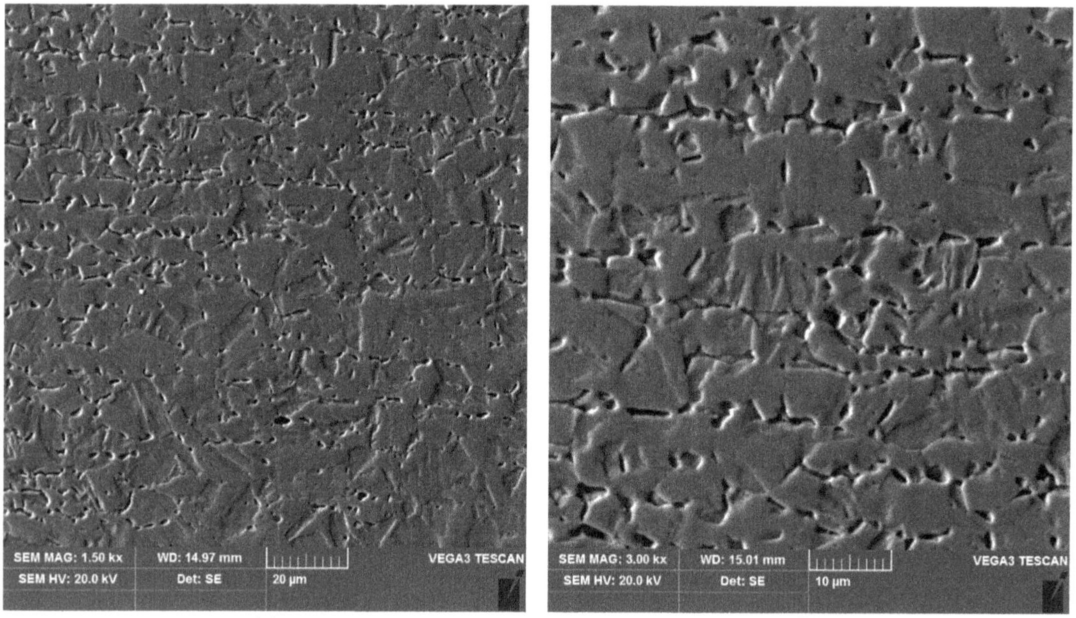

(**a**) (**b**)

Figure 24. Microstructure of the heat affected zone in area B of the test joint made by the automated welding method: (**a**) magnification 1500×, (**b**) 3000×.

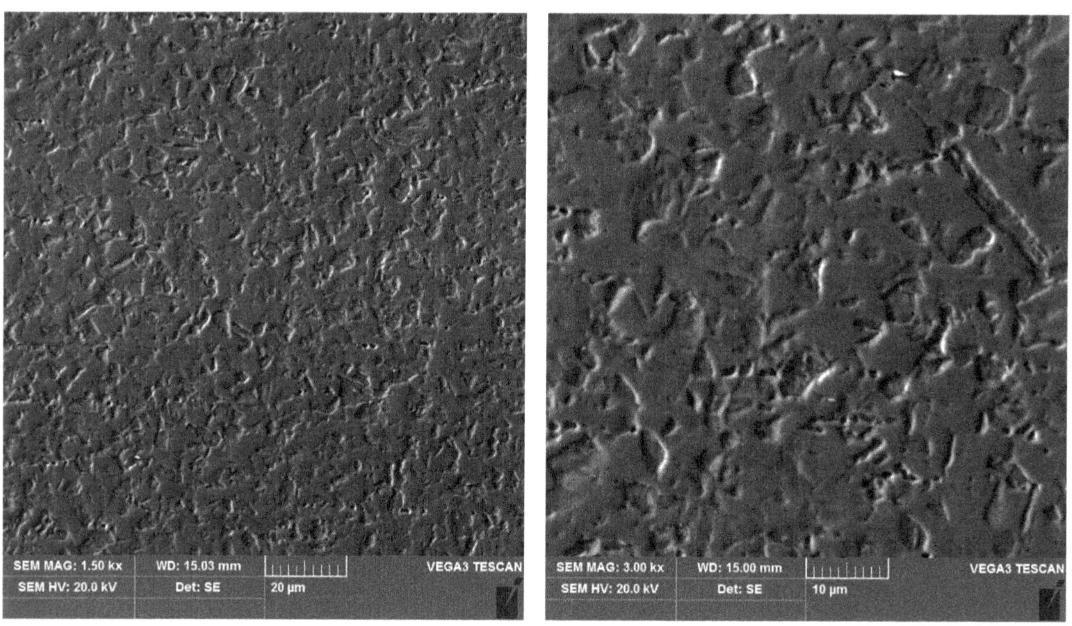

(**a**) (**b**)

Figure 25. Microstructure of the heat affected zone in area C of the test joint made by the automated welding method: (**a**) magnification 1500×, (**b**) 3000×.

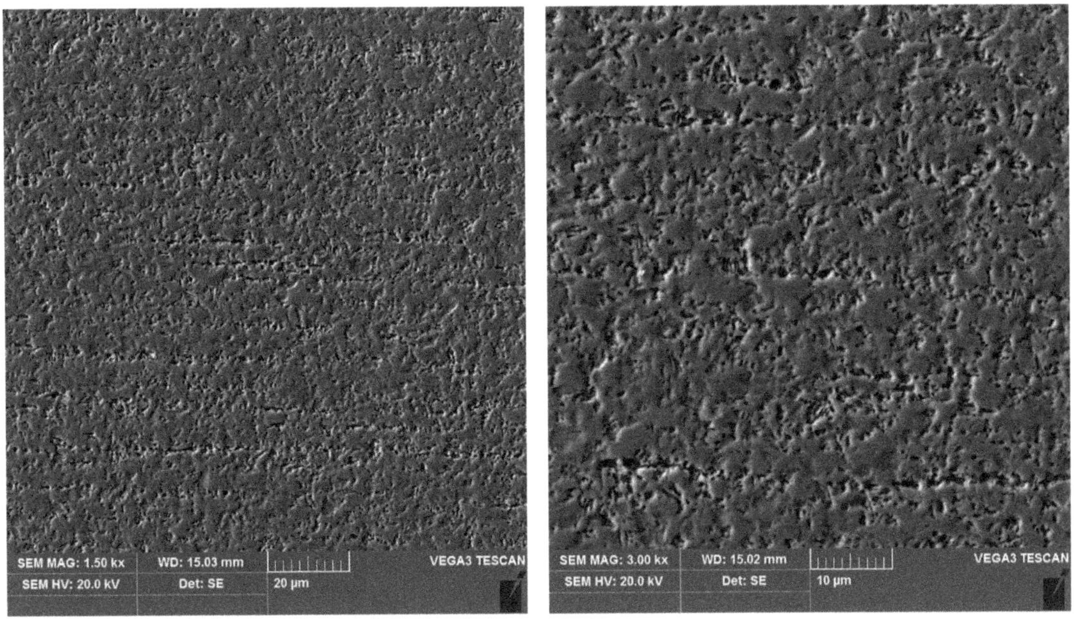

(**a**) (**b**)

Figure 26. Microstructure of the heat affected zone in area D of the test joint made by the automated welding method: (**a**) magnification 1500×, (**b**) 3000×.

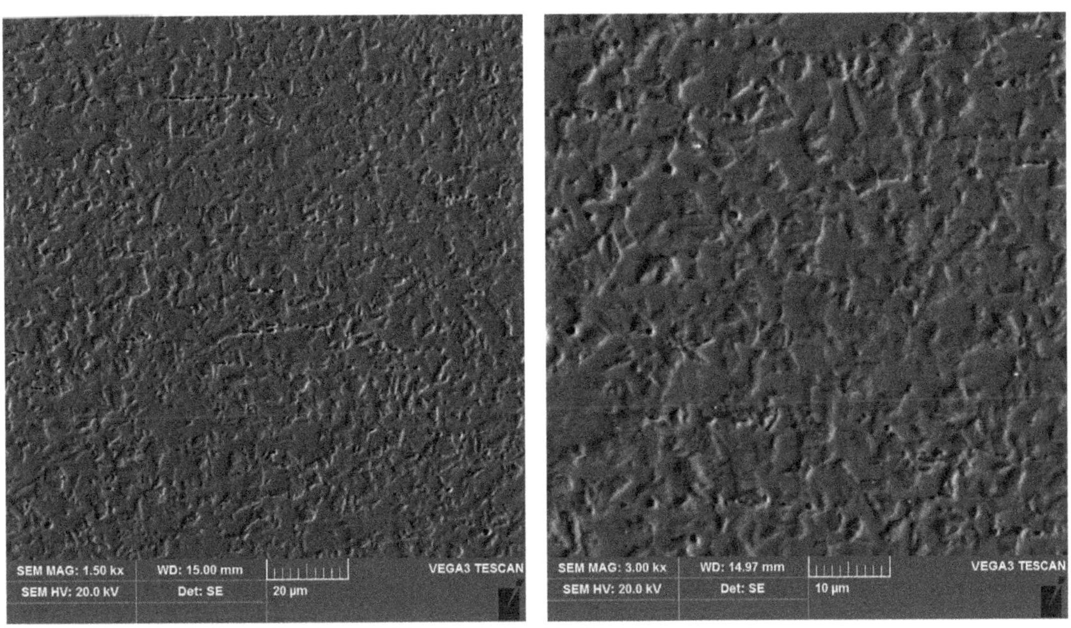

(**a**) (**b**)

Figure 27. Microstructure of the parent material area of the test joint made by the automated welding method: (**a**) magnification 1500×, (**b**) 3000×.

The microstructure of the parent material (Figures 22 and 27), i.e., 17-4 PH steel, is martensitic, with a small number of very fine, almost invisible carbides located at grain boundaries.

Analysis of the test welded joints made manually and automatically showed a great similarity in the effect of the welding thermal cycle on the microstructure of their respective areas.

The weld, in both joints, reveals precipitates of ferrite δ and martensite. However, the weld of the manually welded joint showed a large number of fine carbides. It can be assumed that the thermal cycle of welding is responsible for the carbide release process. The manually welded joint was made with a back-step technique, i.e., sectional weld. After making the first section, the thermal cycle of the next section caused heating of the previously made weld which resulted in the process of carbide release in the weld from the supersaturated solution of carbon in iron α. In this way, laying each successive weld section intensified the secretion processes in the previously laid sections.

Upon examining the microstructure of the heat affected zone (HAZ), ferrite δ and martensite are immediately adjacent to the fusion line (area A). Areas B and C exhibit the typical structure of quenching products (martensite and tempered martensite). As we move away from the weld toward the parent material, there are small amounts of tempered martensite in the D area. This is attributable to the lower intensity of the impact of the thermal cycle of the welding process.

Carbides were observed in the heat affected zone of the manually welded joint (as in the weld) (Figures 22–25). The intensity of the carbide release process in individual areas A, B, C and D varied depending on the intensity of the impact of the thermal cycle of the welding process. The highest amount of carbides was witnessed in areas closer to the weld.

The results of X-ray microanalysis of the chemical composition of the carbides are presented in Table 4.

Table 4. Results of chemical composition analysis of carbides (Figure 18b).

Point	Chemical Composition [%wt.]				
	C	Cr	Ni	Cu	Fe
1	2.87	16.69	4.30	2.76	73.39
2	2.17	16.64	4.54	3.10	73.55
3	1.73	16.83	4.53	2.94	73.97

From the obtained results, it was found that the identified carbide phases mainly contain chromium, nickel and copper. The iron content results from carbides that are too small and electron beam scanning of areas next to these carbides. Even though the determination of carbon content by X-ray microanalysis is not accurate (carbon contamination phenomenon), Table 4 provides the carbon content to confirm that the phases analyzed are carbides.

The analysis of the welded joint made by the automated welding method shows a similar effect of the thermal cycle of the welding process on the microstructure of its individual areas, with the exception that carbides were not found in such large numbers in both the weld and the heat affected zone. This indicates the smaller extent and lower intensity of the impact of the thermal cycle of the welding process. This is further supported by measurements of the width of the entire heat affected zone and its individual areas, as shown in Table 5. The total width of the heat affected zone for the joint made by automated welding is almost half that of the test joint made by hand.

Table 5. Results of measurements of the width of individual areas of the heat affected zone of test joints made manually and by automated welding.

Welding Method	Measurement Point	Width of the Heat Affected Zone Area [μm]
manually	A	120–150
	B	90–105
	C	1800
	D	2000
automatically	A	90–100
	B	80–90
	C	900
	D	1100

3.4. Hardness Measurements

The hardness measurements results of test welded joints are presented in Table 6 and in Figure 28.

Table 6. HV5 hardness measurements results.

Welding Method	Hardness Measurement Point (HV5)												
	1	2	3	4	5	6	7	8	9	10	11	12	13
Manual	330	336	366	338	338	344	340	338	340	330	336	366	364
average value	334 in weld			338 in area A		342 in area B		339 in area C		333 in area D		365 in BM	
Automated	350	348	355	354	358	356	354	342	340	366	368	368	366
average value	351 in weld			356 in area A		355 in area B		341 in area C		367 in area D		367 in BM	

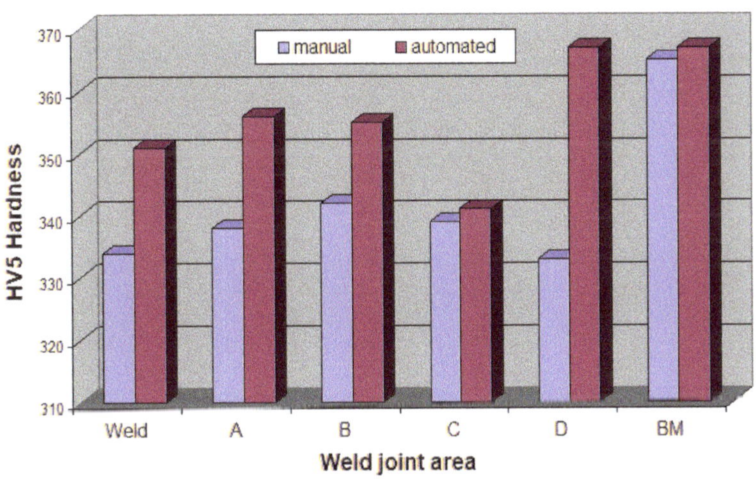

Figure 28. Hardness of the test joint made by the automated welding method.

The obtained results indicate differences in hardness values in individual areas of welded joints made manually and automatically. In the case of an automatically produced joint, due to the less intensive impact of the heat cycle of the welding process, the hardness distribution in individual areas of the heat affected zone, compared to the weld and the base material, is more stable than in a manually made joint. However, in both joints, area C of the heat affected zone witnessed a large decrease in hardness, dropping to about 340 HV5. This indicates that the most significant microstructural changes occurred in area

C due to the thermal cycle of the welding process. A more detailed explanation of these changes, given the thematic scope of this publication, will be the subject of a subsequent publication. The publication will employ the results of additional research, including color metallography results, to reveal the percentage of martensite or other hardening products in individual HAZ areas, which form due to the thermal cycle of the welding process.

4. Conclusions

Not only does the type of welding method used exert a significant impact on the size and distribution of welding deformations, but so does the method of implementation. Replacing manual welding of thin-walled 17-4 PH steel elements with automatic welding led to a reduction in the amount of heat introduced into the joined elements (lower linear energy of the welding process). Consequently, the deformation of the automatically welded joint was less than that of the manually welded joint. In the case of a manually welded joint, significant ripples appeared in the areas extending from the weld to the parent material.

The positive effect of automatic welding on weld quality, especially on the condition of the weld face and the weld root surface, was confirmed.

The microstructure test results of welded joints corroborate the lower intensity of the heat cycle of the welding process in an automatically welded joint. In this joint, a heat affected zone of a similar structure was found as in a manually welded joint, but with a significantly smaller HAZ width. Both welded joints contain areas with a similar microstructure, featuring quenching and tempering products resulting from heating in the thermal cycle of the welding process and subsequent rapid cooling. Furthermore, the presence of carbides was observed in the microstructure of the manually welded joint.

The welding method profoundly influences the hardness distribution in individual areas of welded joints. In automatic welding, the HV5 hardness distribution is more stable compared to the hardness distribution in a manually welded joint, reaffirmed by the lower intensity of the thermal cycle of the welding process on the microstructure of the heat affected zone in the automatically welded joint. The hardness values in the HAZ of an automatically welded joint are closer to the hardness of the weld and the parent material, compared to the HAZ of a manually welded joint. It can be deduced that, in the case of an automatically welded joint, the intensity of various types of hardening and tempering processes caused by the thermal cycle of the welding process is lower compared to a manually welded joint.

Automated TIG welding is recommended for manufacturing 17-4 PH thin-walled components. This is particularly critical when such components are used in the construction of aircraft engines.

Author Contributions: Conceptualization, M.M. and B.K.; Methodology, B.K., P.R. and S.O.; Software, B.K.; Validation, M.M.; Formal analysis, M.M. and B.K.; Investigation, M.M., B.K., P.R. and S.O.; Resources, P.R.; Writing—original draft, M.M. and P.R.; Writing—review & editing, M.M.; Supervision, M.M. All authors have read and agreed to the published version of the manuscript.

Funding: This research received no external funding.

Institutional Review Board Statement: Not applicable.

Informed Consent Statement: Not applicable.

Data Availability Statement: Data is contained within the article.

Conflicts of Interest: The authors declare no conflict of interest.

References

1. Zhang, H.; Zhan, M.; Zheng, Z.; Li, R.; Ma, F.; Cui, X.; Chen, S.; Lei, Y. Forming dependence on spin roller paths for thin-walled complex components from 2195 Al-Li alloy TWBs. *Int. J. Adv. Manuf. Technol.* **2022**, *120*, 3113–3122. [CrossRef]
2. Zhang, C.; Li, H.; Yang, J.; Lu, H.; Su, P. Research on Model Slice and Forming Method of Thin-walled Parts. In *E3S Web of Conferences*; EDP Sciences: Les Ulis, France, 2021; Volume 233. [CrossRef]

3. Zhang, C.; Gümmer, V. High temperature heat exchangers for recuperated rotorcraft powerplants. *Appl. Therm. Eng.* **2019**, *154*, 548–561. [CrossRef]
4. Abdolvand, H.; Faraji, G.; Besharati Givi, M.K.; Hashemi, R.; Riazat, M. Evaluation of the microstructure and mechanical properties of the ultrafine grained thin-walled tubes processed by severe plastic deformation. *Met. Mater. Int.* **2015**, *21*, 1068–1073. [CrossRef]
5. Ma, J.; Atabaki, M.M.; Liu, W.; Pillai, R.; Kumar, B.; Vasudevan, U.; Kovacevic, R. Laser-based welding of 17-4 PH martensitic stainless steel in a tubular butt joint configuration with a built-in backing bar. *Opt. Laser Technol.* **2016**, *82*, 38–52. [CrossRef]
6. Zhu, X.K.; Chao, Y.J. Effects of temperature-dependent material properties on welding simulation. *Comput. Struct.* **2002**, *80*, 967–976. [CrossRef]
7. Lahtinen, T.; Vilaça, P.; Peura, P.; Mehtonen, S. MAG Welding Tests of Modern High Strength Steels with Minimum Yield Strength of 700 MPa. *Appl. Sci.* **2019**, *9*, 1031. [CrossRef]
8. Haslberger, P.; Holly, S.; Ernst, W.; Schnitzer, R. Microstructure and mechanical properties of high-strength steel welding consumables with a minimum yield strength of 1100 MPa. *J. Mater. Sci.* **2018**, *53*, 6968–6979. [CrossRef]
9. Liu, Y.; Wang, P.; Fang, H.; Ma, N. Mitigation of residual stress and deformation induced by TIG welding in thin-walled pipes through external constraint. *J. Mater. Res. Technol.* **2021**, *15*, 4636–4651. [CrossRef]
10. Rubio-Ramirez, C.; Giarollo, D.F.; Mazzaferro, J.E.; Mazzaferro, C.P. Prediction of angular distortion due GMAW process of thin-sheets Hardox 450® steel by numerical model and artificial neural network. *J. Manuf. Process.* **2021**, *68*, 1202–1213. [CrossRef]
11. Mróz, M.; Czech, R.; Kupiec, B.; Dec, A.; Spólnik, M.; Rąb, P. Numerical and Physical Simulation of MAG Welding of Large S235JRC+N Steel Industrial Furnace Wall Panel. *Materials* **2023**, *16*, 2779. [CrossRef]
12. Solfronk, P.; Sobotka, J.; Bukovská, Š.; Bradáč, J. Experimental and Numerical Analysis of the Residual Stresses in Seamed Pipe in Dependence on Welding and Metal Forming. *Materials* **2023**, *16*, 2256. [CrossRef]
13. Tien, D.N. Numerical simulation for determination of temperature field and residual stress of stainless steel butt joints with and without clamping. *Vietnam. J. Sci. Technol.* **2022**, *60*, 713–725. [CrossRef] [PubMed]
14. Ramos, H.M.E.; Tavers, S.M.O.; de Castro, P.M.S.T. Numerical modelling of welded T-joint configurations using SYSWELD. *Sci. Technol. Mater.* **2018**, *30*, 6–15. [CrossRef]
15. Lima, T.R.; Tavares, S.M.O.; de Castro, P.M.S.T. Residual stress field and distortions resulting from welding processes: Numerical modelling using Sysweld. *Ciênc. Tecnol. Mater.* **2017**, *29*, 56–61. [CrossRef]
16. Capriccioli, A.; Frosi, P. Multipurpose ANSYS FE procedure for welding processes simulation. *Fusion Eng. Des.* **2009**, *84*, 546–553. [CrossRef]
17. Unni, A.K.; Vasudevan, M. Computational fluid dynamics simulation of hybrid laser-MIG welding of 316 LN stainless steel using hybrid heat source. *Int. J. Therm. Sci.* **2023**, *185*, 108042. [CrossRef]
18. Satyanarayana, G.; Narayana, K.L.; Rao, B.N. Incorporation of Taguchi approach with CFD simulations on laser welding of spacer grid fuel rod assembly. *Mater. Sci. Eng. B* **2021**, *269*, 115182. [CrossRef]
19. Chou, A.P.; Shi, G.; Liu, C.; Zhou, L. Residual stress and compression buckling of large welded equal-leg steel angles. *J. Constr. Steel Res.* **2023**, *201*, 107756. [CrossRef]
20. Saternus, Z.; Piekarska, W.; Kubiak, M.; Domański, T.; Goszczyńska-Króliszewska, D. Numerical estimation of temperature field in a laser welded butt joint made of dissimilar materials. In *MATEC Web of Conferences*; EDP Sciences: Les Ulis, France, 2018; Volume 157, p. 02043. [CrossRef]
21. Adewuyi, R.A.; Aweda, J.O. Modeling and simulation of welding temperature fields in cr-mo steel bar. *Eur. J. Mater. Sci. Eng.* **2021**, *6*, 3–18. [CrossRef]
22. Horajski, P.; Bohdal, L.; Kukielka, L.; Patyk, R.; Kaldunski, P.; Legutko, S. Advanced Structural and Technological Method of Reducing Distortion in Thin-Walled Welded Structures. *Materials* **2021**, *14*, 504. [CrossRef] [PubMed]
23. Li, Y.; Li, Y.; Ma, X.; Zhang, X.; Fu, D.; Yan, Q. Study on Welding Deformation and Optimization of Fixture Scheme for Thin-Walled Flame Cylinder. *Materials* **2022**, *15*, 6418. [CrossRef]
24. Stanisz, D.; Machniewicz, T.; Parzych, S.; Jeż, G.; Dvorkin, L.; Hebda, M. Microstructure and Mechanical Properties of Joints Depending on the Process Used. *Materials* **2022**, *15*, 5171. [CrossRef]
25. Machniewicz, T.; Nosal, P.; Korbel, A.; Hebda, M. Effect of FSW Traverse Speed on Mechanical Properties of Copper Plate Joints. *Materials* **2020**, *13*, 1937. [CrossRef]
26. Balajaddeh, M.B.; Naffakh-Moosavy, H. Pulsed Nd: YAG laser welding of 17-4 PH stainless steel: Microstructure, mechanical properties, and weldability investigation. *Opt. Laser Technol.* **2019**, *119*, 105651. [CrossRef]
27. Tuz, L.; Sokołowski, Ł.; Stano, S. Effect of Post-Weld Heat Treatment on Microstructure and Hardness of Laser Beam Welded 17-4 PH Stainless Steel. *Materials* **2023**, *16*, 1334. [CrossRef] [PubMed]
28. Lin, Z.M. Making aviation green. *Adv. Manuf.* **2013**, *1*, 42–49. [CrossRef]
29. Federal Aviation Administration. *Airplane Flying Handbook*; Skyhorse Publishing Inc.: Oklahoma City, OK, USA, 2011.
30. Gurrappa, I.; Yashwanth, I.V.S. Design and Development of Smart Coatings for Gas Turbines. In *Gas Turbines*; Intech Open: Rijeka, Croatia, 2010; pp. 65–78.
31. Uliasz, M.; Ornat, A.; Burghardt, A.; Muszyńska, M.; Szybicki, D.; Kurc, K. Automatic Evaluation of the Robotic Production Process for an Aircraft Jet Engine Casing. *Appl. Sci.* **2022**, *12*, 6443. [CrossRef]

32. Biao, Z.; Wenfeng, D.; Zhongde, S.; Jun, W.; Changfeng, Y.; Zhengcai, Z.; Jia, L.; Shihong, X.; Yue, D.; Xiaowei, T.; et al. Collaborative manufacturing technologies of structure shape and surface integrity for complex thin-walled components of aero-engine: Status, challenge and tendency. *Chin. J. Aeronaut.* **2023**. [CrossRef]
33. Pfennig, A.; Kranzmann, A. Corrosion and Fatigue of Heat Treated Martensitic Stainless Steel 1.4542 Used for Geothermal Applications. *Matter. Int. J. Sci. Technol.* **2019**, *5*, 138–158. [CrossRef]
34. Wu, J.H.; Lin, C.K. Tensile and fatigue properties of 17-4 PH stainless steel at high temperatures. *Metall. Mater. Trans. A* **2002**, *33*, 1715–1724. [CrossRef]
35. Arisoy, C.F.; Başman, G.; Şeşen, M.K. Failure of a 17-4 PH stainless steel sailboat propeller shaft. *Eng. Fail. Anal.* **2003**, *10*, 711–717. [CrossRef]
36. Ziewiec, A.; Czech, J.; Tasak, E. Welded Joint Cracking in Martensitic Stainless Steel Precipitation-Sthregthened with Copper. *Arch. Metall. Mater.* **2012**, *57*, 1055–1061. [CrossRef]
37. Ponnusamy, P.; Sharma, B.; Masood, S.H.; Rashid, R.R.; Rashid, R.; Palanisamy, S.; Ruan, D. A study of tensile behavior of SLM processed 17-4 PH stainless steel. *Mater. Today Proc.* **2021**, *45*, 4531–4534. [CrossRef]
38. Trzepieciński, T.; Pieja, T.; Malinowski, T.; Smusz, R.; Motyka, M. Investigation of 17-4 PH steel microstructure and conditions of elevated temperature forming of turbine engine strut. *J. Mater. Process. Technol.* **2018**, *252*, 191–200. [CrossRef]
39. Bhaduri, A.K.; Sujith, S.; Srinivasan, G.; Gill, T.P.S.; Mannan, S.L. Optimized postweld heat treatment procedures for 17-4 PH stainless steels. *Weld. J.* **1995**, *74*, 153.
40. Hosseini Farzaneh, S.M.; Belbasi, M. Effects of Shielding Gas on the Depth, Width and Hardness of 17-4 PH Stainless Steel during TIG Welding. *J. Environ. Friendly Mater.* **2020**, *4*, 25–29.
41. Prabhakar, A.S.; Agilan, M.; Venkateswaran, T.; Sivakumar, D.; Venkatakrishnan, P.G. Effect of Post Weld Heat Treatment on Mechanical Properties of 17-4-PH Stainless Steel Welds. *Mater. Werkst.* **2022**, *53*, 947. [CrossRef]
42. Mousavi, S.A.; Sufizadeh, A.R. Metallurgical investigations of pulsed Nd: YAG laser welding of AISI 321 and AISI 630 stainless steels. *Mater. Des.* **2009**, *30*, 3150–3157. [CrossRef]
43. Liu, W.; Ma, J.; Atabaki, M.M.; Pillai, R.; Kumar, B.; Vasudevan, U.; Sreshta, H.; Kovacevic, R. Hybrid laser-arc welding of 17-4 PH martensitic stainless steel. *Lasers Manuf. Mater. Process.* **2015**, *2*, 74–90. [CrossRef]
44. Ziewiec, A.; Zielińska-Lipiec, A.; Tasak, E. Microstructure of welded joints of X5CrNiCuNb16-4 (17-4 PH) martensitic stainless steel after heat treatment. *Arch. Metall. Mater.* **2014**, *59*, 965–970. [CrossRef]
45. Gholipour, A.; Shamanian, M.; Ashrafizadeh, F. Microstructure and wear behavior of stellite 6 cladding on 17-4 PH stainless steel. *J. Alloys Compd.* **2011**, *509*, 4905–4909. [CrossRef]
46. Hsiao, C.N.; Chiou, C.S.; Yang, J.R. Aging reactions in a 17-4 PH stainless steel. *Mater. Chem. Phys.* **2002**, *74*, 134–142. [CrossRef]
47. Tavares, S.S.M.; Da Silva, F.J.; Scandian, C.; Da Silva, G.F.; De Abreu, H.F.G. Microstructure and intergranular corrosion resistance of UNS S17400 (17-4 PH) stainless steel. *Corros. Sci.* **2010**, *52*, 3835–3839. [CrossRef]

materials

MDPI

Article

Cavitation Erosion Characteristics of the EN AW-6082 Aluminum Alloy by TIG Surface Remelting

Ion Mitelea [1], Ilare Bordeaşu [2], Florin Frant [1], Ion-Dragoş Uţu [1], Corneliu Marius Crăciunescu [1,*] and Cristian Ghera [2]

[1] Department of Materials and Fabrication Engineering, Politehnica University Timisoara, Bulevardul Mihai Viteazul nr.1, 300222 Timisoara, Romania
[2] Department of Mechanical Machines, Equipment and Transports, Politehnica University Timisoara, Bulevardul Mihai Viteazul nr.1, 300222 Timisoara, Romania
* Correspondence: corneliu.craciunescu@upt.ro

Abstract: Components made of aluminum alloys operating under cavitation erosion conditions have low performance and therefore a reduced lifetime. The degradation of these components is a consequence of the repetitive implosion of cavitation bubbles adjacent to the solid surface. In this paper, the effect of the rapid re-melting and solidification modification of the surface microstructure of parts of an Al-based alloy strengthened by artificial ageing on the reduction of material loss through cavitation erosion was investigated. The heat source used was the electric arc generated between a tungsten electrode and the workpiece (i.e., TIG). Local surface melting was performed at different values of linear energy (El = 6600–15840 J/cm), varying the current between 100 A and 200 A, at a constant voltage of 10 V. The obtained results showed an increase in the surface microhardness at values of 129–137 HV0.05 and a decrease in the erosion rate from 0.50 μm/min, characteristic of the artificial ageing heat treatment, to 0.10–0.32 μm/min, specific to TIG re-melted layers. For the study of the cavitational erosion mechanism, investigations were carried out by optical microscopy and scanning electron microscopy. The results showed that the improvement of the cavitational erosion resistance by surface melting was a consequence of the increase in microstructural homogeneity and grains refinement.

Keywords: TIG remelting; aluminum alloy; cavitation erosion

Citation: Mitelea, I.; Bordeaşu, I.; Frant, F.; Uţu, I.-D.; Crăciunescu, C.M.; Ghera, C. Cavitation Erosion Characteristics of the EN AW-6082 Aluminum Alloy by TIG Surface Remelting. *Materials* **2023**, *16*, 2563. https://doi.org/10.3390/ma16072563

Academic Editors: Costica Bejinariu and Jose M. Bastidas

Received: 16 February 2023
Revised: 15 March 2023
Accepted: 20 March 2023
Published: 23 March 2023

1. Introduction

Degradation of component surfaces by cavitation erosion is a result of the rapid formation, growth, and collapse of bubbles in liquids due to strong pressure fluctuations [1]. Cavitation erosion involves different types of interaction between metallurgical, chemical, mechanical and hydrostatic processes [2,3]. This phenomenon causes surface damage and loss of material due to the growth and collapse of bubbles, which are due to local pressure fluctuations within the liquid flowing around the surface of the parts [1,4]. In detail, when there is a localized pressure drop, the liquid reaches the vapor pressure level and undergoes a phase change, forming bubbles (cavities) containing vapor. Cavitation bubbles last only until the low-pressure zone is left. When the fluid returns to the quiet zone, the bubbles immediately implode, forming a shock wave that strikes and erodes the surface of the part it is in contact with.

A series of research papers analyzed the behavior of various types of metallic materials exposed to cavitation during their operation [5–9]. It is well known that among technical alloys, those with an aluminum base have the lowest cavitation erosion resistance [2,4]. In surface engineering, to reduce cavitation wear, various methods are used to prepare coatings applied to a substrate, such as the welding deposition method, thermal spraying, chemical or physical vapor deposition, etc. [1,6]. Li et al. [10] showed that the application of a thermochemical nitriding treatment improved the cavitation erosion behavior of some

Ti-based non-ferrous alloys, which can be explained by the generation of a homogeneous microstructure with a single layer of α-Ti(N) in the diffusion nitrogen-enriched zone. Kwok et al. [11] and Zhang et al. [12] highlighted the advantages offered by the laser technique in the deposition of protective layers against the cavitation erosion of engineering materials.

Man et al. [13] used a laser technique for SiC/Si_3N_4 surface alloying of the AA6061 aluminum alloy. They found that the cavitation erosion resistance was improved three times compared to that of the reference material, while there was no significant improvement in the 100% SiC-alloyed specimen.

However, these surface strengthening methods have some disadvantages, as they result in the presence of pores, relatively low adhesion, inhomogeneous dilution, and a non-uniform microstructure [14–17], which limit the applications in cavitation environments. For these reasons, modifying the surface properties without changing the chemical composition is an innovative approach to avoid the mentioned defects.

Due to the high cost of laser beam operations and the need for vacuum in electron beam equipment, attention has been focused on using TIG welders that are cheap, flexible and easy to handle for surface modification by local melting. The TIG electric arc produces enough thermal energy to perform some surface treatments; it offers some significant advantages, including selective hardening, minimal distortion of the components, controllable depth of the modified layer, and absence of filler material.

The purpose of this paper was to investigate the effect of local TIG surface remelting of a 6xxx series aluminum base alloy—which could be strengthened by aging—in terms of structural transformation mechanism, onto the cavitation erosion resistance. The 6xxx series aluminum base alloys are frequently used for the production of components that undergo severe cavitation erosion phenomena during service, such as diesel cylinder liners, pistons, pumps, hydrofoils, valves, sluice gates, combustion chambers, etc. [1,2,4,6,9,13,18].

2. Experimental Procedure

The material used in the research, type EN AW-6082, (EN AW-AlSi1MgMn according to EN 573) was delivered in the form of sheets with the following dimensions: length L = 300 mm; width l = 150 mm, and thickness g = 30 mm.

The nominal chemical composition of the alloy was: Si = 1.18%, Fe = 0.39%, Cu = 0.065%, Mn = 0.70%, Mg = 1.32%, Cr = 0.10%, Ni = 0.015%, Zn = 0.044%, Ti = 0.011%, Ga = 0.01%, V = 0.023%, Al = Rest%.

The heat treatment was carried out by applying a solution at 535 ± 5 °C/25 min/water, followed by artificial aging at 175 ± 10 °C/8 h/air (Figure 1).

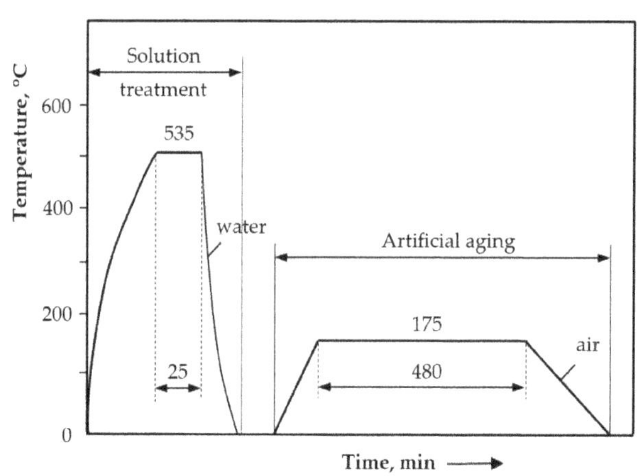

Figure 1. Heat treatment cycle.

Figure 2, the microstructure obtained after this heat treatment was composed of γ solid solution grains with an aluminum base having a polyhedral shape, and inside and on the separation boundaries between them, there were particles of intermetallic phases, consisting of Mg_2Si for the large ones, Al_5FeSi for those in the form of a plate, and $Al_{12}(Fe, Mn)_3Si$ for the spherical ones [18]. This alloy shows the highest values for the mechanical resistance characteristics among all the 6xxx series alloys and is used in the aeronautical, marine, automotive and food industries [7,18].

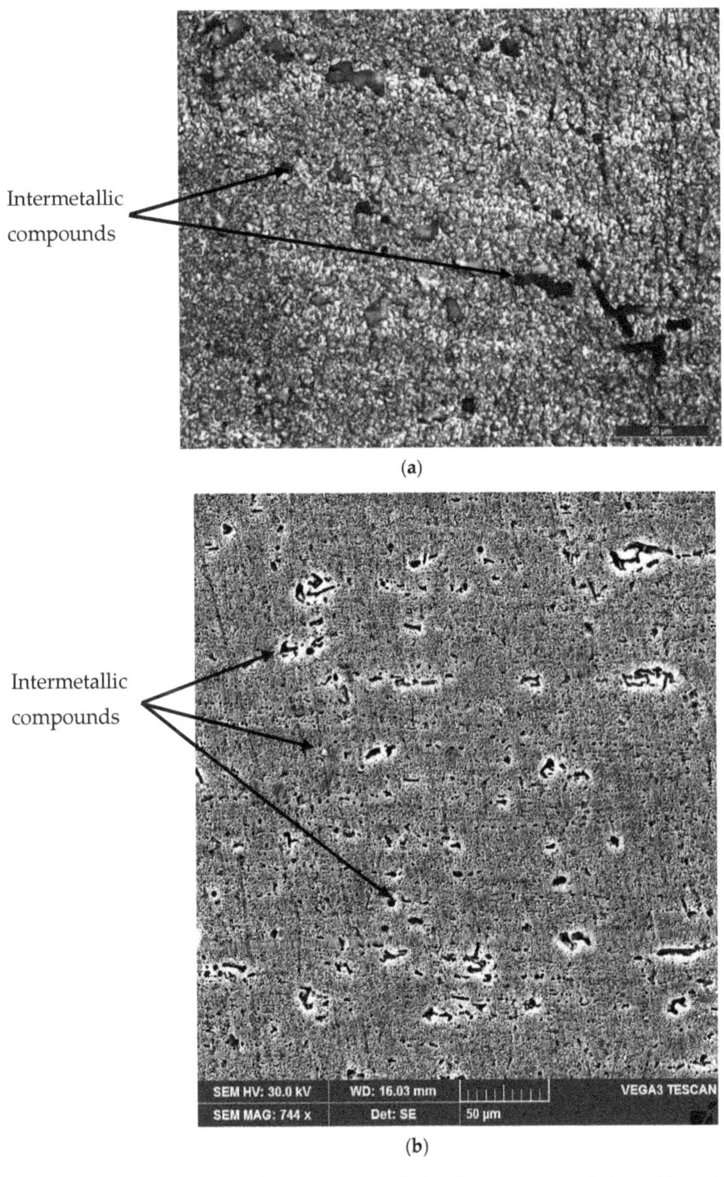

Figure 2. Microstructure of the alloy after solution heat treatment followed by artificial aging: (**a**) MO ×500; (**b**) SEM ×744.

The surface was remelted using a TIG welding equipment (Weld Guru, Buda, TX, USA, Figure 3) in the following working conditions:

- base metal: Al-6082 alloy;
- welding equipment: MAGIC WAVE 300 (Fronius);
- alternating current frequency: 70 Hz;
- nature of current: alternating current;
- balance: 60/40;
- shielding gas: Ar 100%;
- electrode type: EWLa 15;
- electrode diameter: 3.2 mm;
- gas flow: 10 L/min;
- tilt of the gun: 90 degrees.

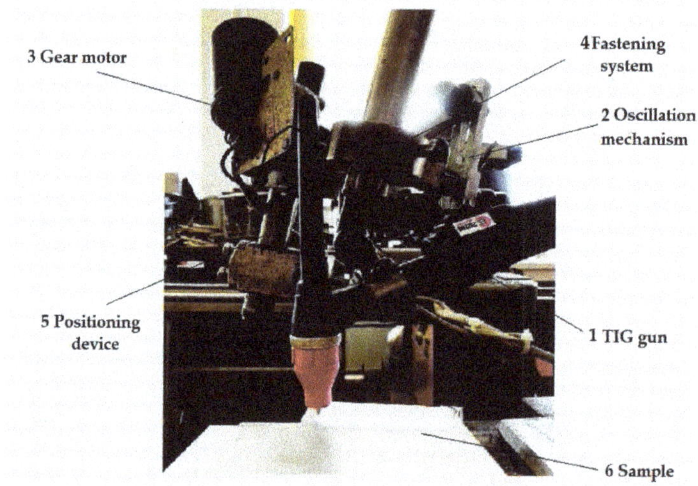

Figure 3. The welding experimental stand.

Further, both from the locally surface remelted plates and from the reference ones, cavitation samples were processed by chipping; their shape and dimensions are shown in Figure 4.

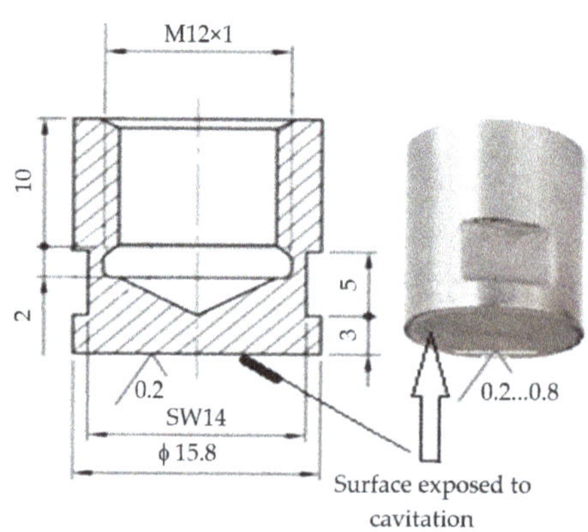

Figure 4. Geometry of the cavitation samples (mm).

The experimental research followed the study of the linear welding energy influence on the improvement of the cavitation erosion resistance of remelted surfaces. For that, three different welding regimes were used. A change in linear energy $E_l = (U_a \times I_s)/v_s \times 60$ [J/cm] was achieved by changing the value of the welding current while keeping the welding speed constant, as follows:

a. Regime 1:
   i. Welding current: Is = 100 A;
   ii. Electric arc voltage: Ua = 11 V (RMS = Root Mean Square);
   iii. Arc length: 2 mm;
   iv. Welding speed: vs. = 10 cm/min;
   v. Linear energy: E_l = 6600 J/cm.

b. Regime 2:
   i. Welding current: Is = 150 A;
   ii. Electric arc voltage: Ua = 12.1 V (RMS);
   iii. Arc length: 2 mm;
   iv. Welding speed: vs. = 10 cm/min;
   v. Linear energy: E_l = 10,890 J/cm.

c. Regime 3:
   i. Welding current: Is = 200 A;
   ii. Electric arc voltage: Ua = 13.2 V (RMS);
   iii. Arc length: 2 mm;
   iv. Welding speed: vs. = 10 cm/min;
   v. Linear energy: E_l = 15,840 J/cm.

According to the technological recommendations, the base material was preheated to a temperature of 100 °C, and the temperature between passes was maintained at values of 120–150 °C. In order to maintain a constant welding speed, the TIG welding head was mounted on a welding installation, the process being mechanized. For the surface remelting of the samples, parallel passages were made, with a step between passages equal to 2/3 of the width of a passage so as to achieve an overlap of the passages of approx. 1/3 of the width of a passage. The remelting details are shown in Figure 5. This made it possible to obtain a smooth molten surface without welding defects (lack of melting or marginal notches). The passages performed aimed at obtaining a width of the remelted area of at least 25 mm to provide cavitation samples with well-specified dimensions, shown in Figure 3.

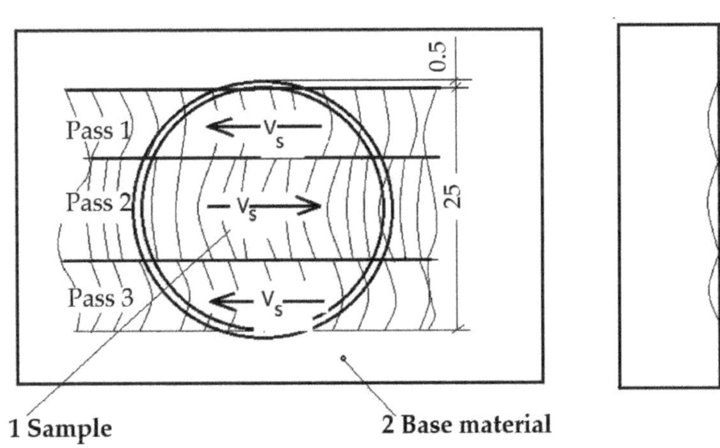

Figure 5. Details of the surface TIG remelting process.

The cavitation tests were conducted on sets of three samples for each remelting regime, using a vibrating device with piezoceramic crystals (Figure 6) [7,8] made in accordance with the requirements of Standard ASTM G32-2016 regarding the indirect testing method [19].

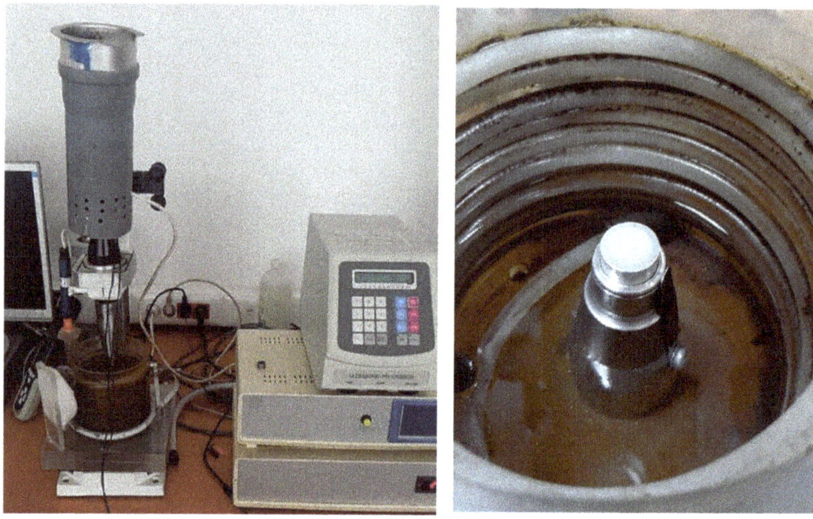

(**a**) (**b**)

Figure 6. Experimental stand: (**a**) overview; (**b**) detail of the sample-fixing system.

The functional parameters of the device were:

- vibration amplitude, 20,000 ± 1% Hz;
- vibration amplitude, 50 μm;
- power of the electronic ultrasound generator, 500 W;
- working environment, potable water having a temperature of 22 ± 1 °C.

A characteristic of this device is the control and constant maintenance of the acoustic and electrical parameters, with the help of a computer, based on a software implemented for this purpose.

Before the cavitation test, the attack surface of each sample was polished on a Buehler Phoenix Beta machine (Spectrographic Ltd., Shipley, UK) to a roughness Ra = 0.2 ÷ 0.8 μm. The total testing duration of each sample was 165 min, this being divided into 12 periods (one of 5 min, one of 10 min, and the next 10 periods of 15 min each). At the end of each test period, the sample was cleaned in acetone and air-dried.

Prior to the start of the tests and at the end of each intermediate test period, the cavitation exposed surfaces were examined under an optical Olympus SYX7 (Olympus, Zhengzhou, China) microscope and photographed with a high-resolution Canon Power Shot A 480 camera (Canon, Tokyo, Japan) to monitor the surface damage exposed to the cavitation attack. The samples were weighed before starting the tests and at the end of each intermediate period. The weighing was carried out with an analytical Zatklady Mechaniki Precyzyjnej WP 1 (Mechaniki Precyzyjnej R&G S.A., Mielec, Poland) balance, whose accuracy was of 5 significant decimals (up to 0.00001 g).

At the end of each intermediate test period, "i", the corresponding mass loss was determined, Δm_i.

The eroded mass was established according to the relation:

$$m_i = \sum_{i=1}^{12} \Delta m_i \tag{1}$$

Later, based on the mass losses, the values of the cumulative mean penetration depth of erosion, $MDE_{\Sigma i}$ and the values of the erosion penetration rate related to the period "i" $MDER_i$ were determined [20,21]:

$$MDE_{\Sigma i} = \sum_{i=1}^{12} \Delta MDE_i = \sum_{i=1}^{i=12} \frac{4\,\Delta m_i}{\rho\,\pi\,d_p^2}\ [\mu m] \tag{2}$$

$$MDER_i = \Delta MDE_i / \Delta t_i\ [\mu m/min] \tag{3}$$

where:

i is the testing period;

Δm_i is the mass of material lost through erosion, in period i, in grams;

P is the material density, in grams/mm^3;

Δt_i is the duration of cavitation corresponding to period "i" (5 min, 10 min or 15 min);

d_p is the diameter of the sample surface subjected to cavitation attack (d_p = 15.8 mm);

ΔMDE_i is the value of the mean penetration depth of erosion, achieved by cavitation during the period Δt_i.

At the end of the test, the cavitation exposed surface of the samples was examined by an optical Leica DM 2700 M microscope (Leica Microsystems, Madrid, Spain) and by the scanning electron microscope TESCAN VEGA 3 LMU Bruker EDX Quantax (Bruker Corporation, Billerica, MA, USA).

3. Results and Discussion

3.1. Cavitation Erosion Curves

In Figures 7 and 8, the characteristics of the cavitation erosion curves are shown comparatively and indicate the variation of the parameters MDE (mean penetration depth of erosion) and MDER (mean penetration rate of erosion) with the duration of the vibrating cavitation attack, for the three structural states obtained after local surface remelting as well as for the reference material subjected to the solution-based heat treatment followed by artificial aging. It is specified that the experimental values in these diagrams represent the mean values of the three tested sets of samples, for each type of material processing.

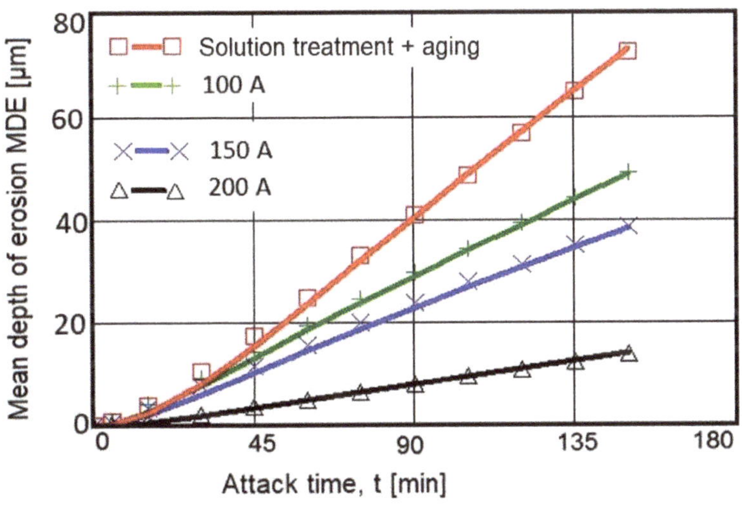

Figure 7. Evolution of the mean penetration depth of erosion with the cavitation attack time.

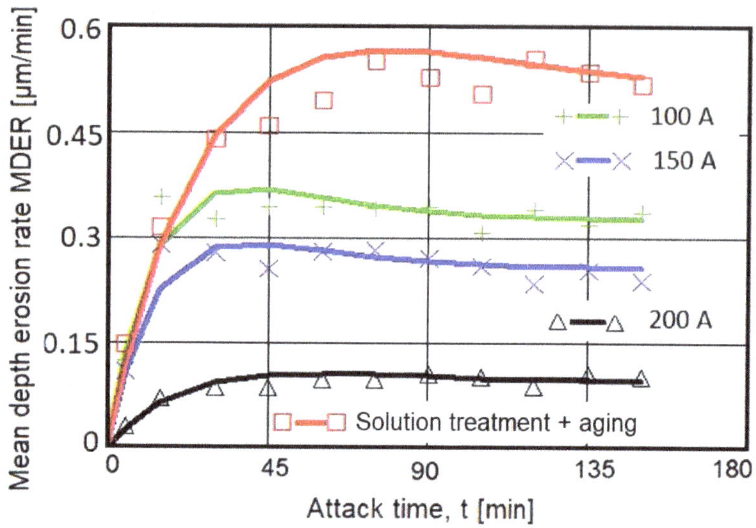

Figure 8. Evolution of the mean erosion rate with the cavitation attack time.

The following observations resulted from the analysis of these graphs:

- TIG remelting of the considered alloy surface at the currents Is = 100 A, 150 A, and 200 A caused an increase in the cavitation erosion resistance, Rcav, from 1.5 times to 5 times (where Rcav. = 1/MDER, t = 165), compared to the conventional heat treatment specific for this material;
- the use of welding currents of 200 A provided the lowest values of the MDE and MDER;
- during the stabilization period, until the end of the test duration (165 min), the erosion rates of the remelted TIG surfaces acquired values of approx. 0.10–0.32 µm/min, and those specific to the age hardening heat treatment were approx. 0.50 µm/min.

3.2. *Macro- and Micrographic Examinations*

3.2.1. Macrographs of the Surfaces Tested for Cavitation

With the help of a Canon Power Shot A480 camera, images of the cavitation exposed surfaces were obtained at each test time. Figure 9 shows these characteristic images for each value of melting current with respect to the linear energy. The energetic impact generated by the collapse of the cavitation bubbles made the samples surface uneven. Up to 30 min of attack, no surface degradation was observed. At longer times, the formation of a ring from the surface periphery together with more and more microcraters (pinches) with variable sizes and uneven distribution was revealed. Thus, for cavitation attack times greater than 90 min, the remelted samples surfaces, especially with Is = 100 A and Is = 150 A, became very rough, which implies a strong damage to them. On the contrary, the degree of surface damage of samples remelted with Is = 200 A was lower, which proved that their cavitation erosion resistance was significantly improved. The main reason for this phenomenon can be the existence of fine grains and a microstructure with a uniform distribution of the constituent phases. After cavitation erosion for 165 min, the pinches on the material surface developed into an irregular group of cavitation craters.

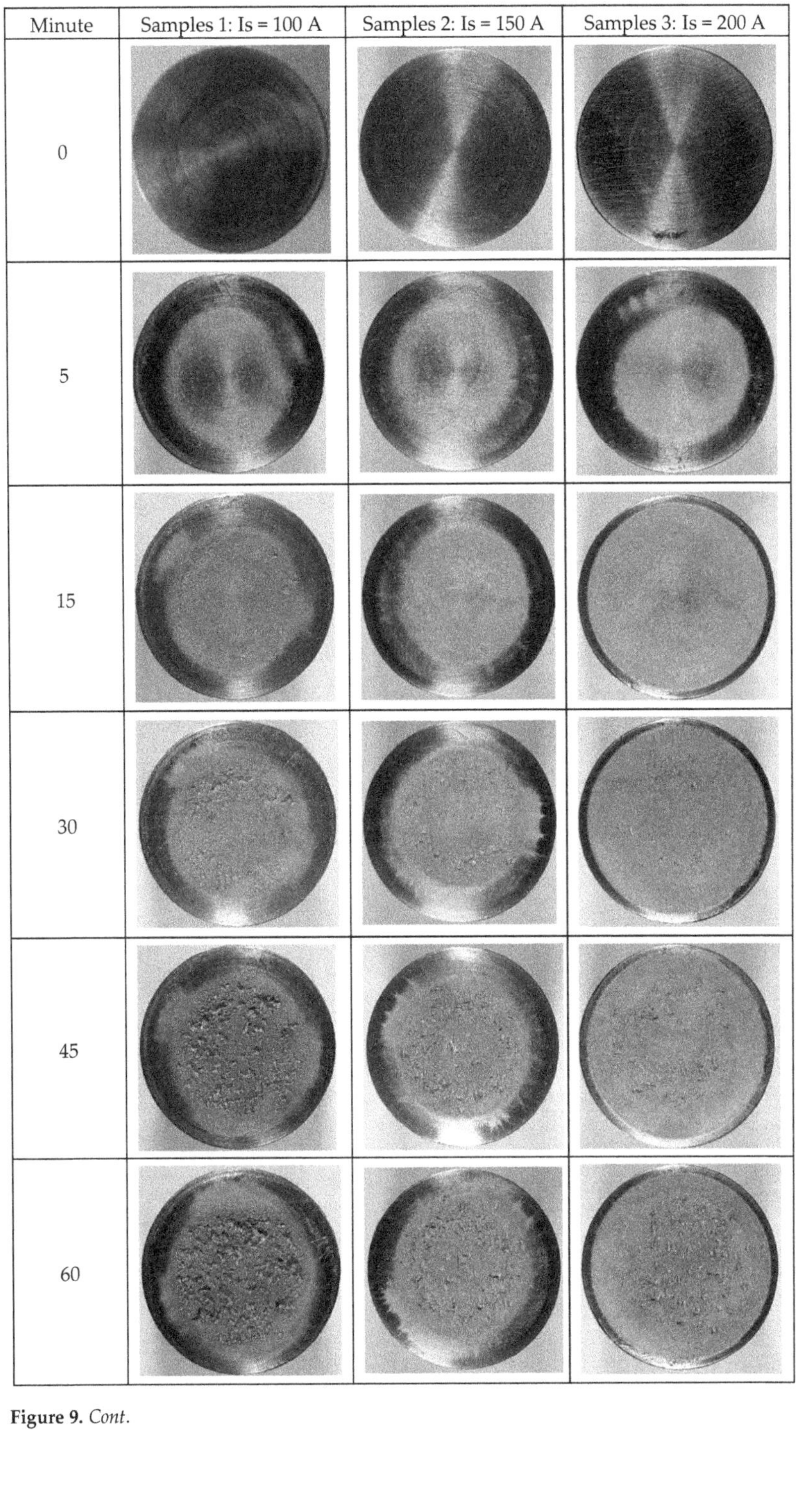

Figure 9. *Cont.*

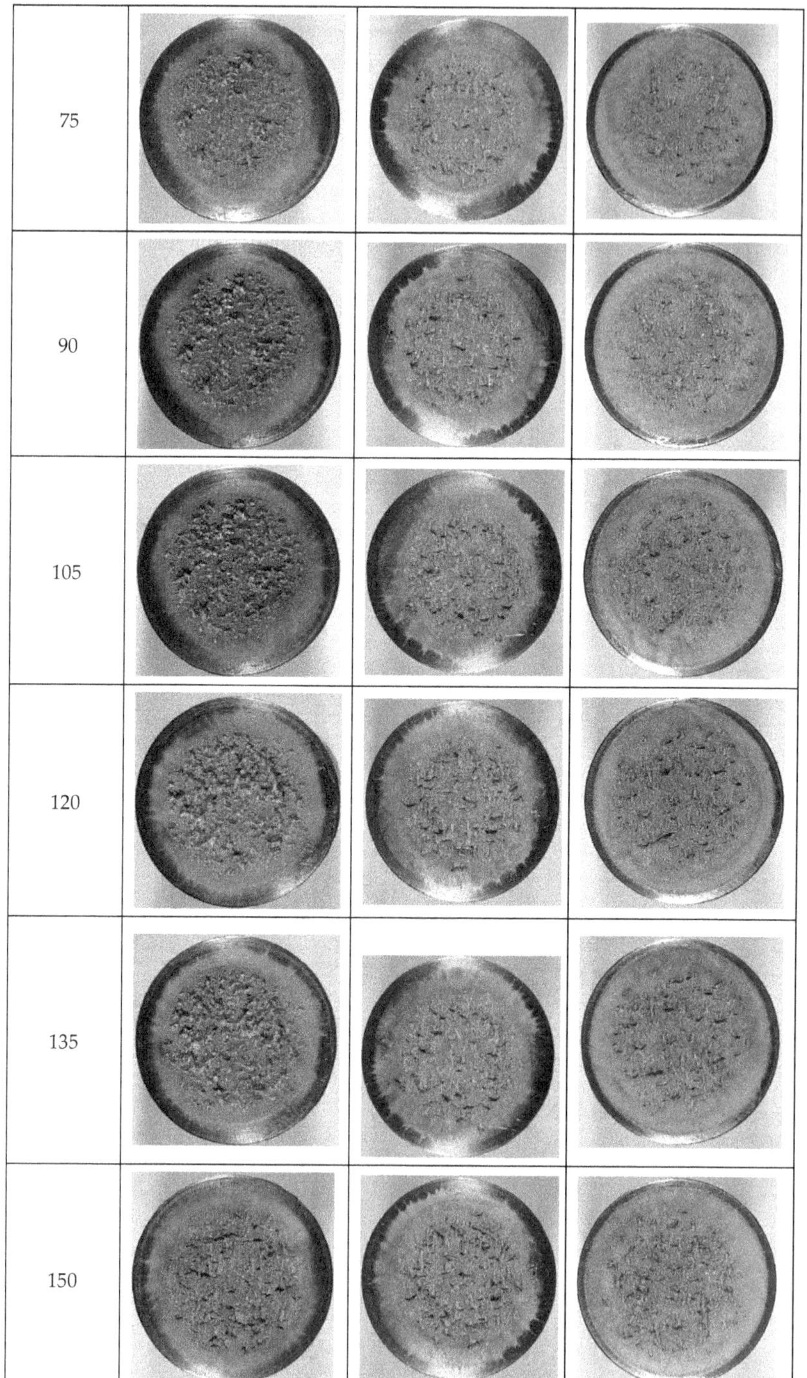

Figure 9. *Cont.*

165			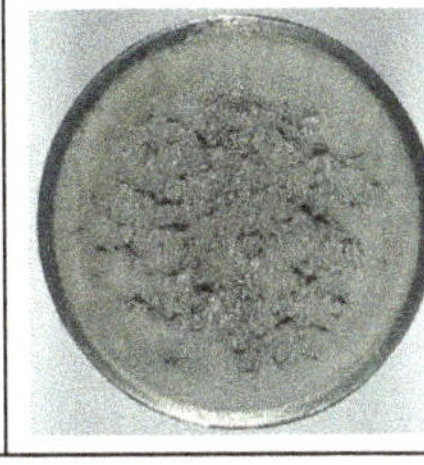

Figure 9. Macrographic images of the surface in cavitation tests of variable durations.

3.2.2. Micrograph of the TIG Remelted Layer

After the TIG remelting process, the surface of the material solidified at a higher degree of subcooling, with respect to that at a lower real temperature than the equilibrium temperature, so that the dimensions of the germs were smaller, their number was larger, and the microstructure obtained was finer. In Figure 10a,b the microstructure of the layer obtained at a current I = 200 A and a linear energy E_l = 15,840 J/cm is shown. It was observed that it contained small, equiaxial, unoriented crystals of α solid solution with an aluminum base and fine particles of intermetallic compounds [18], a microstructure that provides isotropic mechanical properties and an improvement of the use characteristics due to the strengthening mechanism resulting from the refinement of the granulation. The initiation and subsequent propagation of microcracks in the surface layer occurred at the interface between the particles derived from chemical combinations and the aluminum-based solid solution matrix (Figure 10b). Exposure of the material surface to repeated shock waves and/or repeated microjets led to an accumulation of plastic deformations until micro-damages occurred that caused material loss. With the increase of the cavitation attack time, there was a fragile removal of the phases of the chemical combinations, after which a uniform degradation of the solid solution grains took place, the breaking phenomenon having a ductile character.

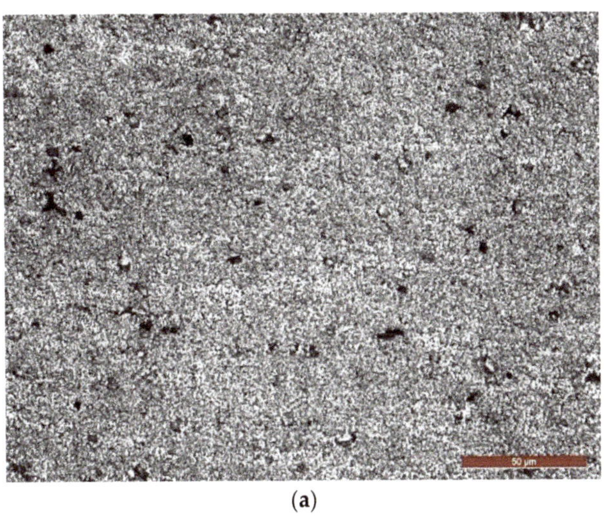

(a)

Figure 10. *Cont.*

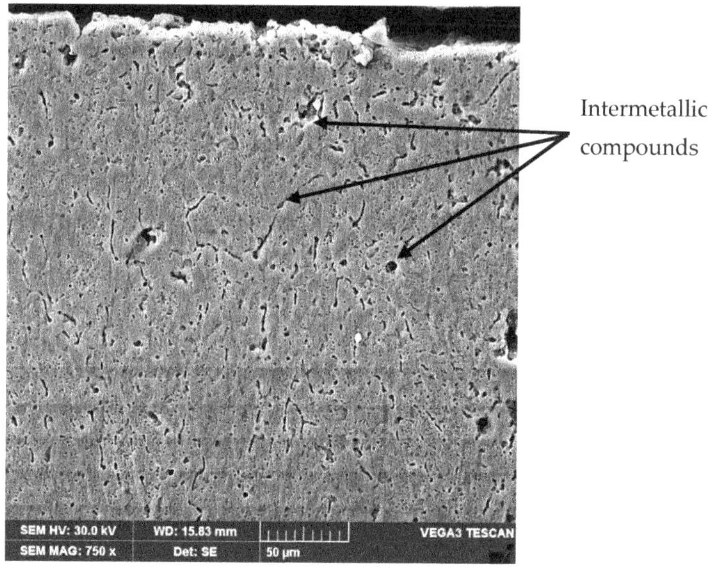

(**b**)

Figure 10. Microscopic images of a cross section through the remelted layer, with Is = 200 A: (**a**) OM ×200; (**b**) SEM ×750.

At melting current values over 200 A and linear energies higher than 15,840 J/cm, the very high thermal shock action induced by the surface melting process caused the appearance of microcracks in the edge layer (Figure 11).

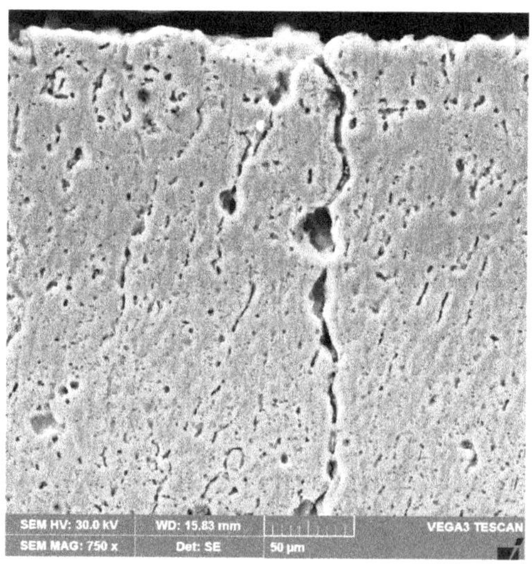

Figure 11. SEM image of a cross section with microcracks through the remelted layer, with Is = 250 A.

The micro-hardness of the alloy subjected to heat treatment with a solution followed by artificial aging showed values of 112–120 HV 0.05, and that of the TIG remelted surfaces reached values of approx. 128–137 HV 0.05.

3.2.3. Topography of Surfaces Eroded by Cavitation

A scanning electron microscope examination of the surfaces tested for cavitation (Figures 12 and 13) demonstrated that the erosion phenomenon occurred in a similar way to that of pure metals with a face-centered cubic lattice, fcc, namely, through plastic deformation followed by a ductile fracture over the entire surface.

(a)

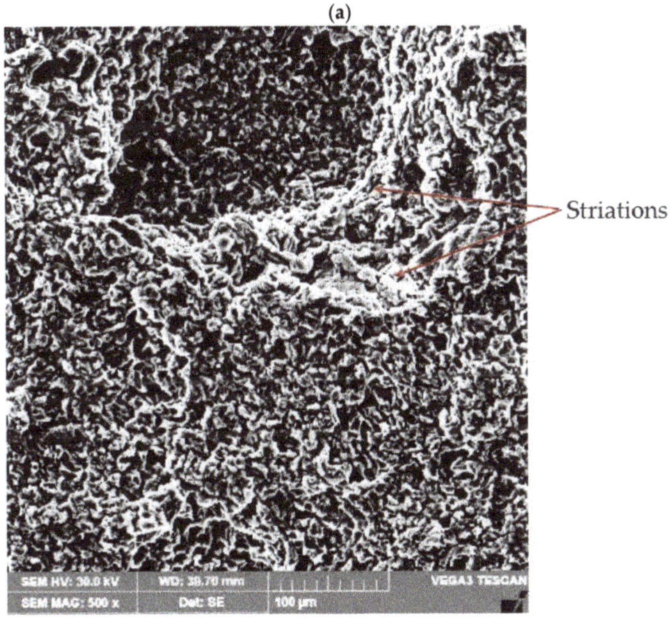

(b)

Figure 12. *Cont.*

(c)

Figure 12. SEM image of the remelted surface, with Is = 100 A, and tested after 165 min for cavitation: (**a**) ×20; (**b**) ×500; (**c**) ×2000.

(**a**)

Figure 13. *Cont.*

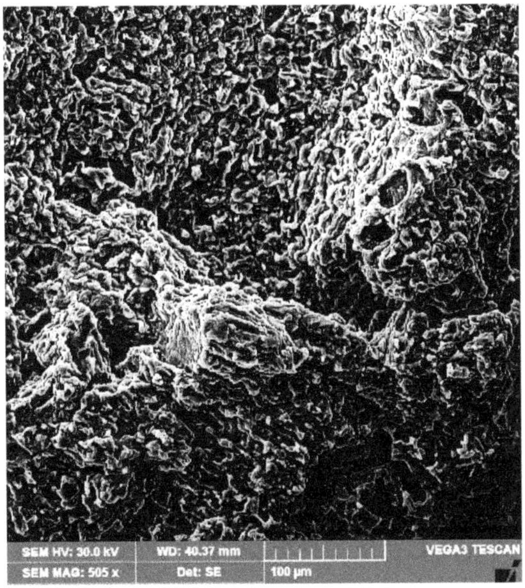

(**b**)

(**c**)

Figure 13. SEM image of the remelted surface, with Is = 200 A, and tested after 165 min for cavitation: (**a**) ×12: (**b**) ×500; (**c**) ×2000.

The results of the metallographic investigations carried out with the optical microscope and the scanning electron microscope, in correlation with the hardness measurements, allowed explaining the mechanism of the cavitation erosion phenomenon. The finishing of granulation and microstructure after the local surface remelting process, beside the hardness increase, were responsible for the cavitation erosion improvement, along with the material removal. The high capacity of mechanical embrittlement (low stacking fault energy) of the solid solution matrix with the fcc crystalline network led to erosion by ductile fracture on the entire surface. The loss of material occurred preferentially in isolated regions, apparently with a random distribution, the formed microcraters showing typical striations similar to the fatigue fractures of aluminum alloys [4,6,9,13].

However, with this alloy containing up to 3% of alloying elements, the material was preferentially lost in isolated regions, apparently randomly, leaving behind striated, flat-bottomed microcraters [1,2,4,7]. The fracture surface of these microcraters was similar to that of fatigue fractures in aluminum alloys. The microcraters expanded laterally to macroscopic sizes without undergoing changes in their topographical characteristics. The diameter of the cavitation microcraters was approximately 20 μm, and some of them reached 50 μm and even 80–90 μm.

4. Conclusions

Based on the investigations regarding the microstructure and cavitation erosion characteristics of TIG remelted surfaces of the EN AW-6082 alloy, the following conclusions can be summarized:

The process of local surface remelting using the TIG electric arc, operated at currents of 100, 150, and 200 A and voltage of 10 V, led to grain and microstructure refinement in the investigated alloy.

It was highlighted that for welding current values Is = 200 A, the cavitation erosion resistance of this alloy increased by a factor of approximately 200%.

The surface microhardness of the structurally modified layers increased from approx. 115 HV0.05 to approx. 134 HV0.05.

The fracture surface of the microcraters formed on the hollow surfaces had a ductile character, similar to the fatigue fractures of aluminum alloys.

Author Contributions: Conceptualization, I.M. and I.B.; methodology, I.M., I.B., F.F., I.-D.U., C.M.C. and C.G.; formal analysis, I.M., I.B. and I.-D.U.; investigation, I.M., I.B., I.-D.U. and C.M.C.; resources, I.M. and C.M.C.; data curation, F.F. and C.G.; writing—original draft preparation, I.M.; writing—review and editing, I.M., I.B., I.-D.U. and C.M.C.; visualization, I.M., I.B., F.F., I.-D.U., C.M.C. and C.G.; supervision, I.M. and I.B. All authors have read and agreed to the published version of the manuscript.

Funding: This research received support from the EEA Grants 2014–2021, under Project contract no. 2/2019 CoDe-PEM (EEA RO-NO-2018-0502).

Institutional Review Board Statement: Not applicable.

Informed Consent Statement: Not applicable.

Data Availability Statement: Not applicable.

Acknowledgments: The research leading to these results has received support from the EEA Grants 2014–2021, under Project contract no. 2/2019 CoDe-PEM (EEA RO-NO-2018-0502).

Conflicts of Interest: The authors declare no conflict of interest.

References

1. Pola, A.; Montesano, L.; Tocci, M.; Marina La Vecchia, G. Influence of Ultrasound Treatment on Cavitation Erosion Resistance of AlSi7 Alloy. *Materials* **2017**, *10*, 256. [CrossRef] [PubMed]
2. Tomlinson, W.J.; Matthews, S.J. Cavitation erosion of aluminium alloys. *J. Mater. Sci.* **1994**, *29*, 1101–1108. [CrossRef]
3. Dybowski, B.; Szala, M.; Kiełbus, A.; Hejwowski, T. Microstructural phenomena occurring during early stages of cavitation erosion of Al-Si aluminium casting alloys. *Solid State Phenom.* **2015**, *227*, 255–258. [CrossRef]
4. Pola, A.; Montesano, L.; Sinagra, C.; Gelfi, M.; La Vecchia, G.M. Effect of globular microstructure on cavitation erosion resistance of aluminium alloys. *Solid State Phenom.* **2016**, *256*, 51–57. [CrossRef]
5. Chen, H.; Li, J.; Chen, D.; Wang, J. Damages on steel surface at the incubation stage of the vibration cavitation erosion in water. *Wear* **2008**, *265*, 692–698.
6. Gottardi, G.; Tocci, M.; Montesano, L.; Pola, A. Cavitation erosion behaviour of an innovative aluminium alloy for Hybrid Aluminium Forging. *Wear* **2018**, *394–395*, 1–10. [CrossRef]

7. Mitelea, I.; Bordeasu, I.; Cosma (Alexa), D.; Utu, I.D.; Craciunescu, C.M. Microstructure and Cavitation Damage Characteristics of GX40CrNiSi25-20 Cast Stainless Steel by TIG. *Materials* **2023**, *16*, 1423. [CrossRef] [PubMed]
8. Mitelea, I.; Bordeasu, I.; Frant, F.; Utu, I.D. Effect of heat treatment on corrosion and ultrasonic cavitation erosion resistance of AlSi10MnMg alloy. *Mater. Test.* **2020**, *62*, 921–926.
9. Tong, Z.; Jiao, J.; Zhou, W.; Yang, Y.; Chen, L.; Liu, H.; Sun, Y.; Ren, X. Improvement in cavitation erosion resistance of AA5083 aluminium alloy by laser shock processing. *Surf. Coat. Technol.* **2019**, *377*, 124799. [CrossRef]
10. Li, H.; Cui ZLi, Z.; Zhu, S.; Yang, X. Effect of gas nitriding treatment on cavitation erosion behavior of commercially pure Ti and Ti-6Al-4V alloy. *Surf. Coat. Technol.* **2013**, *22*, 29–36. [CrossRef]
11. Kwok, C.T.; Man, H.C.; Cheng, F.T.; Lo, K.H. Developments in laser-based surface engineering processes: With particular reference to protection against cavitation erosion. *Surf. Coat. Technol.* **2016**, *291*, 189–204. [CrossRef]
12. Zhang, S.; Wu, S.; Cui, W.; He, S.; Zhang, C.; Guan, M. *Cavitation Erosion Properties of Ni-Based RE Alloy Coating on Monel Alloy by Laser Cladding*; Northwest Institute for Nonferrous Metal Research: Xi'an, China, 2018.
13. Man, H.C.; Kwok, C.T.; Yue, T.M. Cavitation erosion and corrosion behaviour of laser surface alloyed MMC of SiC and Si_3N_4 on Al alloy AA6061. *Surf. Coat. Technol.* **2000**, *132*, 11–20. [CrossRef]
14. Grewal, H.S.; Arora, H.S.; Singh, H.; Agrawal, A. Surface modification of hydroturbine steel using friction stir processing. *Appl. Surf. Sci.* **2013**, *268*, 547–555. [CrossRef]
15. Selvam, K.; Mandal, P.; Grewal, H.S.; Arora, H.S. Ultrasonic cavitation erosion-corrosion behavior of friction stir processed stainless steel. *Ultrason. Sonochem.* **2018**, *44*, 331–339. [CrossRef] [PubMed]
16. Lathabai, S.; Ottmüller, M.; Fernandez, I. Solid particle erosion behaviour of thermal sprayed ceramic, metallic and polymer coatings. *Wear* **1998**, *221*, 93–108. [CrossRef]
17. Wood, R.J.K.; Mellor, B.G.; Binfield, M.L. Sand erosion performance of detonation gun applied tungsten carbide/cobalt-chromium coatings. *Wear* **1997**, *211*, 70–83. [CrossRef]
18. Mrówka-Nowotnik, G.; Sieniawski, J. Influence of heat treatment on the micrustructure and mechanical properties of 6005 and 6082 aluminium alloys. *J. Mater. Process. Technol.* **2005**, *162–163*, 367–372. [CrossRef]
19. *ASTM G 32 Standard*; Standard Test Method for Cavitation Erosion Using Vibratory Apparatus. ASTM International: West Conshohocken, PA, USA, 2016.
20. Bordeaşu, I.; Patrascoiu, C.; Badarau, R.; Sucitu, L.; Popoviciu, M.; Balasoiu, V. *New Contributions in Cavitation Erosion Curves Modeling*; FME Transactions Faculty of Mechanical Engineering, New Series, Nr.1/2006; University of Belgrade: Belgrade, Serbia, 2006; Volume 34, pp. 39–44, ISSN 1451-2092.
21. Bordeasu, I. *Eroziunea Cavitationala a Materialelor*; Editura Politehnica: Timisoara, Romania, 2006.

materials

MDPI

Article

Microstructure and Cavitation Damage Characteristics of GX40CrNiSi25-20 Cast Stainless Steel by TIG Surface Remelting

Ion Mitelea [1], Ilare Bordeaşu [2], Daniela Cosma (Alexa) [1], Ion-Dragoş Uţu [1] and Corneliu Marius Crăciunescu [1,*]

1 Department of Materials and Fabrication Engineering, Politehnica University Timisoara, Bulevardul Mihai Viteazul nr.1, 300222 Timisoara, Romania
2 Department of Mechanical Machines, Equipment and Transports, Politehnica University Timisoara, Bulevardul Mihai Viteazul nr.1, 300222 Timisoara, Romania
* Correspondence: corneliu.craciunescu@upt.ro

Abstract: Cavitation erosion degrades the surface of engineering components when the material is exposed to turbulent fluid flows. Under conditions of local pressure fluctuations, a nucleation of gas or vapor bubbles occurs. If the pressure suddenly drops below the vapor pressure, these bubbles collapse violently when subjected to higher pressure. This collapse is accompanied by the sudden flow of the liquid, which is manifested by stress pulses capable of causing plastic deformations on solid surfaces. Repeating these stress conditions can cause material removal and ultimately failure of the component itself. The present study aims to reduce the negative impact of this phenomenon on the mechanical systems components, using the TIG local surface remelting technique. Cavitation erosion tests were performed in accordance with the ASTM G32-2016 standard on samples taken from a cast high-alloy stainless steel. The alloy response for each melting current value was investigated by measuring mass loss as a function of cavitation attack time and by analyzing the damaged surfaces using optical and scanning electron microscopes. It was highlighted that the TIG remelted layers provide an increase in cavitation erosion resistance of 5–6 times as a consequence of the fine graining and microstructure induced by the technique applied.

Keywords: high Cr-Ni-Si cast stainless steel; cavitation erosion; microstructure; TIG surface remelting

Citation: Mitelea, I.; Bordeaşu, I.; Cosma, D.; Uţu, I.-D.; Crăciunescu, C.M. Microstructure and Cavitation Damage Characteristics of GX40CrNiSi25-20 Cast Stainless Steel by TIG Surface Remelting. *Materials* **2023**, *16*, 1423. https://doi.org/10.3390/ma16041423

Academic Editors: Feng Qiu and Francesco Iacoviello

Received: 11 January 2023
Revised: 3 February 2023
Accepted: 6 February 2023
Published: 8 February 2023

1. Introduction

Cavitation erosion manifests itself by reducing the lifetime of numerous components that belong to mechanical systems. Degradation of parts occurs through the appearance of microcracks that develop into microcraters (pits), and finally through localized material loss [1–4].

A series of studies have been carried out to investigate the specific cavitation phenomena and the relevant physical variables [5–10]. More precisely, under certain conditions of pressure applied to a liquid, vapor bubbles can form, and following their implosion, shock waves and microjets are born that cause damage to adjacent solid surfaces [1,11–15].

Although the physical causes of cavitation erosion are well known, the response of the materials from which engineering components are made is a less complete field, in part because of the wide variety of possible and available materials [16–18].

Cavitation experiments were performed on numerous alloys, comparing and compiling their results in terms of the mass loss of samples related to the time. Correlations between these measurements and some material properties were only partially successful [1,3]. In addition, the initiation and progression of damage mechanisms are not yet fully understood. Therefore, the purpose of this study is to analyze the evolution of cavitation damage for TIG remelted surfaces of a cast, high alloy steel. This process of local surface remelting offers the advantage of simplicity, flexibility and low cost. It has many potential applications in fields such as aerospace, automobiles, nuclear reactors, ships, civil parts, etc. The material GX40CrNiSi25-20 is a cast steel, used for the manufacture of elements used for high mechanical stress and for cavitational or high temperature working environments. Among them are automotive industry: (turbochargers, manifolds, valves) and plant engineering (for oil and gas, valve bodies) [1,19].

2. Experimental Procedure

From cast steel bars, GX40CrNiSi 25–20, (1.4848): EN 10295 (heat treated by annealing for homogenization 1100 °C/8 h/furnace), of the chemical composition C = 0.38%, Cr = 25.20%, Ni = 20.8%, Si = 1.62%, Mn = 1.49%, Mo = 0.34%, p = 0.031%, S = 0.027%, Fe = bal.%, cylindrical samples were made, ϕ 40 mm × 60 mm. Their surface was modified using the local TIG remelting technique, with a non-fusible electrode and using welding currents of different intensities (Is = 100 A; Is = 150 A; Is = 200 A). The other technological parameters were kept constant: electric arc voltage, Ua = 10–11 V; welding speed, vs. = 15 cm/min; electric arc length, L = 2 mm; the step between rows, p = 3 mm. The TIG welding equipment used for the local surface remelting of the front cylindrical samples ϕ 40 mm × 60 mm was a MW-300 inverter source from the Fronius company. Later, according to the model shown in Figure 1, cavitation samples were machined, and the expose surface was polished, on a Buehler Phoenix Beta device to a roughness Ra = 0.2–0.8 μm.

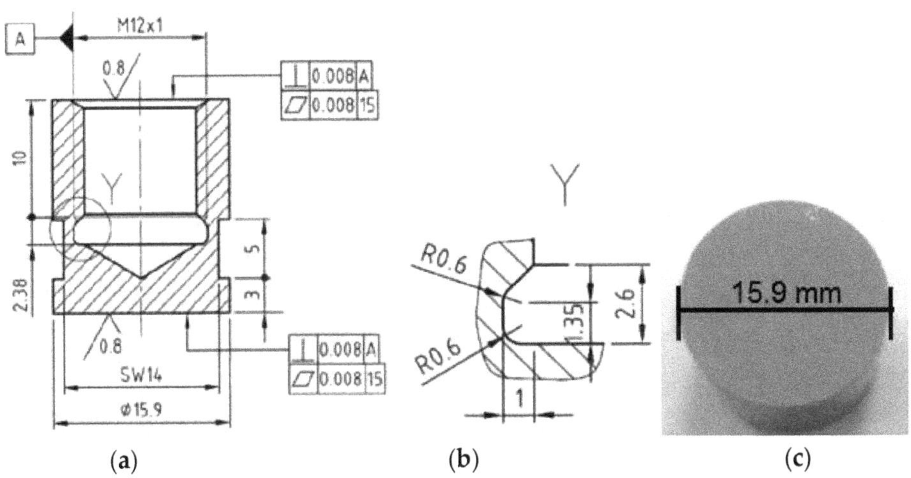

Figure 1. Cavitation sample: manufacturing details: (**a**) axial section through the sample, (**b**) detail of the Y area (thread clearance), (**c**) photo image of the sample surface, before the start of the cavitation attack (dimensions are in mm, according to general tolerance ISO 2768 1&2 standard).

The experimental researches were carried out in the cavitation laboratory of the Timișoara Polytechnic University, on the vibrating device with piezoceramic crystals (Figure 2), which is controlled by a computer and equipped with an automatic control system of the functional parameters that define the cavitation hydrodynamic process. This type of device allows for the testing of different types of materials at intense cavitation erosion regimes, thus considerably reducing the samples attack duration, compared to existing situations, or using other methods of generating the cavitation phenomenon (the case of the tunnel hydrodynamic [20] or of the rotating disk device [21–23]).

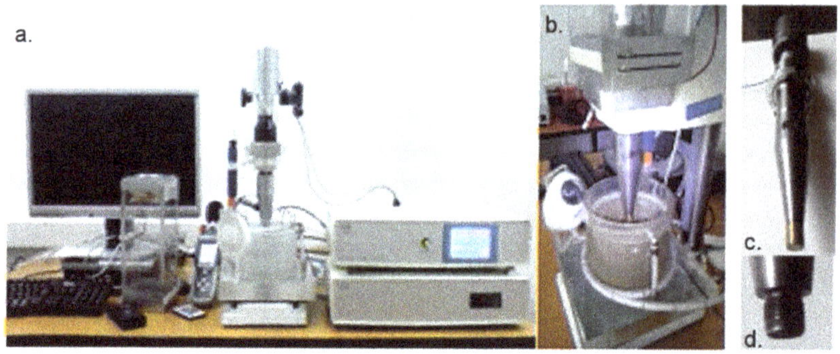

Figure 2. The vibrating device with piezoceramic crystals: (**a**) overview of the device with the computer control system; (**b**) image during the cavitation test (sample immersed in water); (**c**) mechanical vibration system (piezoceramic transducer with amplitude amplification system (buster, sonotrode and cavitation sample); (**d**) threaded end of the sonotrode for fixing the sample.

The functional parameters of the device are:

- power of the electronic ultrasound generator, 500 W;
- vibration frequency, 20,000 ± 2% Hz;
- vibration amplitude, 50 μm;
- sample diameter: 15,9 ± 0.05 mm;
- power supply: 220 V/50 Hz;
- testing liquid: distilled water, having a temperature of 22 ± 1 °C.

The research procedure is the one described by the international standards ASTM G32-2016 [24], and the testing stages specific to the laboratory consisted of:

- weighing each sample and recording the initial mass (first on an electronic balance and then on a precision analytical balance, type Zatklady Mechaniki Precyzyjnej WP 1);
- fixing the sample in the sonotrode by threading;
- fixing the sonotrode in the vibrator support and connecting the piezoceramic transducer to the power source;
- immersing the sample in the vessel with the testing liquid (distilled or potable water from the network), to a depth of 5–10 mm;
- setting up the intermediate cavitation attack duration (5, 10 and 15 min, respectively) and starting the electronic ultrasound generator simultaneously with the cooling water recirculation system from the copper coil in the testing liquid vessel;
- after the end of each time period allocated to the cavitational attack, every sample was washed in acetone solution and dried with a blower, after which it was weighed on the electronic and precision balance to determine the eroded mass;
- after each test period, every sample was photographed using a Canon PowerShot Sx200 IS camera, 12 × Optical Zoom, whose resolution allows the damage expanding on the surface to be highlighted; they were also analyzed using an optical microscope.

According to laboratory procedure, the total duration of the cavitation attack was 165 min, divided into one period of 5 and 10 min, and 10 periods of 15 min each.

Both at intermediate attack durations and at the end of the 165 min on the degraded surfaces, the surface topography was investigated using both the optical stereomicroscope OLIMPUS SYX7 and the scanning electron microscope TESCAN VEGA 3 LMU Bruker EDX Quantax. Following the cavitation tests, the samples were longitudinally sectioned (Figure 3) and metallographically prepared for examination using both the Leica DM2700M optical microscope and the scanning electron microscope of the edge layer where erosion cracks are initiated and propagated by cavitation.

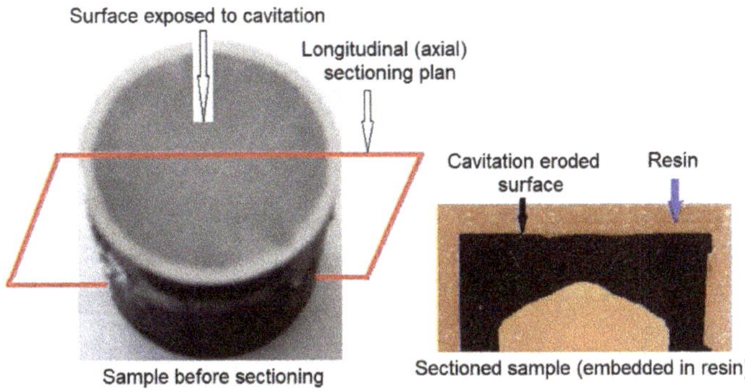

Figure 3. Sample preparation for optical and scanning electron microscopy investigation.

3. Results and Discussion

3.1. Cavitation Curves

The evolution of the mass losses and the erosion rate according to the cavitation attack time for the three values of the surface melting current are shown in Figures 4 and 5. For comparison, Figures 6 and 7 show the same types of curves for the samples that were not surface remelted.

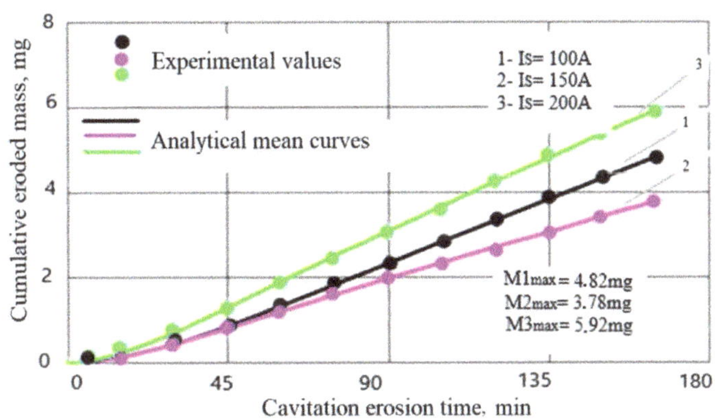

Figure 4. Mass losses of GX40CrNiSi25-20 steel at different remelting current intensities.

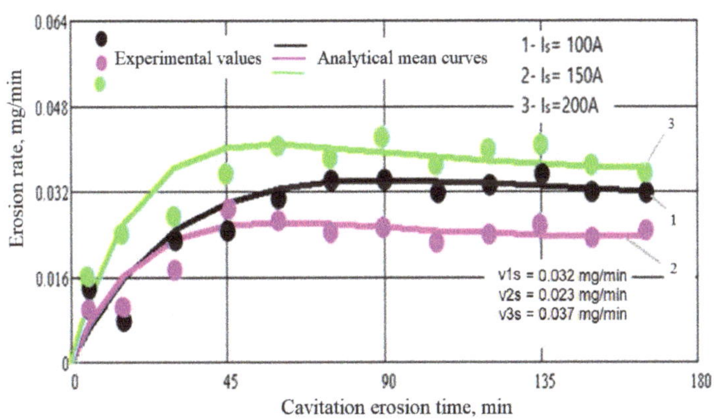

Figure 5. Erosion rate of GX40CrNiSi25-20 steel at different remelting current intensities.

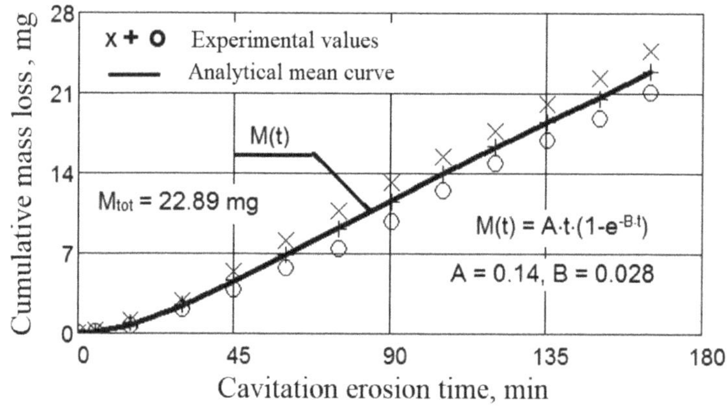

Figure 6. Mass losses of GX40CrNiSi25-20 steel related to the cavitation time (no remelting).

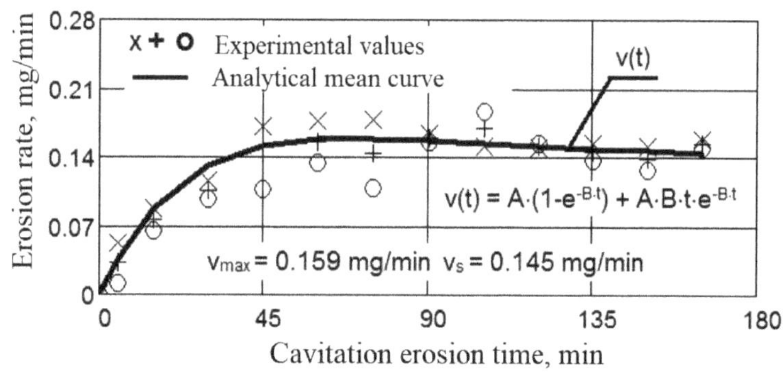

Figure 7. Erosion rate of GX40CrNiSi25-20 steel related to the cavitation time (no remelting).

The TIG process of local surface remelting achieves the modification of the microstructure without adding a new material. Following its application, it was expected to achieve homogenization and finishing of the microstructure and dissolution/redistribution of precipitates or inclusions while the properties of the substrate can be preserved. As can be observed from these diagrams, the GX40CrNiSi25-20 steel with the TIG remelted surface where an intensity current of Is = 150 A was used (samples 2 Figure 4), has the lowest mass loss compared to the same, but unmelted, steel (samples from Figure 6). For values of the remelting current of 150 A, the most favorable hardening situation occurs both through the granulation finishing and through the formation of a solid solution. The heat cycle characteristic for this remelting current value favors the precipitation of fine secondary carbides from the austenite and prevents the formation of eutectic carbides. Implicitly, the same can be said about the erosion rate, which is lower when the steel surface is remelted with a current of intensity Is = 150 A. In the case of the samples where the current of 150 A was used (samples 2 Figure 5), the mass loss is 27% lower than when using a current of 100 A (Sample 1, Figure 4) and 56% lower than when remelting with a current of 200 A (Sample 3, Figure 4). At the same time, compared to the regime with Is = 100 A, the decrease in erosion rate is over 39%, and compared to the regime with Is = 200 A, the decreasing in erosion rate is about 60%. The total cumulative mass loss for the regime with Is = 150 A (Mtot. = 3.78 mg), compared to the initial state (without remelting, Mtot. = 22.89 mg) decreases by more than six times, and the mass loss rate, according to the value of towards which the curve v(t) tends to stabilize, (vs = 0.145 mg/min. for the initial non-remelted surface, respectively vs = 0.029 mg/min. for the remelted surface at a current of 150 A), is reduced by about five times.

The approximation of the experimental values by the curves described by analytical equations, both for the mass losses (Figures 4 and 6) and for the related erosion rates (Figures 5 and 7), was necessary because, according to ASTM requirements G32-2016, they offer a behavior tendency and evaluation, by comparison, of the surface resistance, depending on the parameters values defined by them (Mtot.—after 165 min of cavitation attack, vmax.—the maximum of the curve v(t) and vs—final level value towards which the v(t) curve tends).

The analytical formulas were established by Pătrășcoiu a.o [20], starting from a model developed by Bordeașu et al., in 2006 [25], using a "damped", according to which the loss curve (such as the one for loss volumes—used as an example of Patrăscoiu) is given in the equation:

$$M(t) = A[m_s - f(t)] \tag{1}$$

where m_s represents the maximum value of mass losses, and f(t) is the solution of homogeneous linear differential equations of degree 2 with constant coefficients:

$$\frac{d^2m}{dt} + \frac{dm}{dt} + \beta^2 m \tag{2}$$

which describes "damped" oscillation with "infinite period" and m—represents the eroded mass in time t.

Solving this last equation, considering as a variable the cumulative mass eroded during the experiment (as it is in our case), with the mathematical equation of Pătrășcoiu a.o [20,25], applying the method of least squares, the analytical form of the averinging curve of the experimental values, as follows:

$$M(t) = A \cdot t \cdot (1 - e^{-B \cdot t}) \tag{3}$$

through the derivation in relation to the exposure cavitation time, the equation of mass loss rate (erosion rate) was obtained:

$$v(t) = A \cdot (1 - e^{-B \cdot t}) + A \cdot B \cdot t \cdot e^{-B \cdot t} \tag{4}$$

Coefficients A and B were obtained by the method of least squares.

Comparing the results shown in Figures 4 and 7, it can be seen that the decrease in the cumulative masses from the end of the cavitational attack, in the case of the TIG electric arc remelting process, led to an increase in the cavitation resistance, the erosion rate and the cumulative mass losses being approximately 5–6 times higher in the unmelted samples:

$$\frac{M_{max}}{M_{2max}} = \frac{22.89}{3.78} = 6.05$$

$$\frac{v_s}{v_{2s}} = \frac{0.145}{0.029} = 5$$

M_{2max} și v_{2s} are the values of mass losses and rates shown in the diagrams from Figures 4 and 5. In the above relations Mmax., Mtot., Mtot. and vs. are the values defined by the M(t) and v(t) curves recorded for GX40CrNiSi25-20 steel with a non-remelted surface.

3.2. Macro- and Micrographic Examinations

Using a Canon Power Shot SX200 IS camera, images of the samples surface were taken after each cavitation attack period to track the time evolution of erosion expanding in area and depth. Considering that the samples surfaces, from each set of three, had similar behaviors regarding degradation through cavitation erosion, in Figure 8 macroscopic images are selected, for three significant periods of time, at one of the samples, arbitrarily selected.

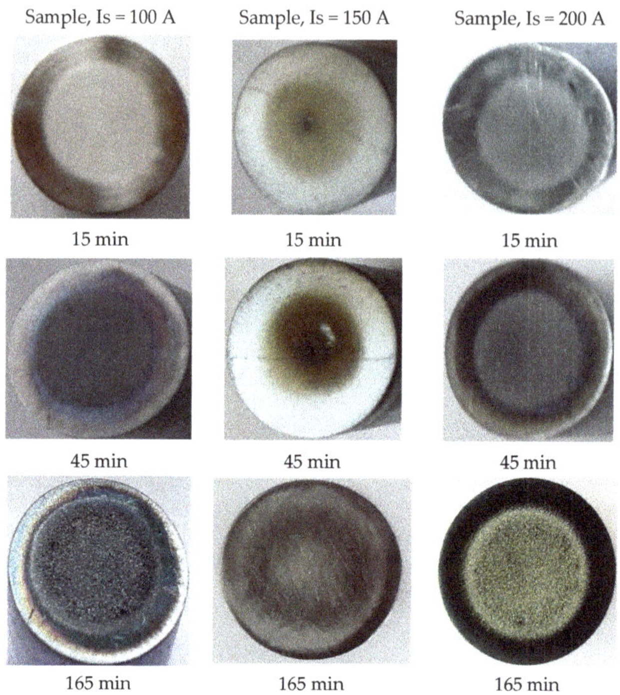

Figure 8. The macrographic image of the TIG surface remelted sample, with different current intensities, after the cavitation attack.

No images from the first two cavitation periods (of 5 and 10 min) were shown because, in this time range, the erosion mechanism consisted of the abrasive dust removal, the destruction of the roughness tip left after the finishing operation, deformations and cracks, without achieving important material expulsions.

The images shown in Figure 8 reveal common elements of behavior and resistance, but also differences in the cavitation surfaces destruction, as a result of the different TIG remelting parameters regime. The main observations that emerge are:

- regardless of the heat adopted regime, the erosion is manifested by the appearance of a peripheral ring, which advances towards the center surface with the increase in the duration of the cavitation attack; the damage is in the form of pitting, as a result of cyclic stresses, created by the impact with the cavitational microjets;
- as cavitation duration increases, the resulting surface roughness for the remelting current Is = 200 A is the most affected, showing that the use of this TIG regime leads to the lowest wear resistance;
- the best resistance to the vibrating cavitation demands is obtained by the remelted surface Is = 150 A, the destruction evolution in terms of area and depth being the most advantageous. The SEM image in Figure 9 proves this fact;
- the use of a remelting current Is = 100 A leads to an intermediate interaction between its other two values;
- the photographic images obtained are in full agreement with the results provided by the specific cavitation curves in Figures 4 and 5.

In Figure 10, for comparison, the photo images are shown, taken at certain times, of the eroded surface for the GX40CrNiSi 25-20 steel samples, tested in the initial state, and of annealing (without remelting the surface). Figure 11 shows the SEM image of the annealed sample surfaces tested for cavitation 165 min. The jagged appearance of the outer area in the cavitated surface is due to the intradendritic microsegregations of the alloying elements present in high concentrations in the steel chemical composition. Comparing the images from Figures 9 and 11 clearly demonstrates that the local TIG remelting leads to a significant improvement in the cavitation erosion resistance.

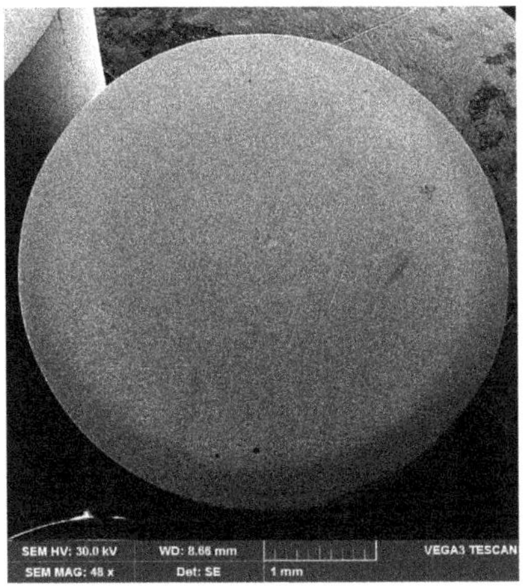

Figure 9. SEM macrograph of theTIG remelted surface with Is = 150 A and tested for cavitation 165 min.

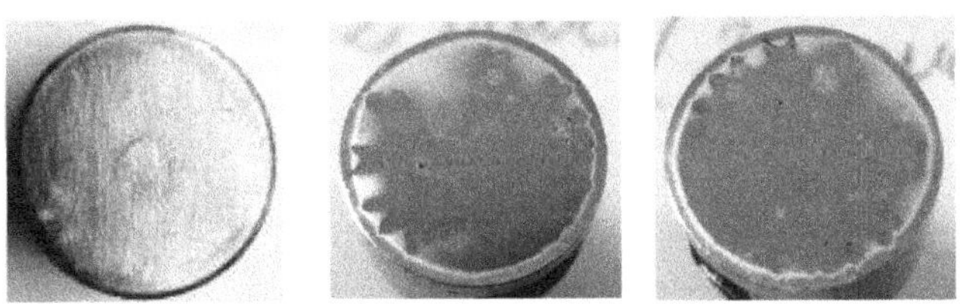

15 min 60 min 165 min

Figure 10. Macrographic image of the unmelted surface of GX40CrNiSi 25-20 sample.

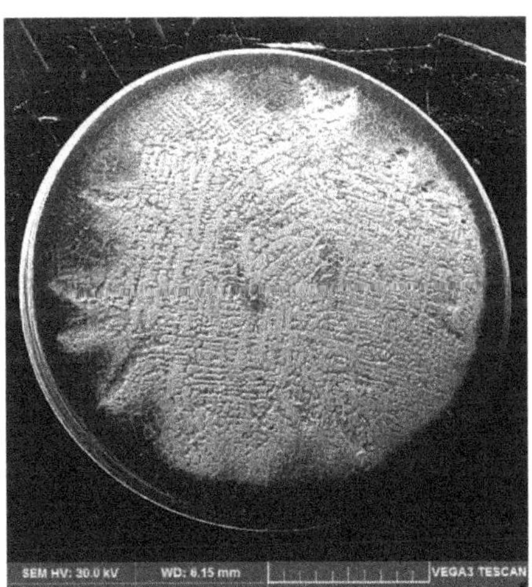

Figure 11. SEM macrograph of the unmelted surface tested by cavitation after 165 min.

The micrographic examination of the longitudinal section through the cavitation samples that were TIG remelted (Figure 12) proves that this surface-ennobling process is manifested by the accentuated grain finishing and implicitly of the microstructure (austenite + carbides), with consequences for the pronounced improvement of the resistance to cavitation erosion degradation.

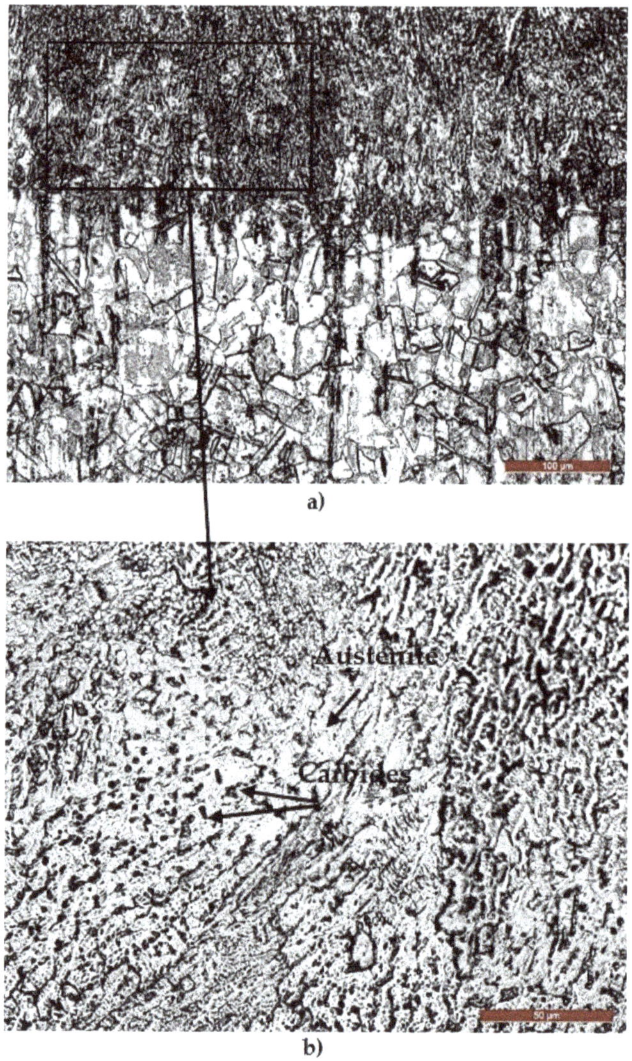

Figure 12. The microstructure of the layer—substrate system during the local TIG surface remelting with Is = 150 A: (**a**)—layer—substrate interface; (**b**)—remelted layer.

Figures 13 and 14 show some representative images that illustrate the material microstructure degradation in the annealed state, respectively, after the TIG surface remelting with a current of Is = 150 A.

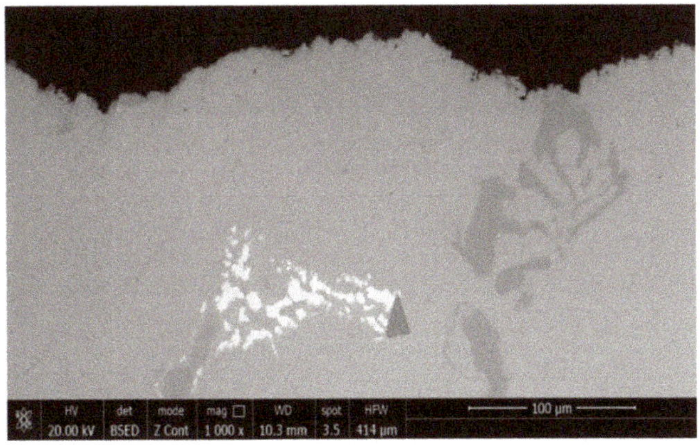

Figure 13. SEM microscopy of an unetched section through the surface of the reference samples (annealed).

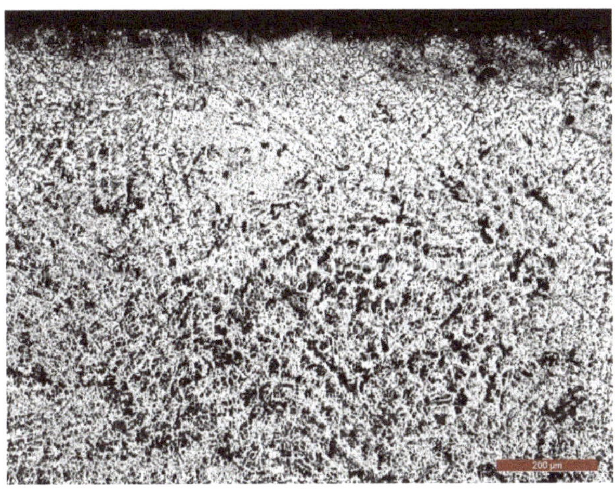

(**a**)

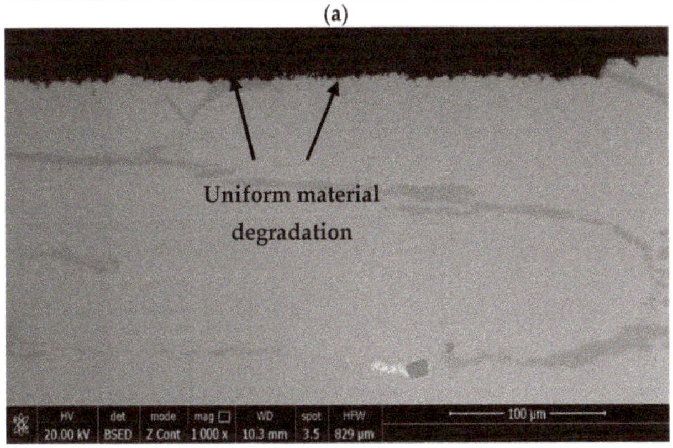

(**b**)

Figure 14. SEM microscopy through the remelted layer with Is = 150 A and after 165 min cavitation time: (**a**)—Villela etched; (**b**)—unetched.

They demonstrate that this form of degradation is predominantly mechanical and is characterized by a surface percussion as a result of the implosion of the gas bubbles in the liquid mass when its temperature is constant and the pressure drops to a certain critical value. Since the steel analyzed is part of a highly alloyed class, cast in pieces, it presents a heterogeneous structure (austenite + carbides) [1]

Both the carbides and the intradendritic microsegregations specific to any high-alloy steel negatively influence the mechanical properties and can constitute stress concentrators, respectively, and crack primers due to cavitational stresses.

From the presented micrographs, it can be seen that the initiation of degradation takes place on the boundaries of the austenite grains, where the carbide particles are present, microstructural constituents with the lowest resistance to cavitation. As erosion progresses, attack occurs in the solid solution matrix γ (austenite). Because of the accentuated granulation finishing and the microstructure through the local surface remelting, the appearance of the cavitation surface is much more uniform compared to the material reference state (Figures 13 and 14b).

3.3. *Topography of the Surfaces Tested for Cavitation*

The typical topographies of the cavitation surfaces for the samples heat-treated through annealing for homogenization and subsequently TIG surface remelted highlight a preferential degradation of the grains boundaries, where the hard and brittle carbide particles are present (Figure 15a,b). This surface modification technique leads to a structure with a high degree of fragmentation and finishing, which favors an increase in hardness and other mechanical resistance characteristics [13–15].

(**a**)

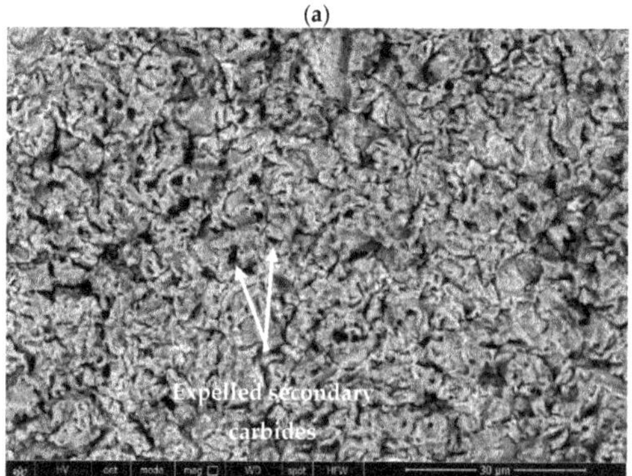

(**b**)

Figure 15. The microfractographic image of the cavition surface for 165 min. (**a**)—annealed structural state; (**b**)—TIG surface remelting with Is = 150 A.

The measured hardness of the annealed surfaces, which were tested in this work, has values of 207~210 HV, and the cavitation erosion rate was 0.145 mg/min. By the remelted TIG samples with Is = 150 A, the surface hardness has values of 386~394 HV, and the erosion rate through cavitation is 0.029 mg/min. Consequently, the cavitation erosion rate of the TIG surface remelted is approx. five times higher than that of the reference material. The removal of chromium carbide particles under the action of cavitation bubbles produces high stress concentrations.

Previous research [1,25–27] reported that in austenitic stainless steels, the interface between chromium carbides and austenite are associated with intense stresses. Interfaces cannot undergo work hardening and consequently become brittle and fragmented through cavitation attack. At the same time, the localized strength of the austenite structural matrix increases with the degree of cold deformation during cavitation erosion. This explains the fact that following the removal of chromium carbide particles under the action of cavitational bubbles, high stress concentrations appear and the boundaries between grains are attacked more compared to solid solution grains γ.

Finishing the grain and limiting the coarse carbide precipitation after the TIG surface remelting increases the energy absorption capacity of the cavitation impact wave, which delays the cracks nucleation.

4. Conclusions

The surface modification of the high alloy steel GX40CrNiSi25-20 using the TIG remelting technique using currents of 100–200 A leads to a significant improvement in the cavitation erosion resistance.

The experimental results proved that for values of the remelting current of 150 A, the total cumulative mass loss (Mtot = 3.78 mg) decreases more than six times compared to the initial state (without remelting, Mtot = 22.89 mg), and the erosion rate, according to the value towards which the v(t) curve tends to stabilize, vs = 0.145 mg/min. for the initial unmelted surface and vs = 0.029 mg/min. for the remelted surface), and decreases by about five times. This hardening phenomenon obtained for remelting values of 150 mA is explained by the accentuated granulation finishing and by the favourable conditions for the formation of solid solutions

The optical and scanning electron microscope investigation of the longitudinal section through the cavitation samples that were TIG remelted proves that this surface-ennobling process causes an increase in the surface hardness from approx. 210 HV at approx. 390 HV, an accentuated granulation finishing and microstructure with consequences for the pronounced increase in the resistance to cavitation erosion degradation.

Author Contributions: Conceptualization, I.M. and I.B.; methodology, I.M., I.B., D.C. and I.-D.U.; formal analysis, I.M. and I.B.; investigation, I.M., I.B., I.-D.U. and C.M.C.; resources, I.M. and I.B.; data curation, D.C.; writing—original draft preparation, I.M.; writing—review and editing, I.M., I.B, I.-D.U. and C.M.C.; visualization, I.M., I.B, D.C., I.-D.U. and C.M.C.; supervision, I.M and I.B. All authors have read and agreed to the published version of the manuscript.

Funding: This research received support from the EEA Grants 2014–2021, under Project contract no. 2/2019 CoDe-PEM (EEA RO-NO-2018-0502).

Institutional Review Board Statement: Not applicable.

Informed Consent Statement: Not applicable.

Data Availability Statement: Not applicable.

Acknowledgments: The research leading to these results has received support from the EEA Grants 2014–2021, under Project contract no. 2/2019 CoDe-PEM (EEA RO-NO-2018-0502).

Conflicts of Interest: The authors declare no conflict of interest.

References

1. Franc, J.-P.; Michel, J.M. *Fundamentals of Cavitation*; Kluwer Academic Publishers: Dordrecht, The Netherlands; Boston, MA, USA; London, UK, 2004.
2. Guiyan, G.; Zheng, Z. Cavitation erosion mechanism of 2Cr13 stainless steel. *Wear* **2022**, *488–489*, 204137.
3. Warren, D.A.; Griffiths, I.J.; Harniman, R.L.; Flewitt, P.E.J.; Scott, T.B. The role of ferrite in Type 316H austenitic stainless steels on the susceptibility to creep cavitation. *Mater. Sci. Eng. A* **2015**, *635*, 59–69. [CrossRef]
4. Xia, D.; Deng, C.; Chen, Z.; Li, T.; Hu, W. Modeling Localized Corrosion Propagation of Metallic Materials by Peridynamics: Progresses and Challenges. *Acta Metall Sin* **2022**, *58*, 1093–1107.
5. Romero, M.C.; Tschiptschin, A.P.; Scandian, C. Low temperature plasma nitriding of a Co30Cr19Fe alloy for improving cavitation erosion resistance. *Wear* **2019**, *426–427*, 581–588. [CrossRef]
6. Mitelea, I.; Bena, T.; Bordeasu, I.; Utu, I.D.; Craciunescu, C.M. Enhancement of Cavitation Erosion Resistance of Cast Iron with TIG Remelted Surface. *Metall. Mater. Trans. A* **2019**, *50A*, 3767–3775. [CrossRef]
7. Wang, Y.; Anp, Y.; Hou, G.; Zhao, X.; Zhou, H.; Chen, J. Effect of cooling rate during annealing on microstructure and ultrasonic cavitation behaviors of Ti6Al4V alloy. *Wear* **2023**, *512–513*, 204529. [CrossRef]
8. Si, C.; Sun, W.; Tian, Y.; Cai, J. Cavitation erosion resistance enhancement of the surface modified 2024T351 Al alloy by ultrasonic shot peening. *Surf. Coat. Technol.* **2023**, *452*, 129122. [CrossRef]
9. Singh, N.K.; Vinay, G.; Ang, A.S.M.; Mahajan, D.K.; Singh, H. Cavitation erosion mechanisms of HVOF-sprayed Ni-based cermet coatings in 3.5% NaCl environment. *Surf. Coat. Technol.* **2022**, *434*, 128194. [CrossRef]
10. Mitelea, I.; Bordeasu, I.; Belin, C.; Utu, I.D.; Craciunescu, C.M. Cavitation Resistance, Microstructure, and Surface Topography of Plasma Nitrided Nimonic 80 A Alloy. *Materials* **2022**, *15*, 6654. [CrossRef] [PubMed]
11. Chen, J.H.; Wu, W. Cavitation erosion behavior of Inconel 690 alloy. *Mater. Sci. Eng. A* **2008**, *489*, 451–456. [CrossRef]
12. Chen, F.; Jianhua Du, J.; Zhou, S. Cavitation erosion behaviour of incoloy alloy 865 in NaCl solution using ultrasonic vibration. *J. Alloy. Compd.* **2020**, *831*, 1547832. [CrossRef]
13. Mitelea, I.; Bordeasu, I.; Belin, C.; Utu Craciunescu, C.M. TIG processing of Nimonic 80A alloy for enhanced cavitation erosion resistance, Metal 2019. In Proceedings of the 28th International Conference on Metallurgy and Materials, Brno, Czech Republic, 22–24 May 2019; pp. 1399–1404.
14. Han, S.; Lin, J.H.; Kuo, J.J.; He, J.L.; Shih, H.C. The cavitation-erosion phenomenon of chromium nitride coatings deposited using cathodic arc plasma deposition on steel. *Surf. Coat. Technol.* **2002**, *161*, 20–25. [CrossRef]
15. Berchiche, N.A.; Franc, J.-P.; Michel, J.M. A cavitation erosion model for ductile materials. *J. Fluids Eng.* **2002**, *124*, 601–606. [CrossRef]
16. Mitelea, I.; Bordeasu, I.; Riemschneider, E.; Utu, I.D.; Craciunescu, C.M. Cavitation erosion improvement following TIG surface-remelting of gray cast iron. *Wear* **2022**, *496–497*, 204282. [CrossRef]
17. Hattori, S.; Kitagawa, T. Analysis of cavitation erosion resistance of cast iron and nonferrous metals based on database and comparison with carbon steel data. *Wear* **2010**, *269*, 443–448. [CrossRef]
18. Mitelea, I.; Dimian, E.; Bordeasu, I.; Craciunescu, C. Ultrasonic cavitation erosion of gas nitrided Ti-6Al-4V alloys. *Ultrason. Sonochemistry* **2014**, *21*, 1544–1548. [CrossRef] [PubMed]
19. Abbasi, M.; Vahdatnia, M.; Navaei, A. Solidification Microstructure of HK Heat Resistant Steel. *Int. J. Met.* **2015**, *9*, 14–26. [CrossRef]
20. Bordeaşu, I.; Patrascoiu, C.; Badarau, R.; Sucitu, L.; Popoviciu, M.; Balasoiu, V. New contributions in cavitation erosion curves modeling. *FME Trans.* **2006**, *34*, 39–43.
21. Steller, J.K. International cavitation erosion test-summary of results. In Proceedings of the ImechE Cavitation Conference, Cambridge, UK, 9–11 December 1992; pp. 95–102.
22. Steller, J.K. *International Cavitation Erosion Test-Test Facilities and Experimental Results*; 2-émés Journées Cavitation: Paris, France, 1992.
23. Steller, J.; Giren, B.G. *International Cavitation Erosion Test; Final Report*; Gdansk: Gdansk, Poland, 2015.
24. *Standard G32-2016*; Standard Method of Vibratory Cavitation Erosion Test. ASTM: West Conshohocken, PA, USA, 2016.
25. Bordeaşu, I. *Eroziunea Cavitaţională a Materialelor*; Editura Politehnica: Timișoara, Romania, 2006.
26. Li, Z.; Han, J.; Lu, J.; Zhou, J.; Chen, J. Vibratory cavitation erosion behaviour of AISI 304 stainless steel in water at elevated temperatures. *Wear* **2014**, *321*, 33–37. [CrossRef]
27. Yucheng, L.; Chang, H.; Guo, X.; Li, T.; Xiao, L. Ultrasonic cavitation erosion of 316L steel weld joint in liquid Pb-Bi eutectic alloy at 550 °C. *Ultrason. Sonochem.* **2017**, *39*, 77–86.

materials

MDPI

Article

Grey Relational Analysis and Grey Prediction Model (1, 6) Approach for Analyzing the Electrode Distance and Mechanical Properties of Tandem MIG Welding Distortion

Hsing-Chung Chen [1,2,*], Andika Wisnujati [1,3], Mudjijana [4], Agung Mulyo Widodo [1,5] and Chi-Wen Lung [6,7,*]

1 Department of Computer Science and Information Engineering, Asia University, Taichung 413, Taiwan
2 Department of Medical Research, China Medical University, Taichung 404, Taiwan
3 Department of Automotive Engineering Technology, Universitas Muhammadiyah Yogyakarta, Yogyakarta 55183, Indonesia
4 Department of Mechanical Engineering, Universitas Muhammadiyah Yogyakarta, Yogyakarta 55183, Indonesia
5 Department of Computer Science, Universitas Esa Unggul, Jakarta 11510, Indonesia
6 Rehabilitation Engineering Lab, University of Illinois at Urbana-Champaign, Champaign, IL 61820, USA
7 Department of Creative Product Design, Asia University, Taichung 413305, Taiwan
* Correspondence: cdma2000@asia.edu.tw or shin8409@ms6.hinet.net (H.-C.C.); cwlung@asia.edu.tw (C.-W.L.)

Abstract: The tandem metal inert gas (MIG) process uses two wires that are continuously fed through a special welding torch and disbursed to form a single molten pool. Within the contact tip of the modern approach, the wires are electrically insulated from one another. This study identified the effect of welding electrode spacing on the distortion of AA5052 aluminum plates and different mechanical properties including hardness and thermal cycle using grey relational analysis. Plate distortion was subsequently predicted using the grey prediction model GM (1, 6). This research used a pair of 400 mm × 75 mm × 5 mm of AA5052 plates and electrode distances of 18, 27, and 36 mm. The welding current, voltage, welding speed, and argon flow rate were 130 A, 23 V, 7 mm/s, and 17 L/min, respectively. The temperature was measured using a type-K thermocouple at 10, 20, 30, and 40 mm from the center of the weld bead. The smallest distortion at an electrode distance of 27 mm was 1.4 mm. At an electrode distance of 27 mm, the plate may reach a proper peak temperature where the amount of heat input and dissipation rate are similar to those for electrode distances of 18 mm and 36 mm. The highest relative VHN of 57 was found in the BM, while the lowest, 46, was found in the WM, showing good agreement with their respective grain sizes. Six parameters were designed using grey relational analysis (GRA) and subsequently employed in the grey prediction model GM (1, 6). Process evaluation results show that predictions for welding distortions are consistent with actual results, thus, the GM (1, 6) model can be used as a predictive model for welding distortions of 5052 aluminum plates.

Keywords: aluminum AA5052; tandem MIG welding; grey relational analysis; grey prediction model GM (1, 6)

Citation: Chen, H.-C.; Wisnujati, A.; Mudjijana; Widodo, A.M.; Lung, C.-W. Grey Relational Analysis and Grey Prediction Model (1, 6) Approach for Analyzing the Electrode Distance and Mechanical Properties of Tandem MIG Welding Distortion. *Materials* **2023**, *16*, 1390. https://doi.org/10.3390/ma16041390

Academic Editors: Cosmin Codrean, Anamaria Feier and Carmen Opriș

Received: 22 December 2022
Revised: 20 January 2023
Accepted: 28 January 2023
Published: 7 February 2023

1. Introduction

Welding is a metal joining method that is widely employed for aluminum and its alloys. Metal inert gas (MIG) welding is a welding process that is generally used to connect aluminum plates [1]. MIG welding uses noble gas (inert gas) to protect the electric arc while the electrode wire is fed continuously by an electric motor so that the welding process can be done semi-automatically or automatically. MIG welding creates the welding arc by continuously regulating metal from the consumable wire electrode. In terms of weld-bead shape and mechanical reliability, welding process factors are also very important in defining weld joint quality [2,3]. Furthermore, Adin discovered that changes in the current and voltage values of MIG welded joints on carbon steel have a significant effect on the tensile strength and elongation values of the joints, increasing the toughness of the MIG welded material [4]. In addition to fabrication, the benefits of the welding process can also be utilized for repairs, such as filling holes in castings, creating a welded layer on tools, thickening worn parts, and a variety of other repairs. The imperative for efficient manufacturing construction has resulted in a new trend in manufacturing design characterized by the use of thin plates. However, the use of welding processes such as MIG for joining thin plates frequently results in severe weld distortion due to local expansion and contraction of the weld metal and its adjoining area during welding [5]. When aluminum alloys are used for welded structures, the situation worsens due to their undesirable thermal material properties, such as a high thermal expansion coefficient, which causes significant distortion.

As a result, welding distortion can result in out-of-tolerance geometry, lowering product quality and having an adverse effect on the buckling strength of welded structures [6–8]. Various MIG welding methods have been developed to improve welding results, one of which is tandem welding. For several decades, the scope of use of tandem MIG welding techniques in construction has been very wide, because it can reduce process time in manufacturing industries, including shipping, bridges, steel frames, pressure vessels, rapid pipes, pipelines, and so on [9–11]. Tandem MIG welding is the best method to increase welding productivity. Tandem MIG welding is considered to have high production effectiveness that can be achieved with high welding speeds [12]. It consists of two independent electrodes (early electrode and delayed electrode) positioned parallel to the weld line and controlled individually under distinct welding conditions (current and voltage) [13]. The main advantages of the tandem MIG welding method are its extremely high efficiency, low heat input, improved weld seam integrity, and extremely high deposition rate. However, it can give rise to distortion known as longitudinal or buckling distortion due to the occurrence of the expansion process during the welding process and shrinkage when the weld begins to cool [14]. Over the last few decades, studies have developed several methods of planning and controlling this distortion [15,16]. In general, control of welding distortion and residual stresses can be accomplished using mechanical effects, thermal effects, or a combination of the two techniques and can be performed either during welding or after [17–19].

Local industry, however, is struggling with a lack of recognition and oversight of the input process factors required to create a quality weld joint that matches the required specifications [20,21]. Traditional processes for determining welding parameter settings currently comprise an empirical method of trial and error, which is a time-consuming and error-based design process. Wu suggested that mechanical and thermal balance techniques be developed during the welding process to control welding distortion by reducing welding thermal gradients and producing a tensile stress field near the welded region. A review of these approaches would be beneficial in order to fully understand the main mechanism of mitigating welding-induced distortion and to select the most appropriate method for minimizing deformation based on realistic fabrication demands [22]. However, welding distortion can be precisely predicted during the manufacturing process while accounting for both local shrinkage and root gap. Mitigating weld distortion to meet product requirements is a difficult task because welding shrinkage cannot be avoided, only controlled. An increase in electrode distance is associated with a decrease in temperature at the center of the welding arc. The determination of the distancing of welding electrodes in tandem MIG welding depends on the type of electrode, the diameter of the electrode core, the material being welded, the geometry of the connection, and the accuracy of the connection [23,24].

Thus, this study proposed using a GRA and GM (1, 6) approach and determined its validity not solely for the analysis of empirical data for MIG welding but also for predicting the distortion of tandem MIG welding in aluminum AA5052 plate. Thus, this research contributes to:

- Determining the influence of mechanical properties parameters and the variable of electrode distance using GRA analysis on the distortion of aluminum AA5052.
- Derive differential equations from empirical data of experimental results to predict plate distortion values of AA5052 using GM (1, 6) based on the parameters resulting from the GRA analysis.
- The value obtained from solving the complete differential equation of the GM (1, 6) prediction model has a value that is broadly similar to the results of the empirical distortion test, so it is useful for the industrial world or users to optimize the parameters that affect the distortion of welding results.

In terms of its relationship to the voltage from the welding current, it can be said that the distance between the welding electrodes and the welding speed are largely independent of the welding voltage but are directly proportional to the welding current so that the effect of the electrode distance on the back bead occurs if the distance is short. In this case, the welding bead will coalesce. The challenge with welded joints is that their mechanical characteristics are inferior to those of the base material. Mechanical properties using microhardness-Vickers, distortion, and thermal cycles in welding temperatures are the output responses evaluated in the study to analyze using the GRA and GM (1, 6) approach based on the effect of variable electrode distance on tandem MIG welding.

2. Related Works

Welding is a manufacturing process in which two or more similar or dissimilar materials are permanently joined by coalescence formation with or without the application of external pressure, heat, or filler material. Weld-bead formation does not require the fusion of the faying surfaces of the base materials [25]. Welding processes are broadly classified into two groups based on whether or not the base material fuses: solid-state welding and fusion welding. Fusion welding occurs when the faying surfaces of the base materials and filler material melt together to form a weld bead. Conversely, solid-state welding occurs when no such melting occurs during the welding process. The following types of welding techniques are available in the manufacturing industry. Each of these welding techniques has its advantages as shown in Table 1.

The material aluminum AA5052 has a process of reduction of 87 percent, resulting in a tensile strength of 325 MPa and a strain of 2.5%. After annealing treatment at 300 °C for 4 h, the elongation is strengthened to 23%, but the tensile strength is reduced to 212 MPa [26,27]. Aluminum AA5052 is an alloy that is not heat treated (non-heat treatable); thus, the strengthening mechanism is carried out by means of a solid solution and cold working. Similarly, aluminum from the AA7075 (a magnesium-zinc alloy) and 6000 series, and AA6082 (a magnesium-silicon alloy), are also commonly utilized in aviation applications due to their excessive specialized strength, hardness, and resistance to corrosion at cleft temperatures [28–30].

Prakash et al. designed the multi-response optimization problem solved by using the GRA method [31]. This is the most often utilized optimization strategy. The parameters are processed, the outputs are observed, and the results are tabulated. This system provides data conditions and facilitates decision-making. Adin claimed that optimizing the welding parameters used in friction welding to join AISI steel bars is quite important. Furthermore, the effects of welding parameters on tensile strength and axial shortening were investigated, and welding parameters were optimized using the Taguchi method to achieve high-quality weld joints [4,32]. The GRA approach is used to solve complex problems. In this paper, we will obtain a GRG through the GRA process, which is used for problem evaluation. The GRA approach relates to the calculation of all the effects of multiple aspects and their correlation and is also called the straightening of factor relations. The GRA approach utilizes relevant data from the grey system to quantitatively compare each element according to the level of resemblance and variation between factors in order to determine their relationships [33,34].

Table 1. Different welding techniques.

No	Types of Welding	Advantages	Disadvantages
1	Arc welding	• High strength weld • High weld speed • Low welding equipment	• Low efficiency • Difficult to weld thin materials
2	Gas welding	• Low equipment cost • Portable equipment • Highly skilled welders are not required	• Low weld quality • Not suitable for thick sections
3	Resistance welding	• Low processing cost • Easy to implement • Welding dissimilar materials are possible	• Comparatively low welding strength • Aesthetics are not good near the weld area
4	Solid-State welding	• Weld dissimilar materials and thermoplastic is possible • Weld joints are free from microstructure defects	• Difficult to set up • Surface preparation is required
5	Energy beam welding	• Weld strength up to 95% is achievable • A vacuum environment eliminates impurities	• EBW requires vacuum conditions to prevent dissipation of the electron beam
6	Laser welding	• High-quality welds • Narrow and deep welds are possible • Automation is easy to implement	• High initial equipment and maintenance cost • Sometimes cracking is a concern due to the high cooling rate

Qazi et al. integrated a methodology of GRA combined with primary component analysis (PCA) for the optimization of the shielded metal arc welding (SMAW) technique of steel plates SA 516 grade 70. The authors studied the effect of SMAW specifications on the transformation of mechanical characteristics [35]. They reported that the experiments display acceptable agreement with optimum outcomes. According to Cai et al. [36], the welding position greatly affects the weld pool due to the effect of gravity, where, in a flat position, the surface of the pool becomes concave, while in a vertical position, the pool surface is even more concave due to more flowing filler metal on the welding pool. Previous studies used MS plate (Grade: IS 2062) specimens to solve multi-objective optimization issues in the metal inert gas (MIG) welding process. The specimen was investigated to find the best combination of input elements such as welding current, open circuit voltage, and plate thickness to achieve superior weld strength and bead geometry quality criteria such as tensile strength, bead width, reinforcing, and penetration. Sahu et al. used GRA and PCA to look at how multiple objectives turned into specific feedback. The goal was to find the best way to set the relationship between the input factors [37]. GRA is a method for calculating the degree of correlation within sequences and combining several output responses into a single output response by applying a GRG to each output response [38,39]. Researchers [40–44] used GRA to optimize grey-based Taguchi analysis (GTA) welding process variables for the multi-objective response by GRG. Taguchi's GRA multi-objective optimization approach was used in this study to optimize gas tungsten arc welding (GTAW) characteristics of various AA6061 and AA2014 alloys. Chavda et al. [45] studied the MIG welding process as well as Taguchi's DOE Method. They concluded that the main factors in the GMAW process are current, voltage, speed, inert gas, gas flow rate, wire feed rate, wire diameter, and so on, and that these variables will affect the various weld properties. Kulkarni et al. [46] investigated the welding parameters of MIG welding by GRA employing an L32 orthogonal array. Several performance factors, including torch angle, wire feed rate, standoff distance, welding speed, and welding current, were optimized using a Taguchi L32 array and grey relational analysis. GRA is used for optimization because it is the most effective strategy for multi-response optimization. Material removal rate (MRR) and surface finish are influenced the most by feed rate, according to the literature [47,48].

3. The Proposed Method and Materials

3.1. Data Acquisition

The input data for this method are the welding electrode distance and mechanical properties of a MIG welding experiment. For conducting the experiments, electrode distances of 18, 27, and 36 mm, and a welding speed of 7 mm/s, are chosen as input parameters. The welding speed was selected by trial and error using dummy specimens, whereas the electrode distances were based on Mudjijana et al. [49,50]. The mechanical properties used in this study are Vickers hardness, distortion, and thermal cycle, which are used to determine the distortion effect of the AA5052 aluminum plate. Figure 1 depicts a semi-automatic welding tool that can be used to place specimens and adjust the welding speed, the distance between the electrode and the specimen, the welding gun position, and the welding gun movement. Our method is illustrated in Figure 2, which shows the whole system architecture.

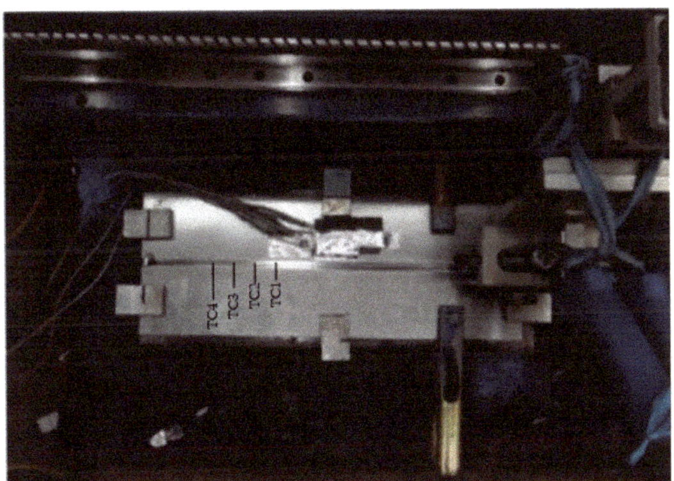

Figure 1. Semi-automatic welding tool.

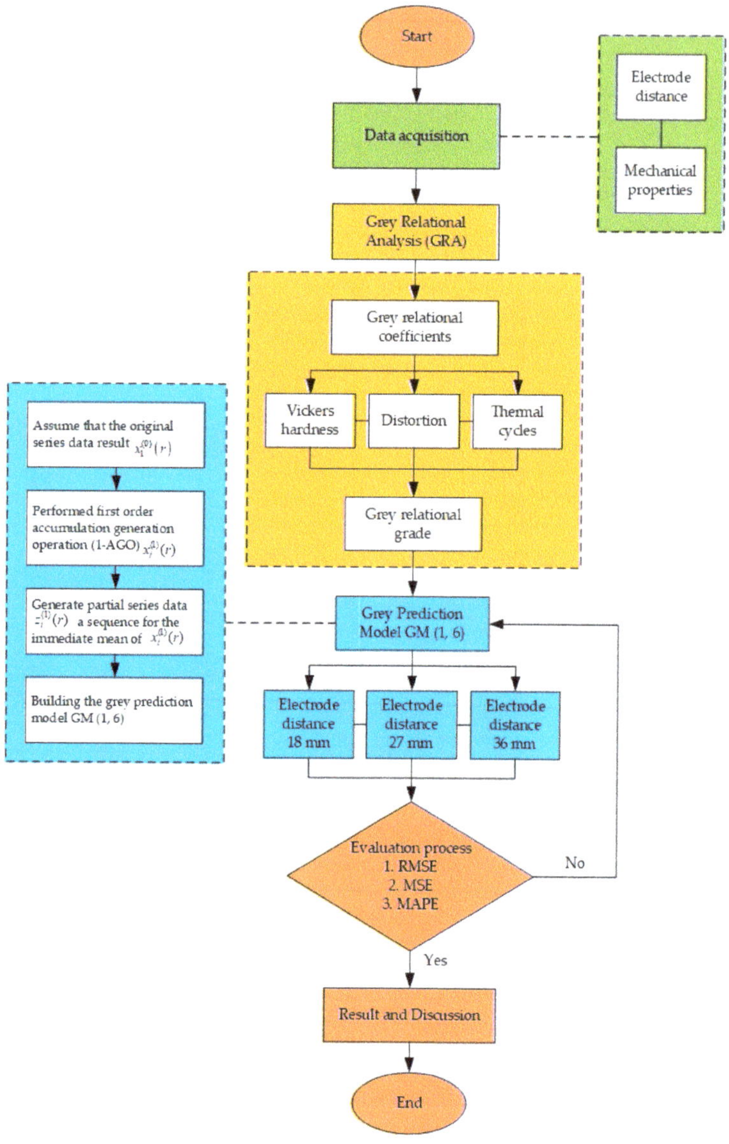

Figure 2. System architecture.

3.2. Materials

In this study, a pair of aluminum AA5052 materials 400 mm × 75 mm × 5 mm in size, with a V (70°) grove, and 2 mm root were welded using MIG tandem welding, and 0.8 mm diameter ER5356 electrodes. Aluminum AA5052 series aluminum alloys provide good weight-to-strength ratios, corrosion resistance, weldability, and recycling possibility. This alloy has excellent workability, excellent corrosion resistance, excellent weldability, and moderate strength. As a result, it is used in aviation fuel or oil lines, gasoline tanks, various modes of transportation, sheet metal work, appliances and lighting, wire, and rivets [51]. Welding was done with a Tenjima 200S MIG welding machine with leading and trailing torches tilted at an angle of 80°, a current of 130 A, a voltage of 23 V, a feeding rate of

25 mm/s, an arc length of 10 mm, and a flow rate of 17 L/min of argon gas. The tandem MIG welding process does not use external cooling. This is because the increase in cooling causes a change in the weld fusion line [52].

The output of this MIG welding is measurements of the distortion of the welding joints on the aluminum AA5052 plate. As illustrated in Figure 3, distortion was evaluated after welding was completed using a dial indicator with 0.01 mm accuracy. The welding temperatures of the thermal cycle (TC), TC1, TC2, TC3, and TC4, were measured at 10, 20, 30, and 40 mm from the center of the weld bead using a type-K wire thermocouple, an ADAM-4561 data acquisition module, and verified using a computer.

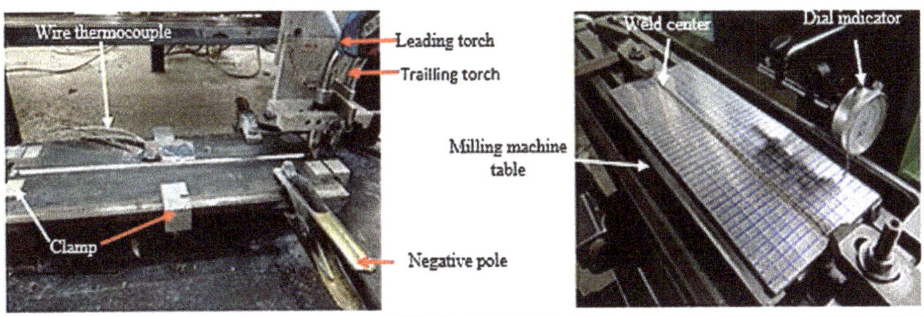

Figure 3. Setup for tandem MIG welding AA5052.

3.3. Grey Relational Analysis Procedure

GRA can handle multi-response optimization problems in the presence of incomplete and unclear information. GRA is used to obtain the value of GRG to evaluate multiple responses, allowing the optimization of complex multiple responses to be transformed into the optimization of a single response with the GRG [36]. The welding process includes several responses, and welding quality is heavily dependent on maximizing all of these at the same time. As a result, researchers typically use GRA to analyze diverse responses at the same time [18,37]. GRA is part of the grey system theory and aims to analyze the grey relational degree between each factor in the grey system [53].

The detailed steps of the GRA method are as follows:

Step 1: Standardized data transformation.

GRA functions as a discovery concept, assembling known and unknown components to achieve the best level of reaction. Grey relational coefficients (GRC) and GRG are calculated with the help of value normalization.

- Benefit-type factor

A factor with a greater value than the original data yields better quality characteristics (larger-the-better) as shown in Equation (1).

$$x_i^*(k) = \frac{x_i^{(0)}(k) - \min x_i^{(0)}(k)}{\max x_i^{(0)}(k) - \min x_i^{(0)}(k)} \tag{1}$$

- Defect-type factor

A factor with a lower value than the original data yields higher-quality characteristics (smaller-the-better) as shown in Equation (2).

$$x_i^*(k) = \frac{\max x_i^{(0)}(k) - x_i^{(0)}(k)}{\max x_i^{(0)}(k) - \min x_i^{(0)}(k)} \tag{2}$$

- Medium-type or nominal-the best

Factors that have the same value or are closest to the specified value standard show better quality characteristics, as shown in Equation (3).

$$x_i^*(k) = 1 - \frac{\left|x_i^{(0)}(k) - x_{tv}^{(0)}\right|}{\max\left(\left(\max x_i^{(0)}(k) - x_{tv}^{(0)}\right), \left(x_{tv}^{(0)} - \min x_i^{(0)}(k)\right)\right)} \quad (3)$$

where $x_i^*(k)$ is the data after grey relational generation, $\max x_i^{(0)}(k)$ is the maximum value of the original sequence factor, $\min x_i^{(0)}(k)$ is the minimum value of the original sequence factor, and $x_{tv}^{(0)}$ is the target value.

Data standardization is a stage in grey relational generation, where the experimental results are normalized to a value on a scale of 0 to 1 due to different units of measurement. Data pre-processing converts original sequences into a set of comparable sequences [53].

Step 2: Calculating the deviation sequence.

Using the following formula, the deviation sequence calculation in Equation (4) attempts to estimate the actual difference between the compared series and the reference series.

$$\Delta_{0i}(k) = |x_0^*(k) - x_i^*(k)| \quad (4)$$

where $\Delta_{0i}(k)$ is the deviation sequence, $x_0^*(k)$ is the reference sequence, and $x_i^*(k)$ is the comparability sequence.

Step 3: Calculating the grey relational coefficient

The calculation of the GRC in Equation (5) is carried out to determine the sequence with the lowest deviation using a discriminating coefficient between 0 and 1. In general, the discriminating coefficient is 0.5. The sequence with the lowest deviation will produce the GRC with the highest value, which is "1".

$$\xi_i(k) = \frac{\Delta_{\min} + \xi\Delta_{\max}}{\Delta_{0i}(k) + \xi\Delta_{\max}} \quad (5)$$

where $\xi_i(k)$ is the grey relational coefficient, ξ is the distinguishing coefficient (0.5), $\Delta_{\min}$ is the minimum deviation sequence, and $\Delta_{\max}$ is the maximum deviation sequence.

Step 4: Calculating the relative grey relational grade

GRG represents the degree of correlation between the reference and comparative sequences in Equation (6). The higher the value of GRG, the stronger the correlation between the reference and comparability sequences. GRG calculation is done using the following formula.

$$r_i = \sum_{i=1}^{n} (w(k) \times \xi_i(k)) \quad (6)$$

where r_i is the grey relational grade and $w(k)$ is the quantity *of the number k influence factor.*

3.4. Existing Grey Prediction Model GM (1, N)

The accuracy of the current grey prediction model GM (1, N) cannot be predicted because of the solution's inaccuracy. It is improper to solve the existing GM (1, N) with the assumption that the related series first-order accumulated generating operation data are constants. GM type (1, N) is an extension of GM (1, 1). In this study, we proposed a new grey prediction model GM (1, 6). This model is extremely useful when historical data is unavailable. However, the model requires at least four periods of time of historical data.

The steps to generating the GM (1, 6) model are described below.

Step 1: Assume that the original series of data came from determining how welding distortion changed over time as a series of dependent variables.

Build the original data series in chronological order in Equation (7) for the dependent variables and Equation (8) for the independent variables.

$$x_1^{(0)}(r) = \left\{x_1^{(0)}(1), x_1^{(0)}(2), \ldots, x_1^{(0)}(k)\right\}; r = 1,2,3,\ldots,k \tag{7}$$

All features or dimensions used as independent variables of the system are formed into a sequence according to Equation (8).

$$x_i^{(0)}(r) = \left\{x_i^{(0)}(1), x_i^{(0)}(2), \ldots, x_i^{(0)}(r)\right\}; i = 1,2,3,\ldots,N \text{ and } r = 1,2,3,\ldots,k \tag{8}$$

Step 2: To eliminate the uncertainties in the original data, we generate $x_i^{(0)}(r)$ using the accumulating generation operation (AGO).

The first-order accumulation generation operation (1-AGO) is performed from $x_i^{(0)}$; $i = 1,2,3,\ldots,N$ as in Equation (9).

$$x_i^{(1)}(r) = \sum_{j=1}^{k} x_i^{(0)}(j); k = 1,2,3,\ldots,r; i = 1,2,3,\ldots,N \tag{9}$$

Step 3: Evaluate the background value of $z_i^{(1)}(r)$ constructed by the generation method based on the average rate of two adjacent $x_i^{(1)}(r)$ datasets in Equation (10).

$$z_i^{(1)}(r) = 0.5\left(x_i^{(1)}(r-1) + x_i^{(1)}(r)\right), r = 1,2,3,\ldots,k \tag{10}$$

Step 4: Building a grey equation generation model in Equation (11).

$$x_i^{(1)}(r) = \left\{x_i^{(1)}(1), x_i^{(1)}(2), \ldots, x_i^{(1)}(k)\right\}; r = 1,2,3,\ldots,k; i = 1,2,3,\ldots,N \tag{11}$$

Furthermore, for each pair of values, $x_1^{(0)}(r), z_i^{(1)}(r)$ and $x_i^{(1)}(r)$ are formed to apply the grey differential equation in GM (1, N). However, before forming GM (1, N) the meaning of the grey differential equation GM (1, N) must be known as in Equation (12).

$$\frac{dx_1^{(1)}(t)}{dt} + b_1 x_1^{(1)}(t) = b_2 x_2^{(1)}(t) + b_3 x_3^{(1)}(t) + \ldots + b_N x_N^{(1)}(t) \tag{12}$$

where coefficient b_1 is the coefficient of grey development, and b_2, b_3, …, b_N are the respective coefficients corresponding to the corresponding series. The coefficients b_1, b_2, …, b_N are model parameters to be estimated.

The grey derivative for the first-order grey differential equation with 1-AGO is conventionally represented as in Equations (13) and (14).

$$\frac{dx_1^{(1)}(t)}{dt} = \lim_{\Delta t \to 0} \frac{x_1^{(1)}(t+\Delta t) - x_1^{(1)}(t)}{\Delta t} \tag{13}$$

$$\frac{dx_1^{(1)}(t)}{dt} = \frac{\Delta x_1^{(1)}(t)}{\Delta t} = x_1^{(1)}(t+1) - x_1^{(1)}(t) = x_1^{(0)}(t+1) \tag{14}$$

when $\Delta t \to 1$.

Step 5: Background values from $\frac{dx_1^{(1)}(t)}{dt}$, $x_1^{(1)}(t)$ are taken as the mean of $x_1^{(1)}(t)$ and $x_1^{(1)}(t+1)$ respectively, while $x_j^{(1)}(t)$, j = 2, 3, …, n is taken as $x_j^{(1)}(t)$, j = 2, 3, …, n.

The least squares solution for the model parameters of GM (1, N) in Equation (12) where t from 1 to r is in Equation (15).

$$\begin{bmatrix} b_1 & b_2 & \ldots & b_N \end{bmatrix}^T = \left(B^T B\right)^{-1} B^T Y_N \tag{15}$$

where

$$B = \begin{bmatrix} -\frac{1}{2}\left(x_1^{(1)}(1) - x_1^{(1)}(2)\right) & x_2^{(1)}(2) & x_3^{(1)}(2) & \ldots & x_N^{(1)}(2) \\ -\frac{1}{2}\left(x_1^{(1)}(2) - x_1^{(1)}(3)\right) & x_2^{(1)}(3) & x_3^{(1)}(3) & \ldots & x_N^{(1)}(3) \\ \vdots & \vdots & \vdots & \ldots & \vdots \\ -\frac{1}{2}\left(x_1^{(r-1)}(1) - x_1^{(1)}(r)\right) & x_2^{(1)}(r) & x_3^{(1)}(r) & \ldots & x_N^{(1)}(r) \end{bmatrix} \tag{16}$$

and $Y_N(r) = \begin{bmatrix} x^{(0)}(2) & x^{(0)}(3) & \ldots & x^{(0)}(k) \end{bmatrix}^T$. Then, the modeling value of the series prediction in Equation (16) is obtained as in Equation (17).

$$\hat{x}_1^{(1)} = \left[x_1^{(0)}(1) - \frac{1}{b_1} \sum_{i=1}^{n} b_i x_i^{(1)}(t) \right] e^{-b_1(t-1)} + \frac{1}{b_1} \sum_{i=1}^{n} b_i x_i^{(1)}(t) \tag{17}$$

From Equation (17), and by the inverse first-order accumulation generation operation (1-IAGO) of $\hat{x}_1^{(1)}$, the modeling value of $\hat{x}_1^{(0)}$ can be reduced in Equations (18) and (19).

$$\hat{x}_1^{(0)}(1) = \hat{x}_1^{(1)}(1) = x_1^{(0)}(1) \tag{18}$$

$$\hat{x}_1^{(0)}(t) = \hat{x}_1^{(1)}(t) - \hat{x}_1^{(1)}(t-1); t = 2, 3, \ldots \tag{19}$$

where t is a time (s), $\hat{x}_1^{(0)}$ is a prediction value, $\hat{x}_1^{(1)}$ is a first-order accumulation generation operation (1-IAGO).

4. Results and Discussion

4.1. Effect of Electrode Distance on Welding Distortion

Measurement of the distortion of the AA5052 aluminum plate aims to determine the curvature of the plate after experiencing the tandem MIG welding process caused by the uneven temperature distribution that occurs during the welding process. In the welding process, the weld metal will experience thermal problems. The results of out-of-plane distortion measurements for all welded plates versus electrode distance are depicted in Figure 4. During the welding process, the temperature of the weld metal will rise, but the amount of temperature increase varies from the highest temperature, namely the melting point of the metal, which occurs at the center of the weld, to the lowest temperature, which occurs at the edge of the metal, which is influenced by the ambient temperature. The temperature distribution, especially the peak temperature that occurs for each part of the weld metal and the time it takes to reach this temperature, greatly affects the properties of the weld metal. The properties of the weld metal that are affected by the temperature distribution include the shape of the macrostructure and the mechanical strength of the weld metal, such as hardness. In this study, the thermal cycle on the AA5052 aluminum plate occurs as a result of the magnitude of the temperature distribution in the weld metal, which is affected by the distance between the electrode and the center of the weld per unit time during the welding process.

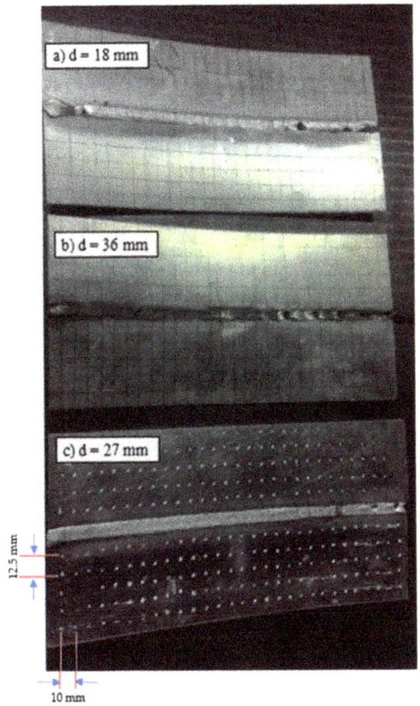

Figure 4. Specimen distortion in the welded plates.

The results of this study indicate that, at an electrode distance of 27 mm, torch interaction in tandem MIG welding will form a different thermal cycle than multi-run, which has two peaks, because the torch distance is very far. This phenomenon indicates that the interaction between torches in tandem MIG welding is very influential on the heat input that occurs; the closer the distance between the torches, the greater the heat input that occurs, so that the welding temperature becomes very high and the distortion will also be greater.

Based on Figure 4, showing the effect of the distance between the peak temperature of tandem MIG welding and the center of the weld, it is known that the closer the distance of the metal part of the weld to the center of the weld, the greater the temperature achieved, so that the part that reaches a higher peak temperature will experience a faster cooling process when the part that is reaching a lower peak temperature is still undergoing a heating process. The greater the peak temperature reached, the greater the cooling rate.

4.2. Vickers Hardness

The smallest distortion at an electrode distance of 27 mm was 1.4 mm. At an electrode distance of 27 mm, the plate may reach a proper peak temperature where the amount of heat input and dissipation rate are to those of electrode distances of 18 mm and 36 mm, resulting in greater distortion. Thus, this distance may be recommended for MIG tandem welding of a 5 mm thick plate of AA5052 material using ER5356 electrodes. Ghosh et al. [54] reported that the welding distortion of the V grove is commonly larger than that of an I grove. Therefore, in order to minimize welding distortion, it can be done by employing double-sided arc welding.

The Vickers hardness number (VHN) values of AA5052 tandem MIG welding with electrode distances of 18, 27, and 36 mm at the welding speed of 7 mm/s are shown in Table 2. The average VHN values in the base metal (BM), heat-affected zone (HAZ), and weld metal (WM) regions are presented. The highest relative VHN of 57 was found in the BM, while the lowest, 46, was found in the WM, showing good agreement with their respective grain sizes [51]. Finer grain size results in higher strength or hardness, with strength being proportional to $d^{-1/2}$, but according to Doksanovic et al. [55], their difference is not significant. Thus, this welding quality can be considered good.

Table 2. Average Vickers Hardness Number (VHN) of tandem MIG welding.

Welding Speed (mm/s)	Electrode Distance (mm)	Average Vickers Hardness Number (VHN)		
		BM	WM	HAZ
7	18	56	49	53
	27	57	48	53
	36	55	46	52

The weld area is divided into three main parts, namely the weld metal, the heat-affected zone (HAZ), and the base metal, which is not affected by welding. The metal being welded and the HAZ zone will go through a number of temperature cycles during the welding operation. The thermal cycle will have an effect on the microstructure of the weld metal and HAZ, as the weld metal will go through a series of phase changes during the cooling process.

4.3. Thermal Cycle

Detailed results of the thermal cycle measurements during the welding processes for a welding speed of 7 mm/s are shown in Figure 5, while the peak temperatures TC1, TC2, TC3, and TC4 were presented in Table 3 for the three electrode distances of 18, 27, and 36 mm. The closer the electrode is to the weld center, the higher the peak temperature because heat input is concentrated in a smaller area, resulting in a longer heat dissipation time. The welding process produces a complicated thermal cycle in the weld. This thermal cycle causes changes in the micromaterial structure in the area around the weld (HAZ) and transient thermal stress, which eventually creates residual stress and distortion. Based on Table 3, it is found that the greater the thermal cycle, the greater the thermal stress. However, the resisting stresses tend to be the same for the same material parameters and dimensions. As a result, an increase in the thermal cycle will be accompanied by an increase in thermal stress, resulting in greater arc distortion.

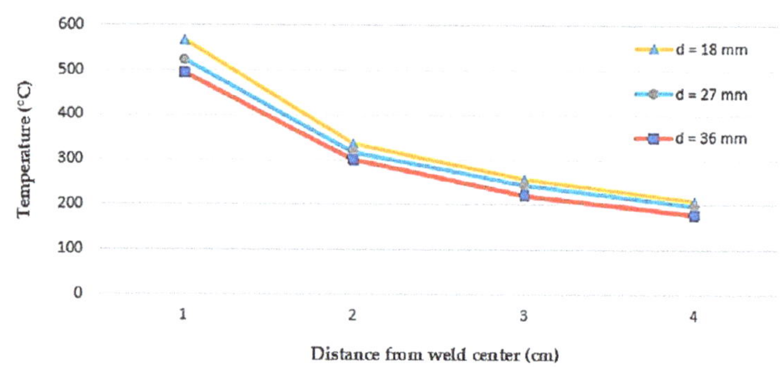

Figure 5. Peak temperature of tandem MIG welding.

Table 3. Peak temperature in the thermal cycle (TC).

Electrode Distance	Peak Temperature (°C) of Tandem MIG Welding at Welding Speed of 7 mm/s			
	TC1	TC2	TC3	TC4
18	567	336	256	208
27	523	317	243	197
36	494	300	220	178

Welding distortion is caused by shrinkage during the cooling process of the AA5052 aluminum plate material, which previously experienced expansion during the welding process. The factors that influence the above distortion include the hardness value and thermal temperature of the connection configuration and the MIG welding method.

The material is subjected to a temperature gradient, and, as long as thermal stress is produced, it tends to expand differently. In the aforementioned procedure, the thermal cycle (TC3) and the effects of distortion happen 30 mm away from the weld's center. This is due to the influence of the optimal welding electrode spacing in this study, which is 27 mm. Within that distance, the welding electrode liquid will flow and settle properly, so that the effect of the weld results will produce welds and translucency on the AA5052 aluminum material. In addition, it also produces a fine sprinkling of slag during the welding process.

According to Hernández et al. [56], the Rosenthal solution to the temperature distribution of a moving point heating element for thin plates in Equation (20) can be used to approximate the four types of weld thermal cycles depicted in Figure 5.

$$T - T_0 = \frac{q_w}{h(4\pi k\rho ct)^{1/2}} \exp\left(-\frac{r^2}{4\alpha t}\right) \tag{20}$$

where q_w is heat input determined by calculating (Q/v), Q is heat energy, v is welding speed, T_0 is the temperature at the beginning of welding, ρc is the specific heat per unit volume, k is thermal conduction, α is thermal diffusivity which is equal to $k/\rho c$, h is the thickness of plate, r is the radial or lateral distance from the weld, and t is time.

The amount of heat input is relatively low, resulting in a low peak temperature, whereas the heat input of tandem welding is close to the total amount of heat input generated by the torches. In a heat sink condition, the cooling system dissipates a significant amount of heat input, resulting in lower peak temperatures.

4.4. Integrated GRA and GM (1, 6)

The grey analysis confirms and finds that the variable parameters that influence distortion from experimental studies are Vickers hardness values and thermal cycles. Both are considered the maximum and average reactions. GRA first normalizes the experimental results and then calculates the grey relational coefficients from the normalized data to reflect the relationship between the desired and actual experimental data, as illustrated in Table 4. GRG can be effectively adapted to solve complex interrelationships among defined performance characteristics.

Table 4. GRG for different electrode distances.

Experiment	Grey Relational Grade			Rank
	18 (mm)	27 (mm)	36 (mm)	
1	0.7030	0.7192	0.6998	2
2	0.5459	0.5863	0.6238	5
3	0.6972	0.6444	0.6138	4
4	0.7579	0.7809	0.6237	1
5	0.6640	0.7513	0.6228	3

(Experiment: 1 = Distortion-VHN; 2 = Distortion-TC1; 3 = Distortion-TC2; 4 = Distortion-TC3; 5 = Distortion-TC4).

This is because, at that distance, a process called thermal cycling happens. This involves heating and cooling a material until it goes through molecular reorganization, which tightens or optimizes the particle structure of the whole material, removes stress, and makes it denser and more uniform so that flaws or imperfections are minimized. This is why the electrode distance of 27 mm at TC3 gives the AA5052 aluminum plate a more even temperature distribution.

The grey prediction model GM (1, 6) is used to anticipate distortion analysis in tandem MIG welding and has a high prediction accuracy [57]. In this research, the modeling values in GM (1, 6) are theoretically the precise solution. There must be a strong correlation between the expected series and the associated series, and the indicator must be highly indicative of the predicted series in order to achieve high prediction accuracy for a dynamic system. Instead of the existing grey relational grade analysis, the upgraded grey relational grade analysis is employed by GM (1, 6) to examine the suitable entries of the projected series and the associated series used for model development.

The following can be made into a row using the results of modeling the welding results at an electrode distance of 27 mm and the distortion value that was determined from the following data:

$$x_1^{(0)}(r) = \{0, 0.14, 0.33, 0.38, 0.51, 0.63, 0.75, 0.87, 0.99, 1.09, 1.24, 1.27, 1.35, 1.43, 1.5, 1.55, 1.6, 1.64, 1.67, 1.69\}$$

Then, here are the results of putting together the list of features or the list of independent variables:

$$x_2^{(0)}(r) = \{48, 50, 49, 48, 48.5, 53, 52, 54, 55, 55, 59, 60, 59, 59, 59, 62, 65, 66, 60, 60\}$$

$$x_3^{(0)}(r) = \left\{ \begin{array}{c} 29.4776, 29.4566, 29.4788, 29.4823, 29.4814, 29.4678, 29.4987, 29.4869, 29.4853, 29.4779, \\ 29.4787, 29.4722, 29.4697, 29.4704, 29.4687, 29.4772, 29.486, 29.4807, 29.4669, 29.4818 \end{array} \right\}$$

$$x_4^{(0)}(r) = \left\{ \begin{array}{c} 29.1908, 29.1754, 29.1899, 29.1936, 29.1791, 29.1904, 29.1958, 29.1903, 29.1961, 29.1888, \\ 29.1954, 29.1835, 29.1886, 29.1888, 29.1932, 29.1844, 29.1987, 29.1987, 29.1919, 29.1849 \end{array} \right\}$$

$$x_5^{(0)}(r) = \left\{ \begin{array}{c} 29.3362, 29.3087, 29.3201, 29.3098, 29.3175, 29.3228, 29.2989, 29.33, 29.3223, 29.3048, \\ 29.3188, 29.3259, 29.3239, 29.3305, 29.3292, 29.3212, 29.3167, 29.3223, 29.3076, 29.3287 \end{array} \right\}$$

$$x_6^{(0)}(r) = \left\{ \begin{array}{c} 29.2944, 29.2655, 29.254, 29.2736, 29.2958, 29.2666, 29.2844, 29.2778, 29.2826, 29.2688, \\ 29.2613, 29.2738, 29.2764, 29.2732, 29.2843, 29.2852, 29.2743, 29.2798, 29.2753.29.2712 \end{array} \right\}$$

The result of the first-order (1-AGO) accumulation generation operation for each variable is

$$X_1^{(1)} = [0\ 0.14; 0.47; 0.85; 1.36; 1.99; 2.74; 3.61; 4.6; 5.69; 6.93; 8.2; 9.55; 10.98; 12.48; 14.03; 15.63; 17.27; 18.94; 20.63]$$

and

$$X_i^{(1)} = \left[\begin{array}{cccccccccccccccccccc} 48 & 98 & 147 & 195 & 243.5 & 296.5 & 348.5 & 402.5 & 457.5 & 512.5 & 571.5 & 631.5 & 690.5 & 749.5 & 808.5 & 870.5 & 935.5 & 1001.5 & 1061.5 & 1121.5 \\ 29.4776 & 58.9342 & 88.413 & 117.8953 & 147.3767 & 176.8446 & 206.3432 & 235.8301 & 265.3154 & 294.7933 & 324.272 & 353.7442 & 383.2139 & 412.6842 & 442.153 & 471.6302 & 501.1162 & 530.5968 & 560.0637 & 589.5455 \\ 29.1908 & 58.3662 & 87.5561 & 116.7496 & 145.9287 & 175.1191 & 204.3149 & 233.5051 & 262.7013 & 291.8901 & 321.0855 & 350.2689 & 379.4575 & 408.6463 & 437.8395 & 467.0239 & 496.2226 & 525.4213 & 554.6132 & 583.7981 \\ 29.3362 & 58.6449 & 87.9649 & 117.2747 & 146.5922 & 175.915 & 205.2139 & 234.5439 & 263.8661 & 293.171 & 322.4897 & 351.8157 & 381.1396 & 410.4701 & 439.7993 & 469.1205 & 498.4372 & 527.7595 & 557.067 & 586.3957 \\ 29.2944 & 58.5599 & 87.8138 & 117.0874 & 146.3833 & 175.6499 & 204.9343 & 234.2122 & 263.4948 & 292.7636 & 322.0249 & 351.2987 & 380.575 & 409.8483 & 439.1325 & 468.4177 & 497.692 & 526.9719 & 556.2472 & 585.5184 \end{array} \right]$$

where $i = 2, 3, \ldots, 20$.

Furthermore, the mean sequence of $x_1^{(1)}(t)$ and $x_1^{(1)}(t+1)$ can be seen in the following matrix.

$$Z_1^{(1)} = [0.07; 0.305; 0.66; 1.105; 1.675; 2.365; 3.175; 4.105; 5.145; 6.31\ ; 7.565; 8.875; 10.265; 11.73; 13.255; 14.83; 16.45; 18.105; 19.785]$$

By using matrix B in Equation (16), we get B:

$$B = \begin{bmatrix} -0.07 & 98 & 58.9342 & 58.3662 & 58.6449 & 58.5599 \\ -0.305 & 147 & 88.413 & 87.5561 & 87.9649 & 87.8138 \\ -0.66 & 195 & 117.8953 & 116.7496 & 117.2747 & 117.0874 \\ -1.105 & 243.5 & 147.3767 & 145.9287 & 146.5922 & 146.3833 \\ -1.675 & 296.5 & 176.8446 & 175.1191 & 175.915 & 175.6499 \\ -2.365 & 348.5 & 206.3432 & 204.3149 & 205.2139 & 204.9343 \\ -3.175 & 402.5 & 235.8301 & 233.5051 & 234.5439 & 234.2122 \\ -4.105 & 457.5 & 265.3154 & 262.7013 & 263.8661 & 263.4948 \\ -5.145 & 512.5 & 294.7933 & 291.8901 & 293.171 & 292.7636 \\ -6.31 & 571.5 & 324.272 & 321.0855 & 322.4897 & 322.0249 \\ -7.565 & 631.5 & 353.7442 & 350.2689 & 351.8157 & 351.2987 \\ -8.875 & 690.5 & 383.2139 & 379.4575 & 381.1396 & 380.575 \\ -10.265 & 749.5 & 412.6842 & 408.6463 & 410.4701 & 409.8483 \\ -11.73 & 808.5 & 442.153 & 437.8395 & 439.7993 & 439.1325 \\ -13.255 & 870.5 & 471.6302 & 467.0239 & 469.1205 & 468.4177 \\ -14.83 & 935.5 & 501.1162 & 496.2226 & 498.4372 & 497.692 \\ -16.45 & 1001.5 & 530.5968 & 525.4213 & 527.7595 & 526.9719 \\ -18.105 & 1061.5 & 560.0637 & 554.6132 & 557.067 & 556.2472 \\ -19.785 & 1121.5 & 589.5455 & 583.7981 & 586.3957 & 585.5184 \end{bmatrix}$$

Then the coefficient value of the independent variables in Equation (15) can be solved and the grey equation generation model is obtained, as shown in Equation (21).

$$\frac{dx_1^{(1)}(t)}{dt} - 0.0462x_1^{(1)}(t) = -0.0077x_2^{(1)}(t) + 3.6606x_3^{(1)}(t) - 1.8749x_4^{(1)}(t) + 1.1569x_5^{(1)}(t) - 2.9587x_6^{(1)}(t) \quad (21)$$

where $b_1 = -0.0462; b_2 = -0.0077; b_3 = 3.6606; b_4 = -1.8749; b_5 = 1.1569; b_6 = -2.9587$.

With the least squares for the model parameters of GM (1, 6) in Equation (21) using Equations (15) and (16). Furthermore, with $Y_5(r) = \begin{bmatrix} 2.8539 & 3.4928 & 3.7879 & 4.1057 \end{bmatrix}^T$ then, the modeling value of the series prediction is obtained as in Equation (22).

$$\hat{x}_1^{(1)} = \left[x_1^{(0)}(1) + \frac{1}{0.0462}\sum_{i=1}^{n} b_i x_i^{(1)}(t)\right]e^{0.0462(t-1)} - \frac{1}{0.0462}\sum_{i=1}^{n} b_i x_i^{(1)}(t), \text{ where } i = 2,3,4,\ldots,12 \quad (22)$$

where

$$\sum_{i=2}^{6} b_i x_i^{(1)}(t) = -0.0077x_2^{(1)}(t) + 3.6606x_3^{(1)}(t) - 1.8749x_4^{(1)}(t) + 1.1569x_5^{(1)}(t) - 2.9587x_6^{(1)}(t)$$

Then, using the inverse-generating operation of the first-order accumulation (1-IAGO) of $\hat{x}_1^{(1)}$, the modeling value of $\hat{x}_1^{(0)}$ can be obtained, as depicted in Table 5 and Figures 6–8.

The GM (1, 6) values are close to the empirical values. This is because the pattern of empirical data follows an upward trend and does not experience fluctuations, so the two models can be used as a reference for looking for random data analysis. The overlapping distortion between the actual data and GM (1, 6) is shown in Figure 7 with a welding electrode distance of 27 mm.

Based on the prediction results using GM (1, 6), a comparison can be made between the actual data and the predicted results using GM (1, 6). A comparison graph of the data pattern with the welding electrode distance variable is shown in Figures 6–8. The distortion value in the actual results is influenced by several factors, including the Vickers hardness and the thermal cycle temperature at a distance of 30 mm from the welding point. Mechanical properties, which consist of Vickers hardness and thermal cycles as independent variables that are controlled by electrode distance as the controlling variable, affect the results of distortion on the AA5052 aluminum plate. The arrangement of the welding electrode distance is able to minimize welding distortion with thin plates.

Table 5. Welding distortion for different electrode distances.

Length (mm)	Distortion in Variable Electrode Distance					
	18 mm		27 mm		36 mm	
	x_1^0	$\hat{x}_1^{(0)}$	x_1^0	$\hat{x}_1^{(0)}$	x_1^0	$\hat{x}_1^{(0)}$
0	0.00	0	0.00	0	0.00	0
10	0.14	0.07	0.08	0.05	0.11	0.14
20	0.33	0.20	0.15	0.14	0.24	0.29
30	0.38	0.69	0.24	0.23	0.36	0.56
40	0.51	0.85	0.32	0.57	0.48	0.72
50	0.63	0.94	0.41	0.63	0.60	0.81
60	0.75	1.06	0.49	0.73	0.73	1.34
70	0.87	1.51	0.56	0.97	0.86	0.88
80	0.99	1.88	0.64	1.18	1.00	1.78
90	1.09	1.69	0.72	1.30	1.13	2.31
100	1.24	2.00	0.81	1.47	1.26	2.43
110	1.27	2.22	0.90	1.62	1.34	2.91
120	1.35	1.89	0.96	1.90	1.51	2.20
130	1.43	1.65	1.13	1.97	1.62	3.16
140	1.50	2.11	1.10	2.29	1.73	2.67
150	1.55	1.21	1.17	1.82	1.82	3.13
160	1.60	1.47	1.23	1.86	1.90	3.33
170	1.64	1.72	1.28	2.01	1.97	2.61
180	1.67	0.73	1.32	2.15	2.02	3.83
190	1.69	0.77	1.35	2.12	2.05	3.55

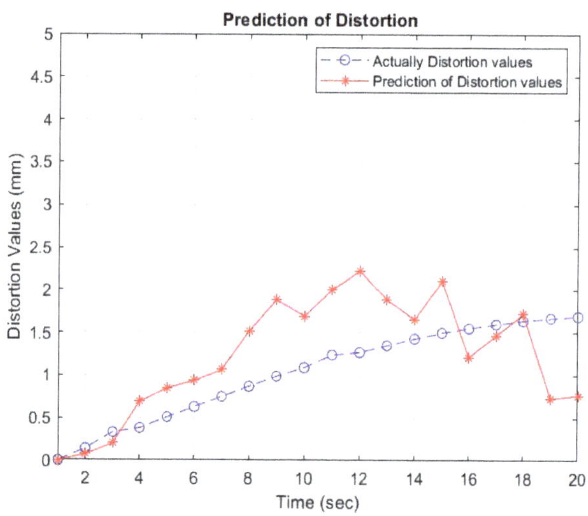

Figure 6. Comparison of predictions and experiments of distortion values at an electrode distance of 18 mm.

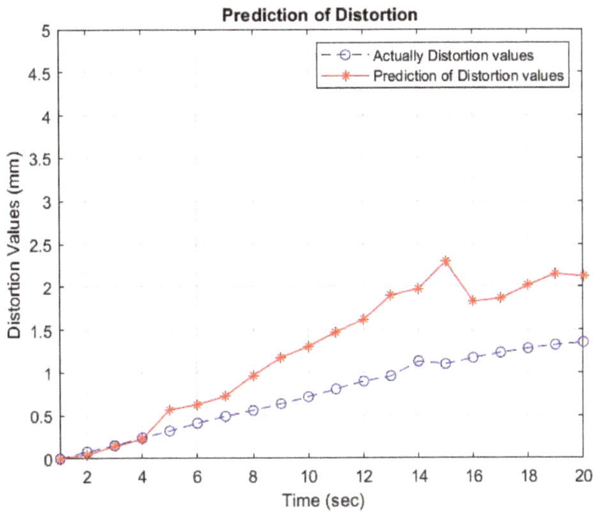

Figure 7. Comparison of predictions and experiments of distortion values at an electrode distance of 27 mm.

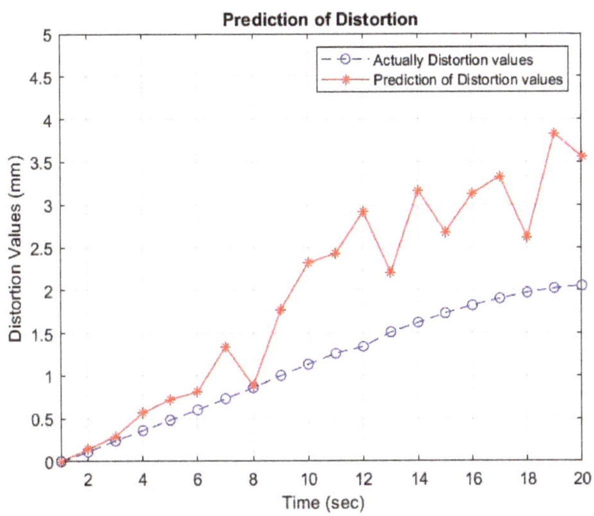

Figure 8. Comparison of predictions and experiments of distortion values at an electrode distance of 36 mm.

4.5. Evaluation Process

The results applying GM (1, 6) on tandem MIG welding indicate the model's effectiveness and prediction accuracy. However, an evaluation is required to measure the error rates between the predicted and the actual values. This is shown in Table 6. The evaluation processes are as follows:

Table 6. Root mean square error (RMSE) for actual and prediction data.

Variable Electrode Distance (mm)	$\sum_{i=1}^{n}\left(x_1^0 - \hat{x}_1^{(0)}\right)^2$	MSE	RMSE	MAPE (%)
18	6.061065231	0.303053	0.550503	0.52
27	7.545394	0.37727	0.614223	1.21
36	19.77073	0.988537	0.994252	1.35

Step 1: Root mean square error (RMSE)

In this research, the root mean square error (RMSE) approach was employed to evaluate the performance of the model GM (1, 6). RMSE is sometimes known as the root mean square deviation (RMSD); its mathematical expression is similar to SD in that RMSE pertains to n data points rather than n^{-1}. The value generated by the RMSE is the average squared value of the number of errors in the prediction model. RMSE is a technique that is easy to implement and has been widely used in various studies related to prediction or forecasting [58,59]. The mathematical equation RMSE is shown in Equation (23).

$$RMSE = \sqrt{\frac{1}{n}\sum_{i=1}^{n}\left(x_1^0 - \hat{x}_1^{(0)}\right)^2} \tag{23}$$

Step 2: Mean square error (MSE)

Calculating the MSE value is similar to calculating the RMSE. MSE is an error calculation method that is calculated by adding up the squared errors and then dividing it by the number of data or periods used [60]. At this stage, the greater the error value, the greater the resulting MSE value shown in Equation (24).

$$MSE = \frac{1}{n}\sum_{i=1}^{n}\left(x_1^0 - \hat{x}_1^{(0)}\right)^2 \tag{24}$$

Step 3: Mean absolute percentage error (MAPE)

The mean absolute percentage error (MAPE) is the average value of the absolute difference between the predicted and actual values, expressed as a percentage of the realized value. The use of the mean absolute percentage error (MAPE) in forecasting results can demonstrate the amount of accuracy of forecasting and realization data. Equation (25) can be used to calculate MAPE value [61].

$$MAPE = \frac{1}{n}\sum_{i=1}^{n}\left(\left|\frac{x_1^0 - \hat{x}_1^{(0)}}{x_1^0}\right| \times 100\%\right) \tag{25}$$

where

x_1^0 = actual value

$\hat{x}_1^{(0)}$ = prediction value

i = order of data in the database

n = datasets

This criterion is similar to the RMSE in its measurement. Nevertheless, it is more reliable than MSE because it is less sensitive to extreme values. All distance measurements (RMSE and MSE) are equivalent and help to quantify the approximated solution's accuracy in comparison to the simulated data between actual and prediction in welding distortion. Low values for these criteria indicate that the estimated model is reasonably close to the true value. The closer the RMSE values are to zero and 1, respectively, the more accurate the model results will be. MAPE provides information on how much the forecast error is compared to the actual value of the series. MAPE represents the average absolute percentage error of each entry in a data set to calculate how accurate the predicted value is compared to the true value. MAPE is a direct metric, with the results of the data above showing that with electrode distances of 18, 27, and 36 mm, the MAPE values are 0.52%, 1.21%, and 1.35%, respectively. It can be interpreted that the value represents the average deviation between the predicted value and the actual value, regardless of whether the deviation is positive or negative.

5. Conclusions

Based on the results of research and testing of tandem MIG welding with electrode distances of 18, 27, and 36 mm on AA5052 aluminum material with a welding speed of 7 mm/s, it can be concluded that the smallest distortion of the AA5052 aluminum plate occurs at a welding electrode distance of 27 mm. This is influenced by the Vickers hardness level in the area, with that of the base metal (BM) being 56.53 VHN, of the heat-affected zone (HAZ) being 52.75 VHN, and of the weld metal (WM) being 48.46 VHN. The level of hardness at that distance is also affected by the thermal cycle temperature at a distance of 30 mm from the weld metal (WM) because the heat input is more concentrated in the WM area. The effect of the distance between the welding electrodes in this research indicated that the heat input is directly proportional to the value of the shrinkage voltage such that, the higher the heat input, the higher the shrinkage voltage. This is consistent with the distortion of large tandem welding results caused by high heat input. By using GRA analysis and the GM (1, 6), the effect of welding distortion is found for an electrode distance of 27 mm with actual and predicted values that coincide with each other. Process evaluation results the predictions for welding distortions do not experience fluctuating numbers; thus, the GM (1, 6) model can be used as a predictive model for welding distortions of 5052 aluminum plates. The results of the evaluation were carried out using the MAPE method to determine the ability of the model used to see the difference in actual and predicted values. The MAPE value at an electrode distance of 18 mm is 0.52%, while at an electrode distance of 27 mm, it is 1.21%, and at 36 mm, it is 1.52%. The optimum value from the experimental results shows that the smallest distortion occurs at the welding electrode distance of 27 mm. However, GM (1, 6) cannot show the optimal value of welding distortion based on electrode distance. In the future, more research can be done on how to use the GM (1, N) model to find the best electrode distance in the welding process to achieve optimal distortion.

Author Contributions: Conceptualization, H.-C.C. and A.W.; Methodology, H.-C.C. and A.W.; Software, A.W. and A.M.W.; Validation, H.-C.C. and A.W.; Formal analysis, A.W. and A.M.W.; Investigation, H.-C.C. and A.W.; Resources, A.W., C.-W.L. and M.; Data curation, A.W. and M.; Writing—original draft preparation, H.-C.C. and A.W.; Writing—review and Editing, H.-C.C.; Supervision, H.-C.C. and C.-W.L.; Project administration, H.-C.C.; Funding acquisition, H.-C.C. and M. All authors have read and agreed to the published version of the manuscript.

Funding: This funding was supported by Universitas Muhammadiyah Yogyakarta (UMY), under grant number 56/R-LRI/XII/2022. It was also supported in part by the Ministry of Science and Technology (MOST), Taiwan, under MOST Grant numbers: 111-2218-E-468-001-MBK, 110-2218-E-468-001-MBK, 110-2221-E-468-007, 111-2218-E-002-037 and 110-2218-E-002-044.

Institutional Review Board Statement: Not applicable.

Informed Consent Statement: Not applicable.

Data Availability Statement: The data used to support the finding of this study are included within the article.

Acknowledgments: This work was supported by Universitas Muhammadiyah Yogyakarta (UMY), under grant number 56/R-LRI/XII/2022. This work was also supported by the Ministry of Science and Technology (MOST), Taiwan, under MOST Grant numbers: 111-2218-E-468-001-MBK, 110-2218-E-468-001-MBK, 110-2221-E-468-007, 111-2218-E-002-037 and 110-2218-E-002-044.

Conflicts of Interest: No conflict of interest.

References

1. Mandal, N.R. *Ship Construction and Welding*; Springer: Berlin/Heidelberg, Germany, 2017; Volume 329.
2. Naik, A.B.; Reddy, A.C. Optimization of tensile strength in TIG welding using the Taguchi method and analysis of variance (ANOVA). *Therm. Sci. Eng. Prog.* **2018**, *8*, 327–339. [CrossRef]
3. Chaki, S.; Shanmugarajan, B.; Ghosal, S.; Padmanabham, G. Application of integrated soft computing techniques for optimisation of hybrid CO_2 laser–MIG welding process. *Appl. Soft Comput.* **2015**, *30*, 365–374. [CrossRef]
4. Adin, M.Ş.; Okumuş, M. Investigation of microstructural and mechanical properties of dissimilar metal weld between AISI 420 and AISI 1018 STEELS. *Arab. J. Sci. Eng.* **2022**, *47*, 8341–8350. [CrossRef]
5. Adin, M.Ş.; İşcan, B. Optimization of process parameters of medium carbon steel joints joined by MIG welding using Taguchi method. *Eur. Mech. Sci.* **2022**, *6*, 17–26. [CrossRef]
6. Abima, C.S.; Akinlabi, S.A.; Madushele, N.; Akinlabi, E.T. Comparative study between TIG-MIG Hybrid, TIG and MIG welding of 1008 steel joints for enhanced structural integrity. *Sci. Afr.* **2022**, *17*, e01329. [CrossRef]
7. Zong, R.; Chen, J.; Wu, C. A comparison of TIG-MIG hybrid welding with conventional MIG welding in the behaviors of arc, droplet and weld pool. *J. Mater. Process. Technol.* **2019**, *270*, 345–355. [CrossRef]
8. Ilman, M.; Muslih, M.; Subeki, N.; Wibowo, H. Mitigating distortion and residual stress by static thermal tensioning to improve fatigue crack growth performance of MIG AA5083 welds. *Mater. Des.* **2016**, *99*, 273–283. [CrossRef]
9. Mohtadi-Bonab, M.; Szpunar, J.A.; Collins, L.; Stankievech, R. Evaluation of hydrogen induced cracking behavior of API X70 pipeline steel at different heat treatments. *Int. J. Hydrogen Energy* **2014**, *39*, 6076–6088. [CrossRef]
10. Akhshik, S.; Behzad, M.; Rajabi, M. CFD–DEM approach to investigate the effect of drill pipe rotation on cuttings transport behavior. *J. Pet. Sci. Eng.* **2015**, *127*, 229–244. [CrossRef]
11. Ilman, M.; Muslih, M.; Triwibowo, N. Enhanced fatigue performance of tandem MIG 5083 aluminium alloy weld joints by heat sink and static thermal tensioning. *Int. J. Lightweight Mater. Manuf.* **2022**, *5*, 440–452. [CrossRef]
12. Liu, G.; Han, S.; Tang, X.; Cui, H. Effects of torch configuration on arc interaction behaviors and weld defect formation mechanism in tandem pulsed GMAW. *J. Manuf. Processes* **2021**, *62*, 729–742. [CrossRef]
13. Wang, J.; Chen, X.; Yang, L.; Zhang, G. Sequentially combined thermo-mechanical and mechanical simulation of double-pulse MIG welding of 6061-T6 aluminum alloy sheets. *J. Manuf. Processes* **2022**, *77*, 616–631. [CrossRef]
14. Zhou, H.; Yi, B.; Shen, C.; Wang, J.; Liu, J.; Wu, T. Mitigation of welding induced buckling with transient thermal tension and its application for accurate fabrication of offshore cabin structure. *Mar. Struct.* **2022**, *81*, 103104. [CrossRef]
15. Yi, J.; Lin, J.; Chen, Z.; Chen, T. Prediction and controlling for welding deformation of propeller base structure. *J. Ocean Eng. Sci.* **2021**, *6*, 410–416. [CrossRef]
16. Huang, H.; Yin, X.; Feng, Z.; Ma, N. Finite element analysis and in-situ measurement of out-of-plane distortion in thin plate TIG welding. *Materials* **2019**, *12*, 141. [CrossRef] [PubMed]
17. Ghafouri, M.; Ahola, A.; Ahn, J.; Björk, T. Welding-induced stresses and distortion in high-strength steel T-joints: Numerical and experimental study. *J. Constr. Steel Res.* **2022**, *189*, 107088. [CrossRef]
18. Li, X.; Hu, L.; Deng, D. Influence of contact behavior on welding distortion and residual stress in a thin-plate butt-welded joint performed by partial-length welding. *Thin-Walled Struct.* **2022**, *176*, 109302. [CrossRef]
19. Li, Z.; Feng, G.; Deng, D.; Luo, Y. Investigating welding distortion of thin-plate stiffened panel steel structures by means of thermal elastic plastic finite element method. *J. Mater. Eng. Perform.* **2021**, *30*, 3677–3690. [CrossRef]
20. Mishra, D.; Roy, R.B.; Dutta, S.; Pal, S.K.; Chakravarty, D. A review on sensor based monitoring and control of friction stir welding process and a roadmap to Industry 4.0. *J. Manuf. Processes* **2018**, *36*, 373–397. [CrossRef]
21. Mishra, D.; Gupta, A.; Raj, P.; Kumar, A.; Anwer, S.; Pal, S.K.; Chakravarty, D.; Pal, S.; Chakravarty, T.; Pal, A. Real time monitoring and control of friction stir welding process using multiple sensors. *CIRP J. Manuf. Sci. Technol.* **2020**, *30*, 1–11. [CrossRef]
22. Wu, C.; Kim, J.-W. Review on mitigation of welding-induced distortion based on FEM analysis. *J. Weld. Join.* **2020**, *38*, 56–66. [CrossRef]
23. Ye, D.; Hua, X.; Xu, C.; Li, F.; Wu, Y. Research on arc interference and welding operating point change of twin wire MIG welding. *Int. J. Adv. Manuf. Technol.* **2017**, *89*, 493–502. [CrossRef]
24. Wang, W.; Yamane, S.; Wang, Q.; Shan, L.; Zhang, X.; Wei, Z.; Yan, Y.; Song, Y.; Numazawa, H.; Lu, J. Visual sensing and quality control in plasma MIG welding. *J. Manuf. Processes* **2023**, *86*, 163–176. [CrossRef]
25. Dwivedi, D.K.; Dwivedi, D.K. Design of Welded Joints: Weld Bead Geometry: Selection, Welding Parameters. In *Fundamentals of Metal Joining Processes, Mechanism and Performance*; Springer: Berlin/Heidelberg, Germany, 2022; pp. 343–351. [CrossRef]
26. Bo, W.; Chen, X.-H.; Pan, F.-S.; Mao, J.-J.; Yong, F. Effects of cold rolling and heat treatment on microstructure and mechanical properties of AA 5052 aluminum alloy. *Trans. Nonferr. Met. Soc. China* **2015**, *25*, 2481–2489.
27. Poznak, A.; Freiberg, D.; Sanders, P. Automotive wrought aluminium alloys. In *Fundamentals of Aluminium Metallurgy*; Elsevier: Amsterdam, The Netherlands, 2018; pp. 333–386.
28. Yelamasetti, B.; Manikyam, S.; Saxena, K.K. Multi-response Taguchi grey relational analysis of mechanical properties and weld bead dimensions of dissimilar joint of AA6082 and AA7075. *Adv. Mater. Process. Technol.* **2021**, *8*, 1474–1484. [CrossRef]
29. Pu, J.; Wei, Y.; Xiang, S.; Ou, W.; Liu, R. Optimization of Metal Inert-Gas Welding Process for 5052 Aluminum Alloy by Artificial Neural Network. *Russ. J. Non-Ferr. Met.* **2021**, *62*, 568–579. [CrossRef]

30. Yelamasetti, B.; Kumar, D.; Saxena, K.K. Experimental investigation on temperature profiles and residual stresses in GTAW dissimilar weldments of AA5052 and AA7075. *Adv. Mater. Process. Technol.* **2022**, *8*, 352–365. [CrossRef]
31. Prakash, K.S.; Gopal, P.; Karthik, S. Multi-objective optimization using Taguchi based grey relational analysis in turning of Rock dust reinforced Aluminum MMC. *Measurement* **2020**, *157*, 107664. [CrossRef]
32. Adin, M.Ş.; İşcan, B.; Baday, Ş. Optimization of welding parameters of AISI 431 and AISI 1020 joints joined by friction welding using taguchi method. *Bilecik Şeyh Edebali Üniversitesi Fen Bilim. Derg.* **2022**, *9*, 453–470.
33. Chafekar, A.; Sapkal, S. Multi-objective optimization of MIG welding of aluminum alloy. In *Techno-Societal 2018*; Springer: Berlin/Heidelberg, Germany, 2020; pp. 523–530.
34. Sefene, E.M.; Tsegaw, A.A. Temperature-based optimization of friction stir welding of AA 6061 using GRA synchronous with Taguchi method. *Int. J. Adv. Manuf. Technol.* **2022**, *119*, 1479–1490. [CrossRef]
35. Qazi, M.I.; Akhtar, R.; Abas, M.; Khalid, Q.S.; Babar, A.R.; Pruncu, C.I. An integrated approach of GRA coupled with principal component analysis for multi-optimization of shielded metal arc welding (SMAW) process. *Materials* **2020**, *13*, 3457. [CrossRef]
36. Cai, X.; Fan, C.; Lin, S.; Yang, C.; Bai, J. Molten pool behaviors and weld forming characteristics of all-position tandem narrow gap GMAW. *Int. J. Adv. Manuf. Technol.* **2016**, *87*, 2437–2444. [CrossRef]
37. Sahu, N.K.; Sahu, A.K.; Sahu, A.K. Optimization of weld bead geometry of MS plate (Grade: IS 2062) in the context of welding: A comparative analysis of GRA and PCA–Taguchi approaches. *Sādhanā* **2017**, *42*, 231–244. [CrossRef]
38. Huang, Y.-F.; Chen, H.-C.; Yen, P.-L. Performance of computer examination items selection based on grey relational analysis. *Int. J. Appl. Sci. Eng.* **2021**, *18*, 2021009. [CrossRef] [PubMed]
39. Sagheer-Abbasi, Y.; Ikramullah-Butt, S.; Hussain, G.; Imran, S.H.; Mohammad-Khan, A.; Baseer, R.A. Optimization of parameters for micro friction stir welding of aluminum 5052 using Taguchi technique. *Int. J. Adv. Manuf. Technol.* **2019**, *102*, 369–378. [CrossRef]
40. Srirangan, A.K.; Paulraj, S. Multi-response optimization of process parameters for TIG welding of Incoloy 800HT by Taguchi grey relational analysis. *Eng. Sci. Technol. Int. J.* **2016**, *19*, 811–817. [CrossRef]
41. Boukraa, M.; Chekifi, T.; Lebaal, N. Friction Stir Welding of Aluminum Using a Multi-Objective Optimization Approach Based on Both Taguchi Method and Grey Relational Analysis. *Exp. Tech.* **2022**. [CrossRef]
42. Sabry, I.; Mourad, A.-H.I.; Thekkuden, D.T. Optimization of metal inert gas-welded aluminium 6061 pipe parameters using analysis of variance and grey relational analysis. *SN Appl. Sci.* **2020**, *2*, 175. [CrossRef]
43. Moi, S.; Rudrapati, R.; Bandyopadhyay, A.; Pal, P. Design Optimization of Welding Parameters for Multi-response Optimization in TIG Welding Using RSM-Based Grey Relational Analysis. In *Advances in Computational Methods in Manufacturing*; Springer: Berlin/Heidelberg, Germany, 2019; pp. 193–203.
44. Wang, Q.; Zeng, X.; Chen, C., Lian, G.; Huang, X. An integrated method for multi-objective optimization of multi-pass Fe50/TiC laser cladding on AISI 1045 steel based on grey relational analysis and principal component analysis. *Coatings* **2020**, *10*, 151. [CrossRef]
45. Chavda, S.P.; Desai, J.V.; Patel, T.M. A review on optimization of MIG Welding parameters using Taguchi's DOE method. *Int. J. Eng. Manag. Res.* **2014**, *4*, 16–21.
46. Kulkarni, S.S.; Konnur, V.S.; Ganjigatti, J.P. Optimization of Mig Welding Process Parameters with Grey Relational Analysis for Al 6061 Alloy. *Weld. Int.* **2022**, *36*, 387–393. [CrossRef]
47. Younas, M.; Jaffery, S.H.I.; Khan, M.; Khan, M.A.; Ahmad, R.; Mubashar, A.; Ali, L. Multi-objective optimization for sustainable turning Ti6Al4V alloy using grey relational analysis (GRA) based on analytic hierarchy process (AHP). *Int. J. Adv. Manuf. Technol.* **2019**, *105*, 1175–1188. [CrossRef]
48. Li, N.; Chen, Y.-J.; Kong, D.-D. Multi-response optimization of Ti-6Al-4V turning operations using Taguchi-based grey relational analysis coupled with kernel principal component analysis. *Adv. Manuf.* **2019**, *7*, 142–154. [CrossRef]
49. Mudjijana, M.; Malau, V.; Salim, U.A. The effect of AA5083H116 2-layer MIG welding speed on physical and mechanical properties. *J. Mater. Process. Charact.* **2020**, *1*, 31–41. [CrossRef]
50. Mudjijana; Himarosa, R.A.; Sudarisman. Macro-Micro Analysis on 2-Layer Semiautomatic MIG Welding of AA5052 Material Using ER5356 Electrode. *Key Eng. Mater.* **2020**, *867*, 204–212. [CrossRef]
51. Callister Jr, W.D.; Rethwisch, D.G. *Fundamentals of Materials Science and Engineering: An Integrated Approach*; John Wiley & Sons: Hoboken, NJ, USA, 2020.
52. Laitila, J.; Keränen, L.; Larkiola, J. Effect of enhanced weld cooling on the mechanical properties of a structural steel with a yield strength of 700 MPa. *SN Appl. Sci.* **2020**, *2*, 1888. [CrossRef]
53. Shih, N.-Y.; Chen, H.-C. An approach for selecting candidates in soft-handover procedure using multi-generating procedure and second grey relational analysis. *Comput. Sci. Inf. Syst.* **2014**, *11*, 1173–1190. [CrossRef]
54. Puh, F.; Jurkovic, Z.; Perinic, M.; Brezocnik, M.; Buljan, S. Optimization of machining parameters for turning operation with multiple quality characteristics using Grey relational analysis. *Teh. Vjesn.* **2016**, *23*, 377–382.
55. Ghosh, A.; Yadav, A.; Kumar, A. Modelling and experimental validation of moving tilted volumetric heat source in gas metal arc welding process. *J. Mater. Process. Technol.* **2017**, *239*, 52–65. [CrossRef]
56. Dokšanović, T.; Džeba, I.; Markulak, D. Variability of structural aluminium alloys mechanical properties. *Struct. Saf.* **2017**, *67*, 11–26. [CrossRef]

57. Hernández, M.; Ambriz, R.; Cortès, R.; Gómora, C.; Plascencia, G.; Jaramillo, D. Assessment of gas tungsten arc welding thermal cycles on Inconel 718 alloy. *Trans. Nonferr. Met. Soc. China* **2019**, *29*, 579–587. [CrossRef]
58. Zeng, B.; Duan, H.; Zhou, Y. A new multivariable grey prediction model with structure compatibility. *Appl. Math. Model.* **2019**, *75*, 385–397. [CrossRef]
59. Chai, T.; Draxler, R.R. Root mean square error (RMSE) or mean absolute error (MAE). *Geosci. Model Dev. Discuss.* **2014**, *7*, 1525–1534.
60. Ebrahimi, M.; Khoshtaghaza, M.H.; Minaei, S.; Jamshidi, B. Vision-based pest detection based on SVM classification method. *Comput. Electron. Agric.* **2017**, *137*, 52–58. [CrossRef]
61. Wu, C.; Wang, C.; Kim, J.-W. Welding sequence optimization to reduce welding distortion based on coupled artificial neural network and swarm intelligence algorithm. *Eng. Appl. Artif. Intell.* **2022**, *114*, 105142. [CrossRef]

materials

MDPI

Article

Effect of Shielding Gas Arc Welding Process on Cavitation Resistance of Welded Joints of AlMg4.5Mn Alloy

Marina Dojčinović [1], Radica Prokić Cvetković [2], Aleksandar Sedmak [2,3], Olivera Popović [2], Ivana Cvetković [2] and Dorin Radu [3,*]

[1] Faculty of Technology and Metallurgy, University of Belgrade, Karnegijeva 4, 11000 Belgrade, Serbia; rina@tmf.bg.ac.rs
[2] Faculty of Mechanical Engineering, University of Belgrade, Kraljice Marije 16, 11000 Belgrade, Serbia; rprokic@mas.bg.ac.rs (R.P.C.); asedmak@mas.bg.ac.rs (A.S.); opopovic@mas.bg.ac.rs (O.P.); icvetkovic@mas.bg.ac.rs (I.C.)
[3] Faculty of Civil Engineering, Transilvania University of Brașov, Turnului Street 5, 500152 Brașov, Romania
* Correspondence: dorin.radu@unibv.ro

Abstract: The effect of the shielding gas arc welding process on the cavitation resistance of the three-component aluminum alloy AlMg4.5Mn and its welded joints was investigated. Welding was performed using the GTAW and GMAW processes in a shielded atmosphere of pure argon. After the welding, metallographic tests were performed, and the hardness distribution in the welded joints was determined. The ultrasonic vibration method was used to evaluate the base metal's and weld metal's resistance to cavitation. The change in mass was monitored to determine the cavitation rates. The morphology of the surface damage of the base metal and weld metal due to cavitation was monitored using scanning electron microscopy to explain the effect of the shielding gas arc welding process on their resistance to cavitation.

Keywords: cavitation rate; weld metal; GTAW; GMAW; aluminum alloys

Citation: Dojčinović, M.; Prokić Cvetković, R.; Sedmak, A.; Popović, O.; Cvetković, I.; Radu, D. Effect of Shielding Gas Arc Welding Process on Cavitation Resistance of Welded Joints of AlMg4.5Mn Alloy. *Materials* **2023**, *16*, 4781. https://doi.org/10.3390/ma16134781

Academic Editors: Cosmin Codrean, Carmen Opriș and Anamaria Feier

Received: 18 May 2023
Revised: 27 June 2023
Accepted: 28 June 2023
Published: 2 July 2023

1. Introduction

The need to reduce vehicle weight and emissions and to make fuel savings has led to an increase in the use of lightweight materials, such as aluminum alloys. Among them, AlMg4.5Mn alloy stands out, used in many areas, such as tanks for storage and pipelines for transportation of liquid gases, for pressure vessels in general, in vehicles, and more recently as the main material for the construction of yachts and ships [1]. It is characterized by high strength and good resistance to corrosion and wear, as well as relatively good weldability [1]. For these reasons, the use of this alloy and the demands for its improvement and the improvement of joining and shaping techniques of this alloy are increasing [2].

The most common problems encountered when welding Al–Mg alloys are degradation of the mechanical properties in the heat-affected zone, reduction in corrosion resistance [3,4], appearance of pores as a result of hydrogen absorption from the air, appearance of cracks, primarily hot as a result of phase transformations that occur in the weld metal and the heat-affected zone, and the appearance of inclusions—most often Al_2O_3 oxides [5]. Other issues include changes in the mechanical properties, which are often related to grain size. During rapid cooling of the fusion zone, intermetallic phase precipitates, such as Mg_2Si and Al_6Mn, can form, increasing the hardness. Further, welding can result in changes in the chemical composition and solidification defects or residual stresses generated during solidification and cooling, which can influence the hardness and susceptibility to fracture. Due to the growth of grains in the HAZ, the strength decreases [6–9]. The addition of elements such as titanium, strontium, and zirconium reduce the mobility of grain boundaries, making the recovery process impossible. In this way, a finer grain is created, which contributes to the quality of the welded joint.

As the most commonly applied joining technique for Al alloys, the gas tungsten arc welding (GTAW) and gas metal arc welding (GMAW) processes are used. The GTAW process is widely used because of the improved control of the heat input. This control is carried out by the correct selection of welding parameters such as the current capacity, welding speed, dimensions and composition of the electrode, flow rate, and composition of the shielding gas [10–12]. In addition to the GTAW process, the GMAW process is also used for welding Al and its alloys; it enables higher welding speeds, a narrower HAZ, excellent cathodic cleaning effect, and welding in all positions [13–15]. However, there are still a few problems with shielded gas arc welding in respect to the mechanical properties and sensitivity to cracking, especially when GMAW is used. As an illustration, significant reduction in the fatigue strength was obtained in an analysis of the fatigue properties and stress concentration of a 6005 aluminum alloy GMAW-welded lap joint [16]. It was shown that the fatigue strength of the welded joint was just 27% of the fatigue strength of the base metal, mostly due to increased sensitivity to cracking and the geometry of the welded joint. Sensitivity to cracking was also analyzed for Al–Mg alloys, used in the shipbuilding industry, welded by GMAW and friction stir welding (FSW) [17]. It was shown that cracking started mainly through decohesion at the matrix–precipitate interfaces and that the cracking mechanism was trans-crystalline ductile.

One of the most important problems when welding aluminum and its alloys by GTAW and GMAW, in addition to the degradation of the mechanical properties, is the appearance of pores. Pores are formed as a result of adsorption, diffusion, and dissolution of gases on the surface and inside the hardened weld metal. Pores are mainly created by hydrogen that dissolves in aluminum. Since hydrogen has a small atomic diameter, it easily diffuses through the metal in the solid state, so porosity can be observed even in the heat-affected zone. Unlike steel, the pores in aluminum are distributed, mostly, in the interior of the seam, they can also be found near its junction with the base material, and sometimes they can also be observed in the heat-affected zone [18]. The effect of the shielding gas (Ar, He, or N_2, individually, or in mixtures) on the porosity in GTAW-made welded joints of AlMg4.5Mn alloy has been investigated by performing a microstructural analysis [19]. It was shown that the best results were obtained with the mixture Ar + 59%He + 0.015%N_2, whereas addition of N_2 alone did not make any difference.

A more comprehensive experimental study was performed on the same GTAW-made welded joints, including metallography, tensile, hardness, and toughness testing to establish the effects of the shielding atmosphere on the mechanical properties and fracture mechanics parameters of the weld metal [20]. It was concluded that the shielding gas mixture significantly affects the impact toughness, slightly affects the resistance to crack initiation, and somewhat more affects crack growth due to fatigue. More recently, fractography was also used to explain higher values of weld metal toughness as evaluated at three different temperatures using instrumental Charpy pendulum impact testing to measure both the crack initiation energy and the crack growth energy [21].

In addition to GTAW, welded joints of AlMg4.5Mn alloy made by the gas metal arc welding (GMAW) process, with different gas shielding atmospheres, were investigated in [1,20], to assess the effects of the gas mixture on the tensile strength, hardness, impact, and fracture toughness, as well as the fatigue crack growth parameters. The main conclusion was that an increased helium content improves the toughness and fatigue crack growth parameters, as shown in Table 1 for the fatigue crack growth rate, whereas its effect on other mechanical properties is not significant [1,22].

In addition to conventional mechanical properties and corrosion, for alloys used in water it is also important to analyze the resistance to corrosion and cavitation [23–28]. Slow strain rate testing (SSRT) was used to study the stress corrosion cracking (SCC) of 5083 Al alloy in a 3.5 percent NaCl solution after superplastic forming and various heat treatments [23], indicating severe SCC and intergranular fracture. Slow strain rate testing (SSRT) was also used to study the effect of the microstructure on the stress corrosion cracking (SCC) susceptibility of Al–Mg alloy sheet containing 6.8% Mg in cold-rolled and fully annealed conditions [24], indicating high SCC susceptibility as well.

Table 1. Data for fatigue crack growth using different shielding gases.

Shielding Gas	ΔK_{th}, $MPa{\cdot}m^{1/2}$	Coefficient C	Coefficient m	da/dN, m/cyc $\Delta K = 7\ MPa{\cdot}m^{1/2}$
Ar	2.88	4.44×10^{-10}	3.86	6.87×10^{-7}
Ar + 0.0307%O_2	2.83	4.17×10^{-10}	3.81	8.12×10^{-7}
Ar + 48%He + 0.0290%O_2	3.02	1.05×10^{-10}	4.07	2.89×10^{-7}

Cavitation is the process of the rapid formation, growth, and implosion of bubbles in a fast-flowing liquid. The implosion causes the formation of shock waves and microjets, which, in contact with the surfaces of the elements in hydro-systems, cause their damage in a short time. In the initial stage, the degree of damage is negligible and this is the incubation period. However, repeated impacts of microjets and shock waves during the cavitation effect leads to material damage [25], causing serious problems in vital components, as described in [26]. Cavitation of Al–Mg alloys is rarely investigated, especially for their welded joints. The correlation between the mechanical properties and cavitation resistance of Al–Mg alloy 5083 was analyzed in [27] to obtain the best combination of heat treatments to apply to cast aluminum products. In ref. [28], the cavitation erosion characteristics of the EN AW-6082 aluminum alloy's surface remelting by GTAW was analyzed. It was shown that the GTAW-remelted layers had cavitation erosion resistance 5–6 times more than the base metal, due to fine graining and microstructure.

In this paper, the cavitation resistance of the welded joints of the three-component aluminum alloy AlMg4.5Mn was investigated by examining the microstructure, determining the hardness distribution, measuring the mass loss, and by fractographic analysis of damaged surfaces. Tests were performed on samples of the base metal and weld metal extracted from the welded joints, made by GTAW and GMAW processes in a shielding atmosphere of pure argon. In this way, the effect of the welding process itself is assessed, whereas the effect of the shielding gas will be analyzed in future research.

2. Materials

For the experimental determination of resistance to cavitation, samples of welded joints of AlMg4.5Mn alloy obtained by the GTAW and GMAW processes were used. The chemical composition and mechanical properties of AlMg4.5Mn alloy are shown in Tables 2 and 3.

Table 2. Chemical composition of aluminum alloy AlMg4.5Mn.

Element	Si	Fe	Cu	Mn	Mg	Zn	Cr	Ti
mass %	0.13	0.21	0.04	0.66	3.95	0.03	0.06	0.025

Table 3. Mechanical properties of aluminum alloy AlMg4.5Mn.

	Tensile Strength R_m (MPa)	Yield Strength $R_{0.2}$ (MPa)	Elongation A (%)	Total Impact Energy (J)
Rolling direction	293	133	25	41
Transverse direction	304	143	24	32

Aluminum alloy plates of AlMg4.5Mn, of dimensions 500 mm × 250 mm × 12 mm, were used for welding, and "Y" grooves were made by milling, as shown in Figure 1, for GMAW. The plates were welded in four passes: one root pass and three filling passes. The appearance of macrographs of GTAW- and GMAW-welded joints are shown in Figure 2. The preheating temperature was above 110 °C for GTAW welding and above 70 °C for GMAW welding. Argon 7.0 with a high purity was used as the shielding atmosphere for both the GTAW and GMAW. The gas flow was 17–19 L/min for the GTAW and 15–16 L/min for the GMAW welding. The welding was performed in horizontal position (PA).

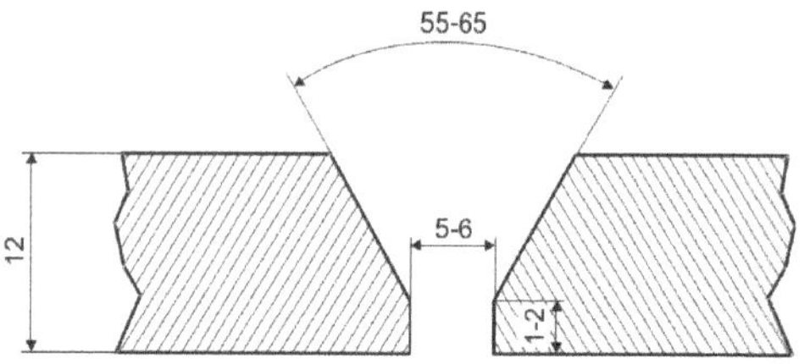

Figure 1. Shape and dimensions of "Y" groove for GMAW.

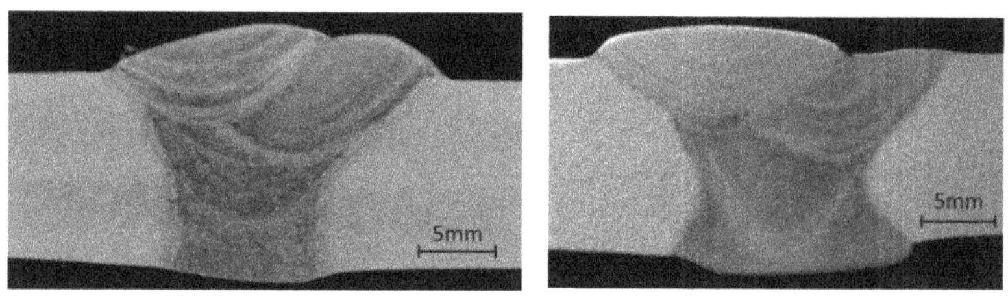

(**a**) GTAW process (**b**) GMAW process

Figure 2. Macrographs of welded joints [17].

For GMAW, welding electrode wire with added zirconium AlMg4.5MnZr in the form of coils weighing 7 kg, Ø1.2 mm was used as a filler metal (Table 4). Zirconium affects the decreasing of the grain size.

For GTAW, welding electrode WZr3 was used and wire made of aluminum alloy AlMg4.5Mn (classifications DIN1732/SG-AlMg4.5Mn or BS2901/5183 or AWS A5.10/ER 5183), Ø5 mm and length 1000 mm, was used as a filler metal (Table 5).

For GTAW, alternative current (AC) was used, whereas direct current reverse polarity (DCEP) was used for the GMAW. The welding parameters, including current, I, voltage, U, welding speed, V_z, and the resulting heat input, $Q = 0.06 \cdot U \cdot I/V_z$, are shown in Table 6.

Table 4. Chemical composition of filler material (GMAW process, remaining is Al).

Element	Si	Fe	Cu	Mn	Mg	Zn	Cr	Ti	Zr
mass %	0.07	0.21	0.01	0.71	4.6	0.02	0.07	0.09	0.11

Table 5. Chemical composition of filler metal (GTAW process, remaining is Al).

Element	Si	Fe	Cu	Mn	Mg	Zn	Cr	Ti	Others	
									Each	Total
mass %	<0.40	<0.40	<0.10	0.5–1.0	4.3–5.2	<0.25	0.05–0.25	0.15	<0.05	<0.15

Table 6. Welding parameters.

Process	I (A)	U (V)	V_z (cm/min)	Q (kJ/cm)
GTAW	215–220	20–22	11–15	19.4–23.5
GMAW	160–170	20–22	27–32	7.0–7.1

3. Testing Methods

Before testing the resistance of the samples to the effect of cavitation, metallographic tests were performed to determine the microstructure of the samples. Samples for microstructural tests were cut from the welded plates. These samples were polished and etched in 10% H_3PO_4.

In addition, the hardness was measured to determine its distribution in the welded joints, using the standard Vickers method HV5, on the polished samples. The hardness was measured in all zones of the weld joint BM, HAZ, and WM, on 3 levels: along the root, mid-thickness, and close to the surface. The distance between the measuring points was 1.5 mm.

The ultrasonic vibration method (with a stationary sample), according to the ASTM G32 standard [29], was used to test the cavitation resistance of the AlMg4.5Mn alloy and welded joints. A test sample and the device on which the tests were performed are shown in Figure 3. The ultrasonic vibration method is based on the creation and implosion of cavitation bubbles on the sample's surface and measuring the mass loss of the sample during the time of exposure to cavitation. The selection of characteristic parameters for testing was performed in accordance with the ASTM G32 standard: frequency of mechanical vibrations was 20 ± 0.2 kHz; amplitude of mechanical vibrations at the top of the concentrator was 50 ± 2 μm; the gap between the test sample and the concentrator was 0.5 mm; water flow was 5–10 mL/s; water bath temperature was 25 ± 1 °C. Before testing, all samples were polished. The change in mass loss was measured with an analytical balance with an accuracy of 0.1 μg. Before measurement, the samples were dried with warm air and kept in a desiccator in order to separate the residual moisture in the material. The samples were measured after 30 min of cavitation exposure for a total test period of 180 min. The results are shown in the mass loss versus time duration exposure diagram. The least squares method was used to construct straight lines whose slope specifies the cavitation rate.

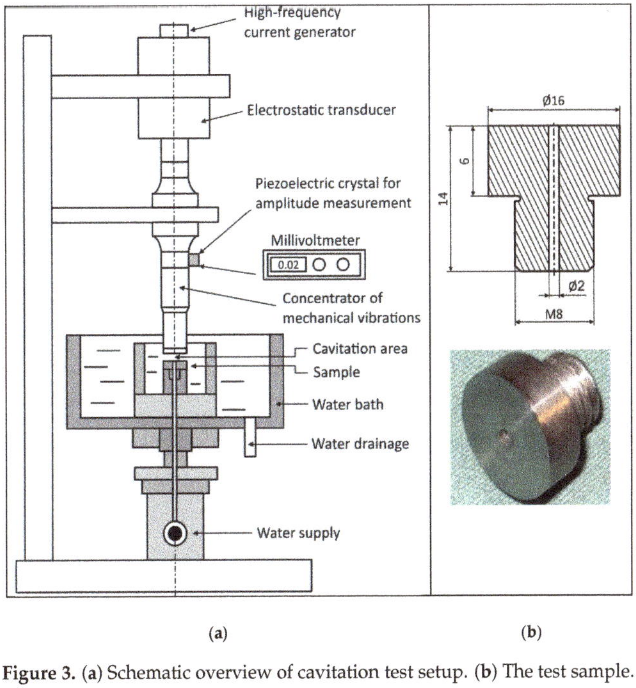

Figure 3. (**a**) Schematic overview of cavitation test setup. (**b**) The test sample.

Scanning electron microscopy (SEM) was used to analyze the morphology of the surface damage of the samples. The FESEM Mira Tescan x3m microscope (Brno, The Czech Republic) was used. The analysis of damage to the sample surface was performed after 60, 120, and 180 min of exposure of the samples to the effects of cavitation.

4. Results and Discussion

4.1. Metallographic Tests

Figure 4a shows the microstructure of aluminum alloy AlMg4.5Mn. The microstructure consists of metal grains of directional orientation, i.e., it is a typical rolling microstructure with a fine Mg_2Al_3 mesh at the grain boundaries. Large dark microconstituents were also observed, which are mostly Mg_2Si and (Fe, Mn)Al_6, as explained in [21]. The pores in aluminum alloys are largely distributed within the weld metal, and sometimes in the HAZ, as shown in Figure 4b, where a pore has a large diameter, almost equal to the width of the HAZ [19]. Figure 5 shows the microstructures of the weld metals obtained by the GTAW (Figure 5a) and GMAW (Figure 5b) processes.

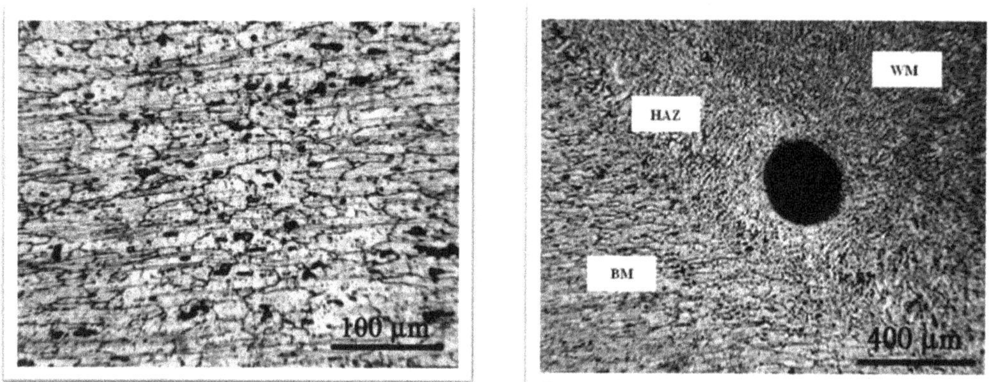

(**a**) (**b**)

Figure 4. (**a**) The microstructure of aluminum alloy AlMg4.5Mn [21]. (**b**) Large pore in the HAZ of alloy AlMg4.5Mn [19].

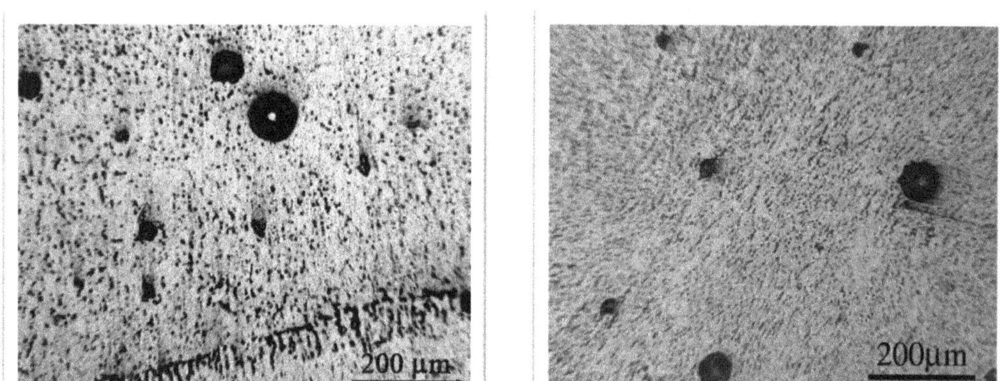

(**a**) (**b**)

Figure 5. Microstructure of the weld metal: (**a**) GTAW [19], (**b**) GMAW [22].

The weld metal in both welding processes has a dendritic structure. Intermetallic phases of the type (Fe, Mn, Cr)Al_6, (Fe, Mn, Cr)$SiAl_{12}$, and Mg_2Al_3 are homogeneously distributed in the weld metal. The porosity is dominantly distributed around the interface between two passes and the interface between the base metal and the weld metal. A somewhat lower porosity and a finer-grained microstructure is obtained with the GMAW process [22]. This is probably the result of the use of a filler material with the small addition of zirconium. It has been found that zirconium may be used to promote a fine grain size, acting as nuclei for the formation of very fine grains during solidification. The grain size has a dominating effect on the mechanical properties in the case of GMAW, while the shielding gas composition effect is dominant for GTAW.

4.2. Hardness Test

The results for the hardness measurements are shown in Table 7. The hardness of the base material ranged from 67.3–74.4 HV. The hardness of the weld metal produced by the GTAW welding process ranged from 67.1–78.5 HV, while that of the weld metal produced by the GMAW process was 70–88.5 HV. The somewhat higher values of the hardness of the weld metal obtained by the GMAW process are the result of a finer grain structure. The hardness of the HAZ was 67.5–76.6 in the GTAW-welded joint, and 71.3–81.5 in the GMAW-welded joint. The hardness values indicate a homogeneous microstructure since the values are in the range of ±7.5%, except in the case of the BM made by GMAW where the range was ±12%. This appears to be the consequence of a harder surface layer obtained by the GMAW process. One can see that the hardness is highest in WM made by GMAW and lowest in BM, with WM made by GTAW closer to BM than to WM made by GMAW. The hardness in the HAZ follows the same trend.

Table 7. Hardness measurement, HV_5.

	GTAW			GMAW		
	Face	Center	Root	Face	Center	Root
BM	74.0			78.6		
	74.4	67.3	71.3	79.8	72.3	77.9
	72.2	70.1	70.6	80.2	78.4	81.6
	71.6	68.5	73.2	76.2	72.7	83.1
	72.0	68.6	72.0	81.1	74.8	82.1
	73.5	74.9	72.2	80.3	73.9	83.3
HAZ	75.5	76.6	70.9	79.9	78.6	81.5
WM	74.7	67.1	69.3	74.0	78.2	81.7
	73.4	76.3	75.1	73.6	82.2	84.3
	74.6	74.4	77.3	82.6	70.8	84.5
	73.4	76.8	71.1	79.1	79.2	79.1
	74.7	78.5	74.1	75.1	77.4	84.1
	73.3	73.4	76.2	70.0	74.3	81.3
	73.3	75.2	74.8	71.2	75.2	88.5
	73.1	71.4	75.1	76.5	77.2	81.8
HAZ	74.3	67.5	72.3	71.3	74.2	79.4
BM	71.1	68.7	71.1	73.1	70.4	78.8
	71.4	70.7	71.8	76.1	71.3	79.6
	71.7	74.3	69.0	79.4	72.3	82.1
	74.2	72.5	73.8	78.5	78.5	82.6
	73.4			76.8		

4.3. Cavitation Rates

The results of measuring the mass loss of the samples (base material AlMg4.5Mn alloy, GTAW weld metal, GMAW weld metal) under the effect of cavitation during the testing time are shown in Figure 6. The mass loss caused by cavitation damage is plotted on the ordinate and the time intervals are given on the abscissa. Using the least squares method, the points of the diagram are approximated by a straight line with slope proportional to the mass loss after the total time of cavitation, which defines the cavitation rate. Therefore, an increase in the slope indicates a reduction in cavitation resistance. The intersection point of the straight line and the abscissa indicates the incubation period, i.e., the period of time in which there is no loss in material mass under the effect of cavitation. The incubation period of all of the samples was about 10 min (Figure 6). In this period, there was no separation of material particles under the effect of cavitation during the test.

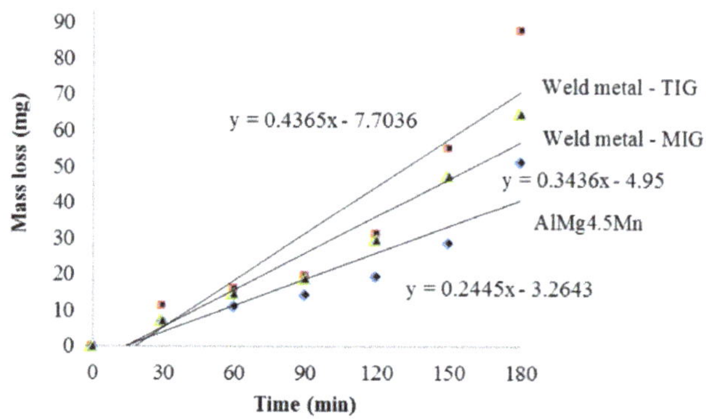

Figure 6. Cumulative mass loss during total exposure time.

Table 8 shows the mass losses of all samples after the cavitation test, which lasted 180 min. The table shows the calculated cavitation rates of all samples. The base material of the AlMg4.5Mn alloy has the highest resistance to the effect of cavitation (the lowest total mass loss and the lowest cavitation rate). The weld metal obtained by the GMAW process has a higher resistance to the effect of cavitation than the weld metal obtained by the GTAW process. This behavior is due to the smaller porosity and finer microstructure of the GMAW WM process. One should note that there is no correlation with hardness (the highest in GMAW WM, lowest in BM and close to that in GTAW WM) since the cavitation rate is the highest in GTAW WM, lowest in BM, and in between in GMAW WM.

Table 8. Total mass losses and cavitation rates of the tested samples.

Sample	Total Mass Loss (mg)	Cavitation Rates (mg/min)
Base metal—AlMg4.5Mn	51.3	0.285
Weld metal—GTAW	87.9	0.488
Weld metal—GMAW	64.5	0.358

4.4. Morphology of Damaged Surfaces

Scanning electron microscopy (SEM) was used to analyze the surface damage of all of the tested samples. Figure 7 shows microphotographs of the surfaces after 60 min of exposure to the effect of cavitation. In all samples, the presence of a uniform undulation of the surface exposed to the effect of cavitation was observed. Small pits appear, but in a matrix that is soft. There is still a pronounced grain boundary network because it forms Mg_2Al_3 which is a hard phase. Dark microconstituents are also present, especially on the surface of the base material sample (Figure 7a), which were observed on the surface of the samples before testing.

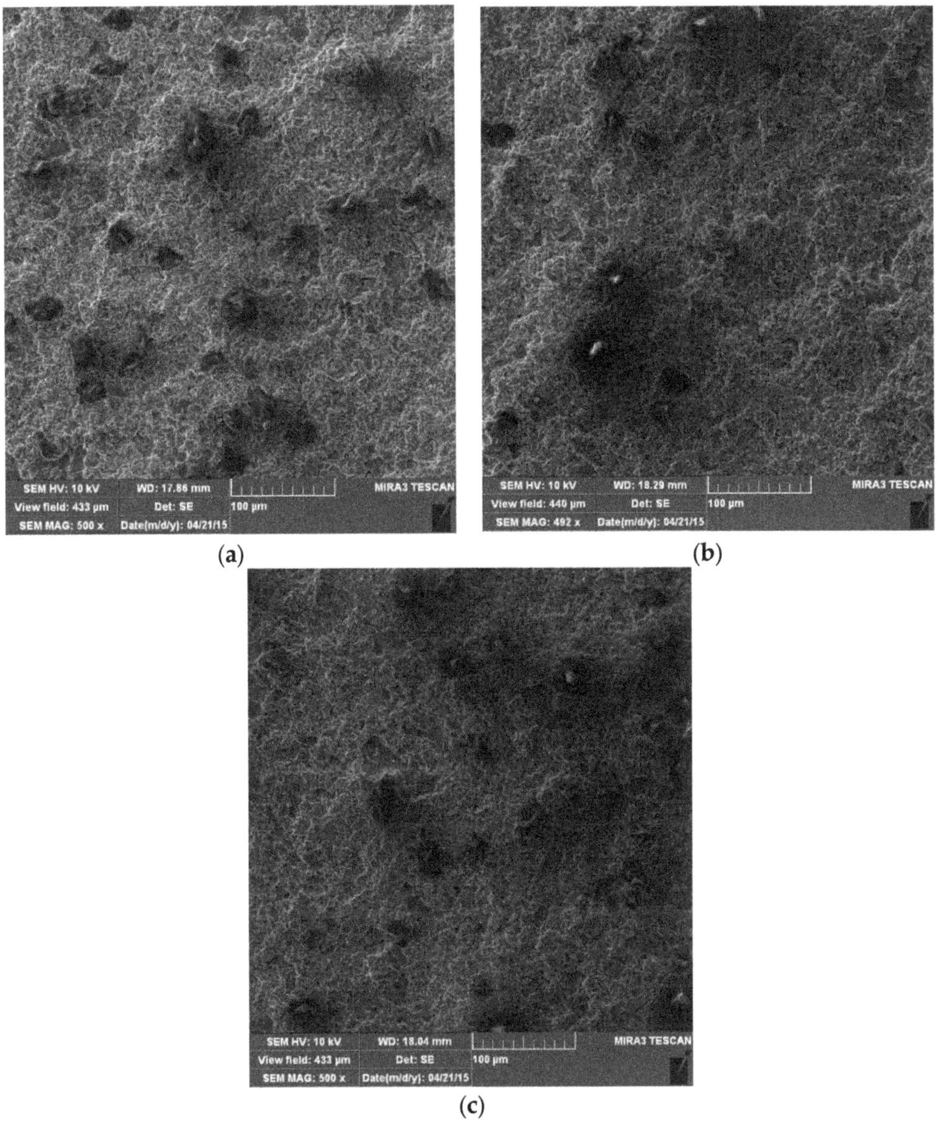

Figure 7. SEM microphotographs of damaged surfaces after 60 min: (**a**) base metal; (**b**) weld metal, GTAW; (c) weld metal, GMAW.

After 120 min of cavitation action, a greater degree of plastic deformation than in the previous test interval was observed in all samples (Figure 8), which is reflected in more expressed surface undulation. Pits created by the removal of microconstituents were observed. The degree of base damage is significantly higher than in the previous test interval, which is in accordance with the mass loss measurements (Figure 6).

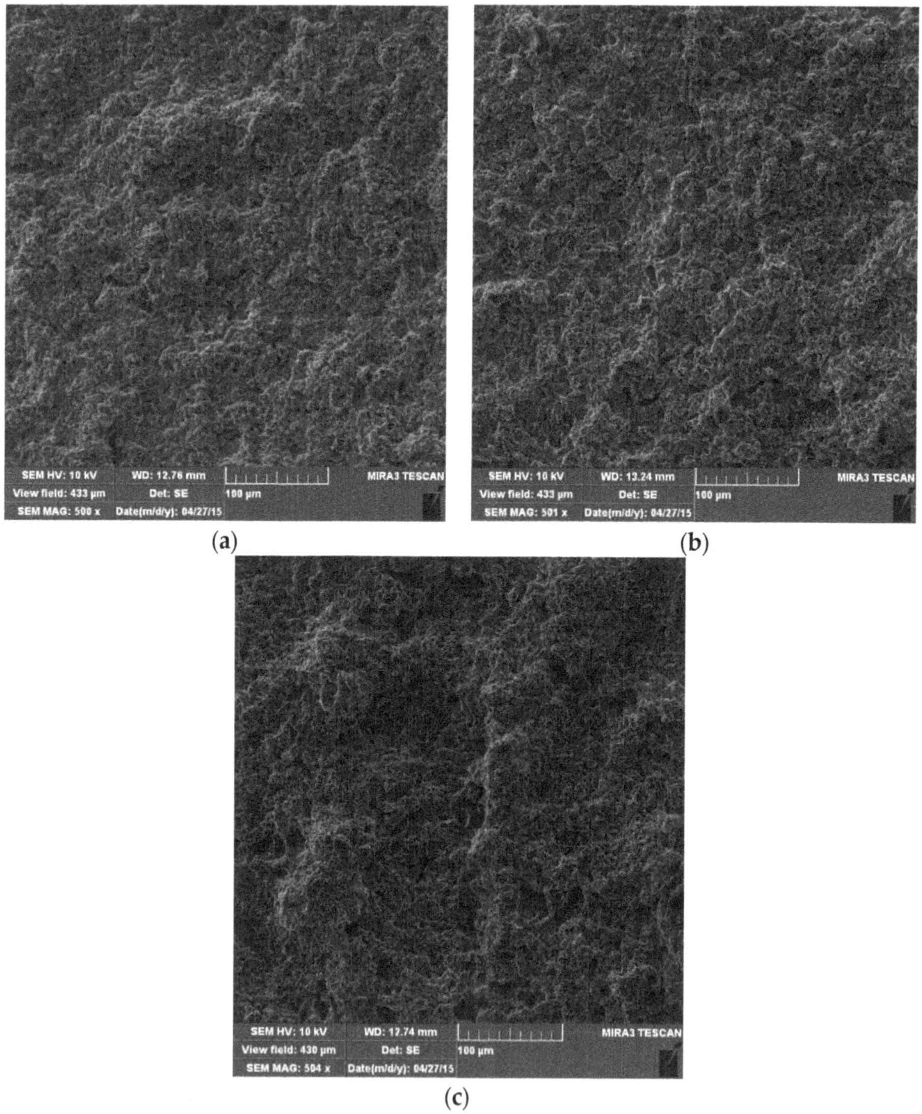

Figure 8. SEM microphotographs of damaged surfaces after 120 min: (**a**) base metal; (**b**) weld metal, GTAW; (**c**) weld metal, GMAW.

After 180 min of testing, it was observed that a large amount of material had been removed in all of the tested samples. That is why priority damaged areas cannot be identified (Figure 9). The surfaces appear distinctly wrinkled, with ridges of irregular shapes, with a morphology reminiscent of a ductile fracture mechanism. Surface damage is so much expressed that pits are not visible because they have merged and formed craters. The appearance of a fatigue-like damage mechanism was observed in all samples, as shown in detail in Figure 10. Craters with pronounced circumferential striations are clearly visible, which is a typical form of fatigue damage. The craters are most expressed in BM (Figure 9a), due to the fatigue damage mechanism, being in accordance with its highest cavitation resistance.

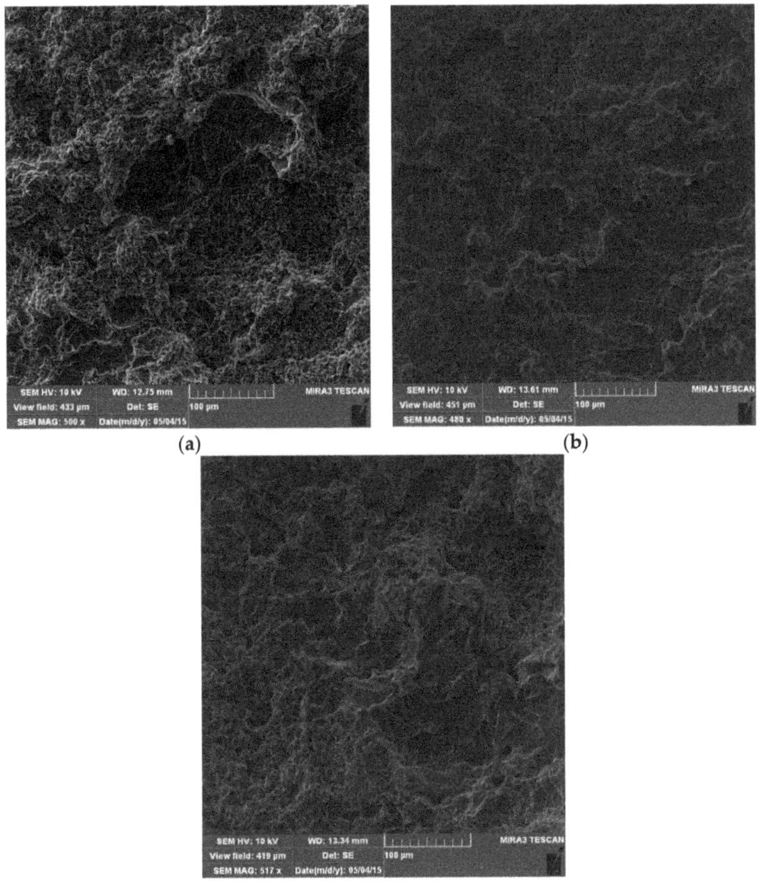

(**a**) (**b**)

(**c**)

Figure 9. SEM microphotographs of damaged surfaces after 180 min: (**a**) base metal; (**b**) weld metal GTAW; (**c**) weld metal, GMAW.

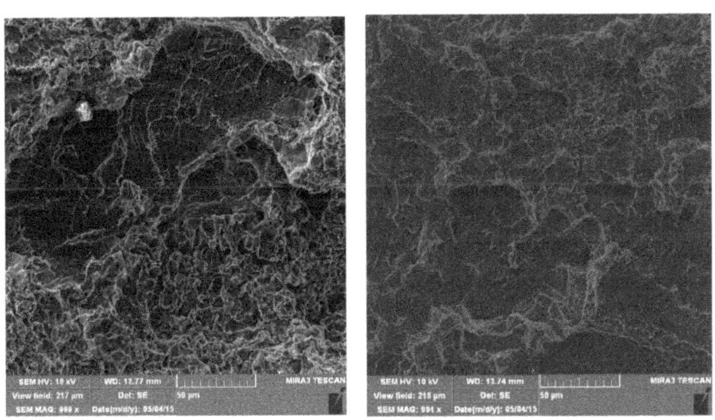

(**a**) (**b**)

Figure 10. *Cont.*

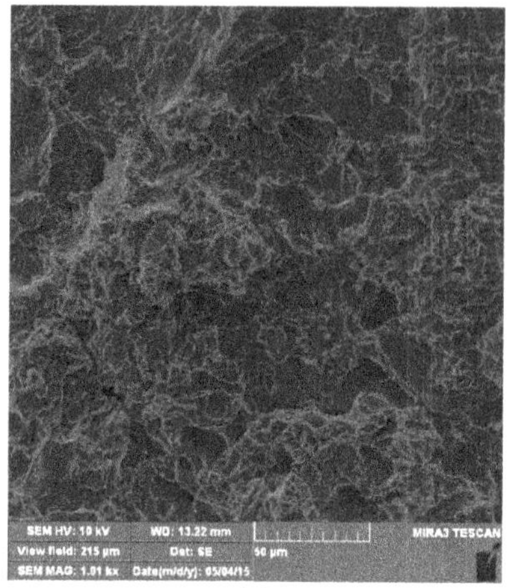

(c)

Figure 10. SEM microphotographs of craters with fatigue-like striations after 180 min: (**a**) base metal; (**b**) weld metal, GTAW; (**c**) weld metal, GMAW.

5. Conclusions

The cavitation resistance of the welded joints of aluminum alloy AlMg4.5Mn obtained by the GTAW and GMAW processes, as well as the base metal, was investigated by examining the microstructure, determining the hardness distribution, measuring the mass loss during the cavitation process, and fractographic examination of the damaged surface.

The microstructure was finer and with less pores in the case of GMAW, as a consequence of using a filler metal with Zr, which creates nuclei for the formation of very fine grains during solidification. One should note that GTAW can produce higher quality welded joints with less pores if shielding gas mixtures are used, but here the focus was on a comparison with GMAW with pure Ar as the shielding gas.

The distribution of hardness in the welded joints was determined and the results showed slightly higher hardness values of the weld metal obtained by the GMAW process compared to the GTAW process, which is a consequence of the finer-grained microstructure of the weld metal. However, hardness is not an important factor here, since it cannot be correlated with the cavitation rates for BM and MW made by GMAW and GTAW.

The base material, AlMg4.5Mn alloy, has the highest resistance to the effect of cavitation (the lowest total mass loss and the lowest value of the cavitation rate). The weld metal obtained by the GMAW process has a higher resistance to the effect of cavitation than the weld metal obtained by the GTAW process. This behavior is a consequence of the lower porosity and finer microstructure of the weld metal obtained by the GMAW process.

Finally, one can conclude from the morphology of the samples' surface damage during the cavitation that all tested samples showed fatigue characteristics. Fatigue behavior should be investigated as the next step to provide a better insight into its effects on cavitation resistance.

The composition of the shielding gas is another very important issue to be investigated in the future, since its effect on the weld metal properties, including cavitation resistance, is expected to be significant.

Author Contributions: M.D., conceptualization, investigation, formal analysis; R.P.C., conceptualization, investigation, formal analysis; A.S., validation, investigation, methodology; O.P., supervision, validation; I.C., visualization, formal analysis; D.R., validation. All authors have read and agreed to the published version of the manuscript.

Funding: This research was funded by the Ministry of Education, Science, and Technological Development of the Republic of Serbia (Contract Nos. 451-03-68/2023-14/200135, 451-03-68/2023-14/200105, and 451-03-68/2023-14/200213).

Acknowledgments: This research was supported by the Ministry of Education, Science, and Technological Development of the Republic of Serbia (Contract Nos. 451-03-68/2023-14/200135, 451-03-68/2023-14/200105, and 451-03-68/2023-14/200213).

Conflicts of Interest: The authors declare no conflict of interest.

References

1. Prokić-Cvetković, R.; Kastelec-Macura, S.; Milosavljević, A.; Popović, O.; Burzić, M. The effect of shielding gas composition on the toughness crack growth parameters of AlMg4,5Mn weld metals. *J. Min. Metall. B Metall.* **2010**, *46*, 193–202. [CrossRef]
2. Jang, K.C.; Kuk, J.M. Welding and environmetal test condition effect in weldability and strenght of Al alloy. *J. Mater. Process. Technol.* **2005**, *164–165*, 1038–1045. [CrossRef]
3. Pickens, J.R.; Gordon, J.R. Green, The effect of loading mode on the stress corrosion cracking of aluminium alloy 5083. *Met. Trans. A* **1983**, *14*, 925–930. [CrossRef]
4. Baer, D.R.; Windisch, C.F. Influence of Mg on the corrosion of Al. *J. Vac. Sci. Technol. A Vac. Surf. Film.* **2000**, *18*, 131–136. [CrossRef]
5. Mathers, G. *The Welding of Aluminium and Its Alloys*; Elsevier: Cambridge, UK, 2003.
6. Zaidi, M.A.; Sheppard, T. Effects of high temperature soak and cooling rate on recrystalization behaviour of two Al-Mg alloys (AA5252 and AA5454). *Met. Technol.* **1984**, *11*, 313–319. [CrossRef]
7. Reddy, G.M.; Mukhopadhyay, A.K.; Rao, A.S. Influence of scandium on weldability of 7010 aluminium alloy. *Sci. Technol. Weld. Join.* **2005**, *10*, 432–441. [CrossRef]
8. Katsas, S.; Dashwood, R.; Grimes, G.; Jackson, M.; Todd, G. Dynamic recrystallization and superplasticity in pure aluminium with zirconium addition. *Mater. Sci. Eng. A* **2007**, *444*, 291–297. [CrossRef]
9. Robinson, J.S.; Liu, T.Y.; Khan, A.K.; Pomeroy, M.J. Influence of processing on the properties of the aluminium alloy 2025 with a zirconium addition. *J. Mater. Process. Technol.* **2009**, *209*, 3069–3078. [CrossRef]
10. Yarmuch, M.A.R.; Patchet, B.M. Variable AC polarity GTAW Fusion Behaviour in 5083 Aluminium. *Weld. J. N. Y.* **2007**, *86*, 196–200.
11. Facts About Aluminium Welding'; AGA Gas AB SKG-203, Sweden, Member of the Linde Gas Group. 1996. Available online: www.aga.se (accessed on 17 May 2023).
12. Aesh, M.A. Optimization of weld bead dimensions in GTAW of aluminium-magnesium alloy. *Mater. Manufact. Process.* **2001**, *16*, 725–736. [CrossRef]
13. Quinn, T.P. Process Sensitivity of GMAW: Aluminium vs. Steel. *Weld. J.* **2007**, *81*, 55–60.
14. Praveen, P.; Kang, M.J. Advancements in pulse gas metal arc welding. *Mater. Process. Technol.* **2005**, *164–165*, 1113–1119. [CrossRef]
15. Kang, B.Y.; Yarlagadda, K.D.V. Prasad, Characteristics of alternate supply of shielding gases in GMA welding. *J. Mater. Process. Technol.* **2009**, *209*, 4716–4721. [CrossRef]
16. Li, Y.; Yang, S.; Peng, Z.; Wang, Z.; Gao, Z. Microstructure, Fatigue Properties and Stress Concentration Analysis of 6005 Aluminum Alloy MIG Welded Lap Joint. *Materials* **2022**, *15*, 7729. [CrossRef] [PubMed]
17. Dudzik, K.; Czechowski, M. The Cracking of Al-Mg Alloys Welded by MIG and FSW under Slow Strain Rating. *Materials* **2023**, *16*, 2643. [CrossRef] [PubMed]
18. Sven-Frithjof Goecke. Auswirkungen von Aktivgaszumischungen im vpm—Bereich zu Argon auf das MIG-impulsschweisen von Aluminium. Ph.D. Thesis, Technische Universität Berlin, Berlin, Germany, 2004.
19. Kastelec-Macura, S.; Prokić Cvetković, R.; Jovičić, R.; Popović, O.; Burzić, M. Porosity of welded joints of AlMg4.5Mn alloy. *Struct. Integr. Life* **2008**, *8*, 114–120.
20. Buyukyildirim, G. Investigation of Effects of Shielding Gas Mixture on Fracture Resistance of AlMg4.5Mn Alloy Welded Joints. Ph.D. Thesis, Univeristy of Belgarde, Beograd, Serbia, 2021.
21. Prokić Cvetković, R.; Popović, O.; Radović Lj Sedmak, A.; Cvetković, I. Fracture Behavior of AlMg4.5Mn Weld Metal at Different Temperatures under Impact Loading. *Sustainability* **2023**, *15*, 1550. [CrossRef]
22. Buyukyildirim, G.; Sedmak, A.; Prokic-Cvetkovic, R.; Popovic, O.; Jovicic, R.; Bulatovic, S. The effect of shielding gas on the toughness of AlMg4.5Mn weld metals made by GMAW. *FME Trans.* **2011**, *39*, 127–132.
23. Chang, J.C.; Chuang, T.H. Stress-corrosion cracking susceptibility of the superplastically formed 5083 aluminum alloy in 3.5 pct NaCl solution. *Metall. Mater. Trans. A* **1999**, *30*, 3191–3199. [CrossRef]
24. Popović, M.; Romhanji, E. Stress corrosion cracking susceptibility of Al–Mg alloy sheet with high Mg content. *J. Mater. Process. Technol.* **2002**, *125–126*, 275–280. [CrossRef]
25. Hammitt, F.G. *Cavitation and Multiphase Flow Phenomena*; McGraw-Hill: New York, NY, USA, 1980.
26. Dojčinović, M.; Arsić, M.; Bošnjak, S.; Murariu, A.; Malešević, Z. Cavitation resistance of turbine runner blades at the hydropower plant 'Djerdap'. *Struct. Integr. Life* **2017**, *17*, 55–60.

27. Istrate, D.; Sbârcea, B.-G.; Demian, A.M.; Buzatu, A.D.; Salcianu, L.; Bordeasu, I.; Micu, L.M.; Ghera, C.; Florea, B.; Ghiban, B. Correlation between Mechanical Properties—Structural Characteristics and Cavitation Resistance of Cast Aluminum Alloy Type 5083. *Crystals* **2022**, *12*, 1538. [CrossRef]
28. Mitelea, I.; Bordeașu, I.; Frant, F.; Uțu, I.-D.; Crăciunescu, C.M.; Ghera, C. Cavitation Erosion Characteristics of the EN AW-6082 Aluminum Alloy by TIG Surface Remelting. *Materials* **2023**, *16*, 2563. [CrossRef] [PubMed]
29. *ASTM G32*; Standard Test Method for Cavitation Erosion Using Vibratory Apparatus. ASTM International: West Conshohocken, PA, USA, 2010.

materials MDPI

Article

Optimisation of the Heterogeneous Joining Process of AlMg3 and X2CrNiMo17-12-2 Alloy by FSW Method

Anamaria Feier [1,*], Ioan Both [2] and Edward Petzek [2,3]

1 Department of Materials and Manufacturing Engineering, Mechanical Faculty, Politehnica University Timișoara, Bl. Mihai Viteazu No. 1, 300222 Timisoara, Romania
2 Department of Steel Structures and Structural Mechanics, 1 Ioan Curea Str., 300224 Timisoara, Romania
3 Department of Steel Structures and Structural Mechanics, SSF Romania, 1 Ioan Curea Str., 300224 Timisoara, Romania
* Correspondence: anamaria.feier@upt.ro; Tel.: +40-256-403612

Abstract: This paper presents experimental investigations on the solid-state joint of 3 mm sheets of AlMg3 alloy with X2CrNiMo17-12-2 stainless steel. The study presents a dissimilar joint that was made in a solid state using a modified milling cutter. The study highlights the possibility of using this type of joint in a naval field. The paper presents all the steps of the joining process, from the technological parameters to the examination and numerical validation of the obtained specimens. A numerical model was defined in Abaqus, considering a Static analysis, and the results demonstrated a good similarity with a small discrepancy observed in the elastic range of the specimen behaviour. In the conclusions, this study will provide some recommendations for the optimisation of this joint and proposals for future studies; the idea for this study started from the dissimilar joints used in the naval field. The article also briefly presents some dissimilar joints made on the same milling machine and in the same laboratory.

Keywords: dissimilar joint; FSW process; milling machine; solid-state joining

Citation: Feier, A.; Both, I.; Petzek, E. Optimisation of the Heterogeneous Joining Process of AlMg3 and X2CrNiMo17-12-2 Alloy by FSW Method. *Materials* **2023**, *16*, 2750. https://doi.org/10.3390/ma16072750

Academic Editor: Tomasz Trzepieciński

Received: 1 January 2023
Revised: 14 March 2023
Accepted: 28 March 2023
Published: 29 March 2023

1. Introduction

In the early 1990s, a new joining technology called friction stir welding (FSW) was invented and patented by TWI Cambridge [1]. The procedure was first applied at an industrial level in Sweden, in the year 1995 [1]. Due to its qualities, this procedure brought interest to some economically powerful countries such as USA and Japan.

In 2015, Lucian A. Blaga and his team had developed a special process on Friction Riveting (FricRiveting) as a new joining technique in GFRP lightweight bridge construction; through this process, it has been possible to make joints from different materials for emergency bridges [1].

A current concern in society is the compromise between the benefits of using lightweight materials and how to integrate these into larger multi-material designs projects. The wider the range of possible joining technologies to perform dissimilar joints, the less compromising or restricted might be the usage of these materials [1]. The more traditional and well-established methods to perform connections between different material classes are known as mechanical fastening, but most recently, the FSW process and FSW-derived processes solved a lot of situations. The recent research on this process demonstrated that it can also be applied in civil engineering, namely bridge construction [2].

Analysing, as an example, the case of emergency bridges [1,2], where joints from different materials were proposed, the joint studied in this work is suitable for certain construction areas.

The schematic illustration of friction stir welding is presented in Figure 1 for a butt joint configuration. Friction stir welding (FSW) is a solid-state process, which means that the objects are joined without reaching the melting point. In friction stir welding (FSW), a cylindrical shouldered tool with a profiled pin is rotated and plunged into the joint area between two pieces of sheet or plate material.

Friction stir welding (FSW) relies on heat derived from the friction between the active element and the materials to be welded, heat that produces the softening of the marginal areas of the base materials. Through the advance movement of the tool along the joint, the plasticised material becomes homogenous, and through cooling, a solid-state joint between the base materials is produced [1,2].

The friction stir welding (FSW) process involves a joint formation below the basematerial's melting temperature. The heat generated in the joint area is typically about 80–90% of the melting temperature [3,4].

There are a variety of friction welding techniques:

Rotary Friction Welding—the most popular type of friction welding and used for parts where at least one piece is rotationally symmetrical such as a tube or bar.

Linear Friction Welding—used for jet engine components, near-net shapes, and more where the limitation on the parts is only based upon the mass of the moving part, not the geometry of the interface.

Friction stir welding—often used for aluminium plates, extrusions, and sheets where seam or butt welds are made between thin components without a restriction on the component length.

There are a variety of types of friction stir welding

- Friction Stir Spot Welding;
- Double-Sided friction stir welding—Stationary Shoulder FSW. [5]

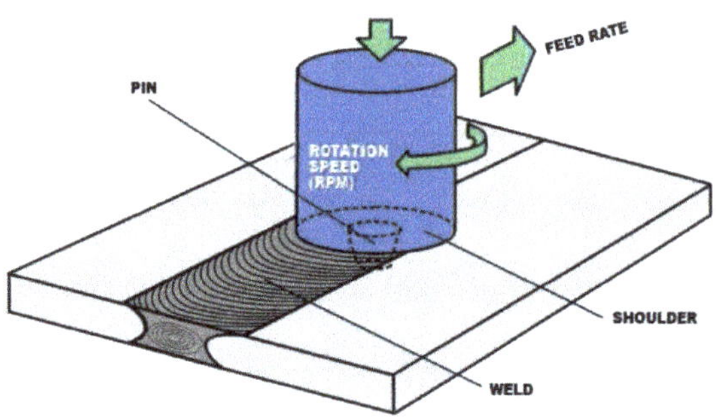

Figure 1. Schematic illustration of friction stir welding [5].

The main features of friction stir welding are:

✓ Friction-based joining procedure.

✓ Solid-state joining process.

Metallurgical advantages of the procedure [5]: There is no melting, reduced deformations, and high dimensional stability and reproducibility. The quantity of alloying elements from the materials does not decrease, there are excellent mechanical properties in the welded joint and fine granulation structure, and there is an absence of the cold cracking phenomenon.

Advantages in terms of environmental protection [5]: There is no need to use protection gas, minimal preparation procedures of the surfaces, and no waste resulted from the operation of polishing; it does not involve the use of solvents and degreasers. There is a low consumption of consumables and absence of harmful emissions.

Energetical advantages [5]: There is more reduced energy consumption compared to laser welding; there is reduced fuel consumption in automotive, naval, and aeronautical applications because of the low weight of the welded parts; the weight reduction results from the use of improved materials.

Disadvantages of FSW [5]: Tool wear/costs. A great part of the tool wear takes place in the plunging phase; the welding speed for single pass welds is lower for some alloys than for other electric arc welding processes; the equipment used for FSW is massive and expensive due to the high pressing forces; high melting point materials such as steel and stainless steel have some limitations in terms of the welding tool; the absence of an additive material leads to difficulties in producing corner welds; there is a presence of a crater (keyhole) at the end of the welding seam

Friction stir welding can be used in the following industrial domains [5]: Naval and offshore constructions, automotive, railway, aerospace, fabrication, and others (electrical, oil and gas, nuclear industry, and construction). In the area of naval constructions, FSW can be used for [5]: Aluminium panels for freezing fish on fishermen, joining extruded elements for bridges, comb-like panels, and panels that are resistant to salty water.

In Figure 2, the influence zones of a welded joint obtained through the friction stir welding technology are represented [5].

These are as follows:

- BM—base metal;
- HAZ—heat-affected zone;
- TMAZ—thermo-mechanically affected zone;
- NZ—nugget zone.

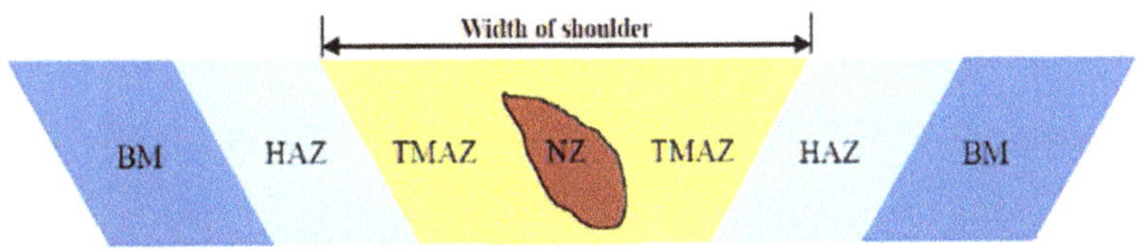

Figure 2. Influence zones of an FSW joint [5].

In the heat-affected zone, the microstructure and mechanical properties are affected by the heat generated during the FSW process, but plastic deformation does not occur.

The thermo-mechanic-affected zone is the one where the material suffers mechanical deformation. It is plastically deformed, and the process is comparable to the hot-metal working [6–9].

The nugget zone is characterised by intense plastic deformation and heating through friction during the FSW process, which leads to the formation of a fine-grain, recrystallised microstructure. This is the zone formerly occupied by the tool's pin. The central nugget contains fine grains and is formed by different thickness layers, such as "onion rings" (also known as "metallurgical strips"). This repetitive macroscopically visible model in the cross and side section of the weld is a unique characteristic which appears during FSW and related processes. As a result, the fine-grain microstructure confers excellent mechanical properties, a good fatigue resistance, high deformability, and very good plasticity [7,8].

Specific defects of FSW welding:

(1) Lack of connection appears when welding dissimilar materials and if there is insufficient diffusion between the materials.
(2) Lack of penetration appears at the inferior base of the sheets, insufficient welding depth, and inadequate welding pin length.
(3) Defects of the closing area— uncleanness or traces on the active rotating element negatively influence the aspect of the weld and also the insufficient pressing forcein-sufficient pressing force.
(4) Structure defects— inadequate thermic regime.

Change of the crystalline grains and the length and pin type.

2. Materials and Methods

The method used in this study is intended to highlight the achievement of a dissimilar joint using a basic method and an upgraded milling machine. The joining of the two materials used was carried out in a solid state without one of the materials used reaching the melting point, which is basically using the principles of the FSW process. The milling machine used was upgraded with a pin, which is a tool specific to the FSW process.

Materials

The materials used in the experiments are:

- ➢ AlMg3 (EN AW 5754);
- ➢ X2CrNiMo17-12-2.

The AlMg3 alloy is a medium-hardness alloy recommended for welded structures in a nuclear environment, the production of boilers and heaters, applications in a marine environment (boat production), and road signs [10].

The X2CrNiMo17-12-2 alloy is characterised by a good intergranular corrosion resistance up to a temperature of 300 °C, good resilience, and heat resistance up to temperatures of 550 °C. It is used in the following industries: pharmaceutical, chemical, food, aerospace, automotive, in construction, and in petrochemistry [11].

The chemical composition and the mechanical characteristics of the AlMg3 alloy are presented in Tables 1 and 2. Table 3 represents the mechanical properties and the content of alloying elements of the X2CrNiMo17-12-2 stainless steel [11,12].

Table 1. Chemical composition of AlMg3 [11].

Charge	S37106411	Min.	
Si	0.25	0.4	
Fe	0.33	0.4	
Cu	0.059	0.1	Obs: 0.10–0.06 Mn + Cr
Mn	0.26	0.5	
Mg	2.8	2.6 0.3	
Cr	0.035	0.05	
Ni	0.01	0.2	Each max.: 0.05
Zn	0.04	0.15	
Ti	0.016	0.05	All max.: 0.15
Ga	0.011	0.05	
V	0.015		
Al	96.17	Difference	

The infrastructure used for the solid-state joining of the two materials AlMg3 + X2CrNiMo17-12-2 was a milling machine on which a pin made of 56SiCr7 steel was mounted. The pin was crafted on some lab from where the milling machine is [13,14].

Table 3 contains a comparison in terms of several criteria between a milling machine such as the one used in the experiments and some equipment that is usually provided in the processes of friction welding with a rotating active element [15,16].

Table 4 shows the chemical composition of the turned pin's material.

Table 2. Mechanical properties of AlMg3 and X2CrNiMo17-12-2 [11].

Mechanical Properties of AlMg3	Value
R_m (MPa)—min. values from Standard (EN 485)	190
R_m (MPa)—max. values from Standard (EN 485)	240
R_m (MPa)—measured values	212
$Rp_{0.2}$ (MPa)—min. values from Standard (EN 485)	80
$Rp_{0.2}$ (MPa)—measured values	128
Elong. % 50 mm—values from Standard	16
Elong. % 50 mm—measured values	26.5
E[MPa]	200,000
Elong. % -mm	45
R_m (MPa)	500
$Rp_{0.2}$ (MPa)	200
ρ (g/cm^3)	8
C%	0.03
Si%	1.00
Mn%	2.00
P%	0.045
S%	0.015
Cr%	16.5
Mo%	2.5
Ni%	10–13
N%	0.11

Table 3. Comparison between the milling machine and the equipment used for FSW welding.

	Milling Machine	FSW Equipment	Parallel Robot	Articulated Robot
Flexibility	Low	Low/Medium	High	High
Cost	Medium	High	High	Low
Stiffness	High	High	High	Low
Work volume	Medium	Medium	Low	High
Setting time	Low	High	Medium	Medium
Number of programs	Low	Medium	High	High
Capacity of producing complex welds	Low	Medium	High	High
Type of control	Movement	Movement/force	Movement	Movement

Table 4. Chemical composition of the pin's alloy % (56SiCr7, in accordance with SR EN 10089:2003 [11].

Chemical Composition	Value
C%	0.52–0.6
Si%	1.6–2.0
Mn%	0.7–1.0
P%	max. 0.25
S%	max. 0.25
Cr%	0.20–0.45
Cu + 10 Sn	<0.6

In Figure 3a,b are shown images illustrating the infrastructure used in the experiments, more specifically, the milling machine and the pin made of 56SiCr7; in Figure 4a,b, the steps of obtaining the joint by the principle of the FSW process are presented [12]. The joint was obtained using a universal milling machine FUS 32. The specifications of the milling machine used for the FSW process are presented in the technology, as shown below. The pin had a cylindrical shape with striations to obtain as much material as possible into the joint.

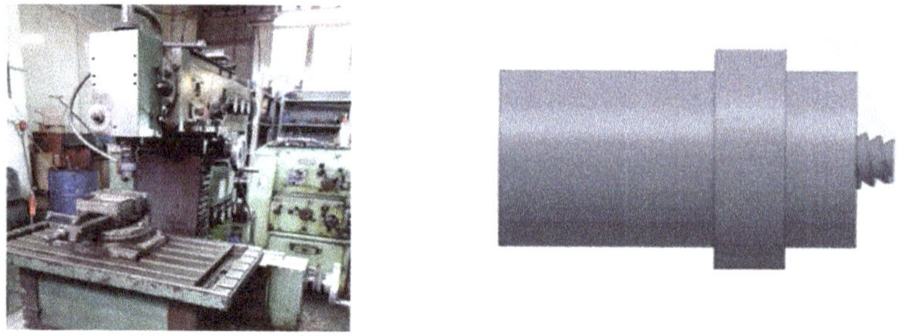

(a) (b)

Figure 3. Equipment used for the FSW experiments, consisting of a mill machine with the following characteristics: Speed control (**a**); FSW tool manufactured for the present work—the length of the pin: 2.62 mm (**b**).

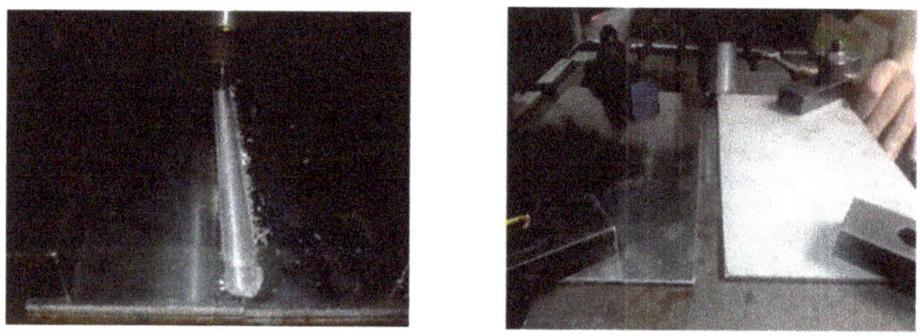

(a) (b)

Figure 4. (**a**,**b**) Images with the steps of the joint making.

The joint was obtained by inclining the pin at an angle of 2 degrees to the aluminium so that the softer material can be brought over the harder material, i.e., aluminium over stainless steel. In obtaining this type of joining, it should be observed that the softer material should be brought over the harder one so that the two different materials mix in the contact area.

3. Results and Discussion

The joint obtained started from the idea of dissimilar joints used in naval fields (namely TRICLAD joints); the joint was obtained from AlMg3 and X2CrNiMo17-12-2. The procedure of the optimisation of the technology took 3–4 weeks; at the beginning, the technology was optimised on the AlMg3 sheets with the same length of pin and after validation of the technology on AlMg3, the study was carried out on the AlMg3 and X2CrNiMo17-12-2.

The study was carried out starting from the infrastructures available in the laboratory and applying the principles of the FSW process. The aim was to demonstrate that even with a machine from the 1960s, slightly upgraded, it is possible to produce high quality and dissimilar joints. The CNC was equipped with a pin, a tool specific to the FSW process, and with the parameters presented in the next subchapter, the joining was obtained and examined both non-destructively and destructively.

3.1. The Process of Joining the AlMg3 + X2CrNiMo17-12-2 Specimens

The research began from the idea of TRICLAD. In the early stages of the research, the same defects appeared in the joint; after the optimisation of the joining technology and the shape of the pin were obtained, the results were presented in this paper.

The dimensions of the parts were 3 × 250 × 350 mm. From the parts (3 × 250 × 350 mm), smaller specimens were sampled for the tensile test and the microscopic tests (can be observed in Section 3.2).

The joining technology optimised used was:

- Milling cutter speed: 1250 rpm;
- Advance speed: 0.40 m/min;
- The angle of inclination of the pin: 2°;
- The length of the pin: 2.62 mm;

The optimisation of the procedure consisted of two stages:

➢ The first phase was the joining on one side of the AlMg3 + X2CrNiMo17-12-2 alloy.
➢ The second phase was the joining on both sides of the AlMg3 + X2CrNiMo17-12-2 alloy.

It was observed that in the first phase, an area of a lack of penetration resulted. The next step was an optimisation of the welding/joining technology; the team decided to join the parts on both sides, as in Figure 5, which means that after making the joint on one side, the parts were turned on the opposite side using the same welding technology. By joining the parts on both sides, the imperfection (area of lack of penetration) disappeared [12].

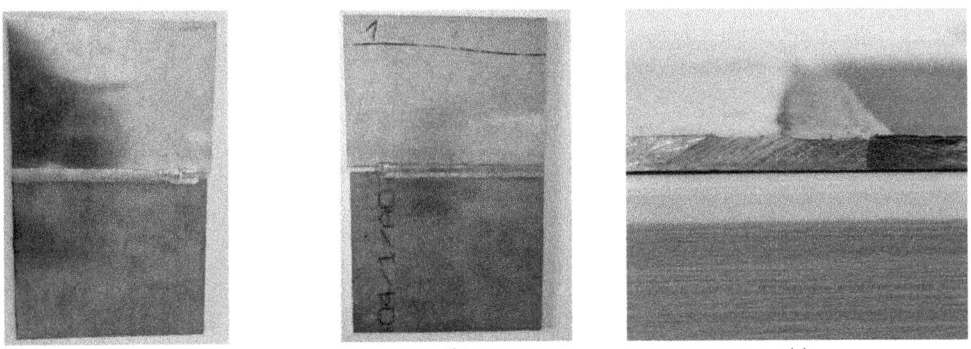

(a) (b) (c)

Figure 5. A joining of the AlMg3 + X2CrNiMo17-12-2 alloy; (**a**) first stage was the joining on one side, (**b**) the root of the joining, and (**c**) section of the joint.

As presented above in this article, a joint between the two materials utilised in the naval industry is TRICLADUL, which is an alloy that consists of two layers of aluminium and one of steel. TRICLAD is a special plating, which is generally assigned to the naval domain and facilitates the joining of aluminium structures with steel structures. It is produced in the form of a board with standard dimensions of 1.5 × 4.0 m, the useful surface of which is 1.3 × 3.8 m. Strips or other forms of semi-finished products can be obtained through cutting from it.

In Figure 6, a part made of TRICLAD and his composition can be observed; this type of material is used often in the naval industry. This material TRICLAD is used as a strip with two sheets positioned vertically [12,17] when a dissimilar joint is needed.

The TRICLAD is used for the idea that the stainless-steel layer will be welded to the underside of the vessel (the area that needs to resist corrosion; a harder, stainless material is required) and the aluminium layer of the TRICLAD composition will be welded to the top of the vessel (the deck—which is made of a lighter material, namely aluminium).

(**a**)

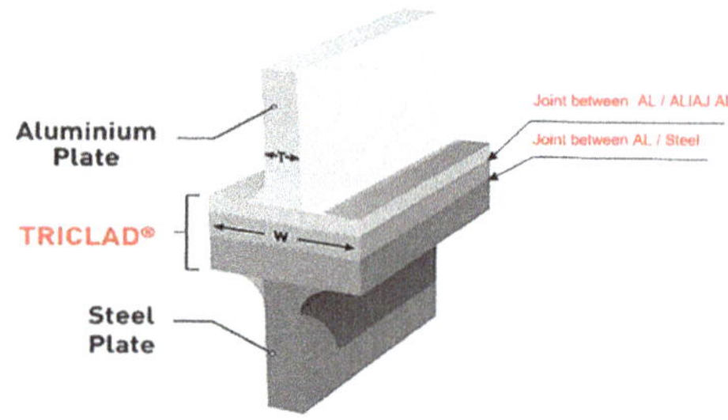

(**b**)

Figure 6. TRICLAD joint used in the naval sector; (**a**) TRICLAD strips; (**b**) TRICLAD composition.

3.2. Non-Destructive and Destructive Examination of Specimens

To evaluate the load-bearing capacity of the obtained joints, non-destructive and destructive tests were required to validate the optimised procedure used. Accordingly, to EN ISO 25239-5, the surface defects can be detected by visual inspection (macroscopic examination), apart from insufficient penetration welding and to validate the load-bearing capacity, tensile tests were needed.

The samples were examined:

- Visual;
- Tensile test;
- Microscopy analysis.

The results of the joining were:

- Excessive surface flash formation.
- Flashes are the excessive expulsion of material on the top surface, leaving a corrugated or ribbon-like effect along the retreating side that is generated under a too hot process condition or a too high weld pitch.

Flashes are caused by excessive overpressure or plunge depth; a thickness mismatch between the advancing side and retreating side [18,19] can be observed in Figure 5a.

An excessive lack of fusion may result in a reduction in terms of mechanical properties and can lead to a thinning of the material thickness. Flashes can be observed upon visual inspection. In the visual examination, the specimens fulfil the requirements, the weld was adequate, and the shape of the joint was appropriate [7,8].

After visual examination, the next step was tensile tests; in Figure 7a,b are the specimens prepared for the tensile test according to the standards.

In the second phase, i.e., the destructive testing, the tests demonstrated that two out of six specimens prepared for tensile testing failed in the joint area. The two specimens that failed are the specimens sampled from the end of the piece of the 3 × 250 × 350 mm plate, the area where the weld closure crater is located (a defect specific to this process). Figure 8 provides all six specimens after the tensile tests, and it can be observed that specimen 1 and 6 failed (at the beginning and the end of the piece).

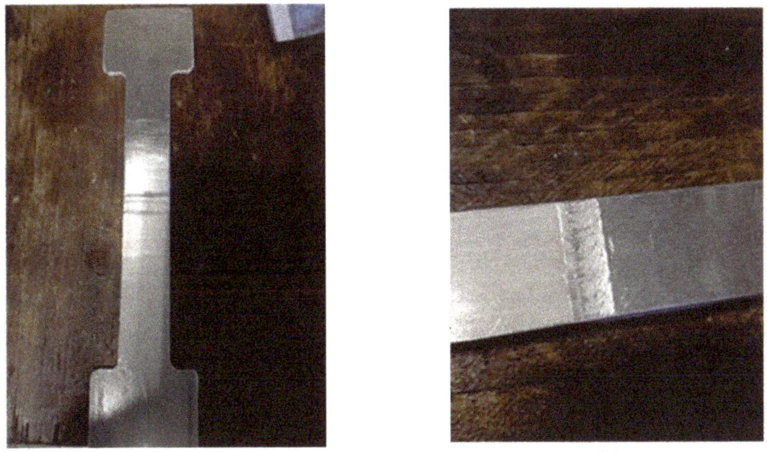

(**a**) (**b**)

Figure 7. (**a**,**b**) Pictures of specimens prepared for tensile testing.

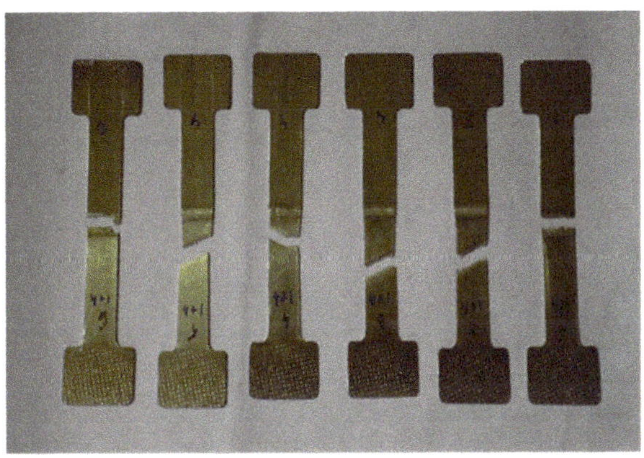

Figure 8. Pictures of the six specimens after the tensile test.

The tensile tests were carried out in the CMMC-UPT department on the TESTWELL/UTS machine. The TESTWELL/UTS tensile testing machine was used, and the machine has the following characteristics:

- Hydraulic pans;
- Tensile tests and compression tests can be performed;
- Maximum force: 250 kN;
- Computer-assisted machine control, acquisition, and post-processing of results.

Table 5 shows the results of the tensile tests performed on the specimens. The lowest yield strength was 227 MPa for specimens 1 and 6. It can be observed that the elongation A_t is also significantly lower in the specimens that failed in the welded joint.

Table 5. Tensile test results.

Specimen Identification Code	$Rp_{0.2}$	R_m	A_t
1 A-I	150.78	227.00	8.218
2 A-I	159.47	230.32	11.43
3 A-I	156.72	229.30	10.90
4 A-I	143.64	228.52	10.70
5 A-I	143.14	227.84	10.25
6 A-I	149.29	227.20	8.73

The usual tensile test was performed; specimens were made according to the tensile standard and tensile test parameters were respected, namely speed, time, specimen grips, etc (Figure 9).

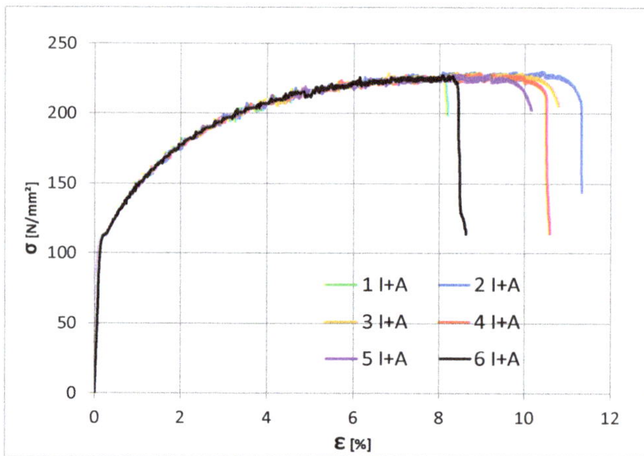

Figure 9. Tensile test curves.

As a complement to the tensile tests, some microscopy analyses were also carried out to observe possible areas with a lack of penetration.

Microscopic structure of specimens:

Figure 10 shows the microstructure of specimens; the microscopic image presents the joint and interface very well. The two materials can be observed distinctly; the lighter coloured material is AlMg3, and the darker coloured material with a higher hardness is X2CrNiMo17-12-2.

The microscopic structure of the specimens can be observed in Figure 10. In Figure 10a, the inclusions of stainless steel in AlMg3 can be observed. All the dark inclusions in the left surface look less like material adhesions and more like material pull-outs due to the rotating tool. This is an indication of a far from optimal joining technology for the material on the left, which may be a type of cold rolled, heavily roughened rolling structure.

The joint was validated because the failure of joint happened due to the final crater of the joint generated by the riveting of the pin at the beginning and end of the weld (hypothesis validated by tensile tests). The line of joint between the two materials is very well highlighted, and the meshing between them and the sealing of the joint with the softer material, namely AlMg3, can be observed

(a)

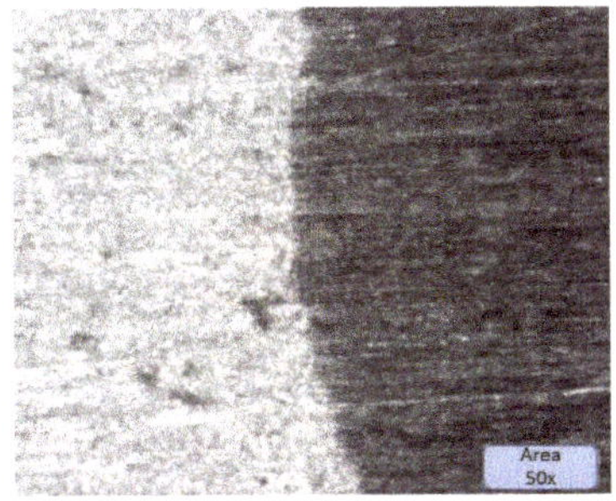

(b)

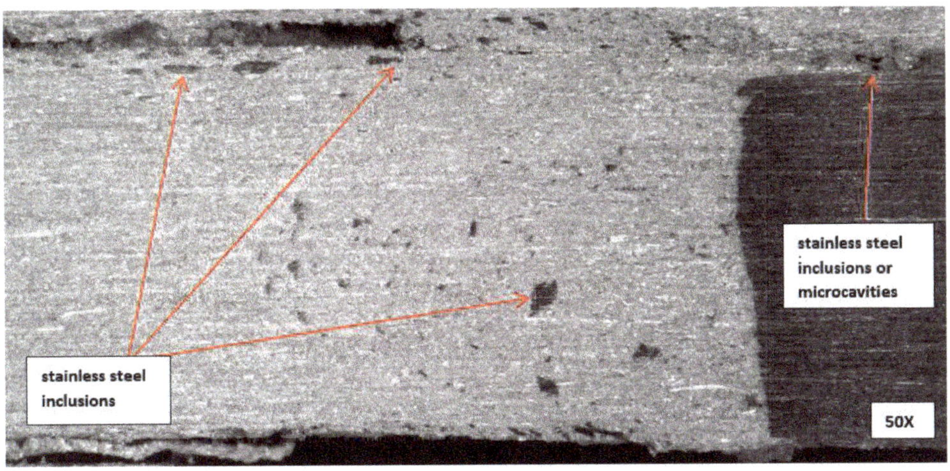

(c)

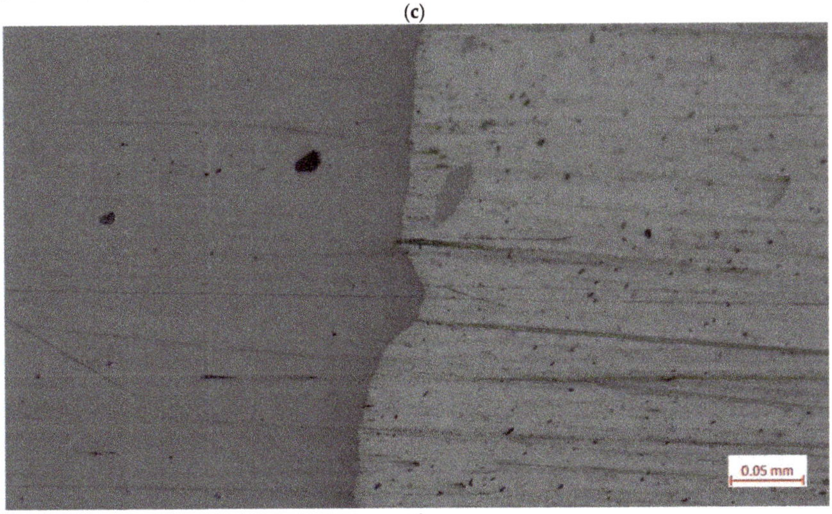

(d)

Figure 10. (**a**) Microscopic image of the aluminium area where the stainless-steel inclusions can be observed, (**b**) microscopic images of the joining area, and (**c**) microscopic image with stainless steel inclusions and microcavities. (**d**) Microscopic image at 200× with stainless-steel inclusions.

From a process point of view, the two materials do not reach the melting point; however, only in the plasticizing zone, the bonding takes place with a small anchoring of the harder material in the softer material. The presence of stainless-steel inclusions in aluminium is justified because the milling speeds were quite high (1250 rpm); other researchers who used lower speeds obtained no inclusions with similar processes (e.g., Sandnes, Lise (NTNU)).

The Interface between the two materials Is very well defined because of the fact that specimens extracted for tensile testing from outside the crater zone have validated the technology, which is also reflected in the microscopic analysis shown in Figure 10b [20–23].

4. FEM Simulation on AlMg3-X2CrNiMo17-12-2 Specimen

A numerical model was defined in Abaqus to assess the response of the entire tensile test specimen. As the tensile tests' failure occurred in the base material, the considered analysis is useful for the assessment of joint elements and not necessarily the joining.

The geometry of the tested specimen is presented in Figure 11, and it was modelled using volumetric elements as two separate identical parts. The two parts were assigned two different material constitutive laws. For the X2CrNiMo17-12-2, the nominal mechanical properties were defined as 200 Mpa, 500 Mpa, and 50% for the yield limit $R_{p0.2}$; the tensile strength—R_m and elongation at fracture—A, respectively. The aluminium alloy, AlMg3, was defined using the properties determined from tests, considering 113 Mpa, 228 Mpa, and 24% for the yield stress $R_{p0.2}$, tensile strength—R_m, and elongation at fracture—A, as shown in Figure 12.

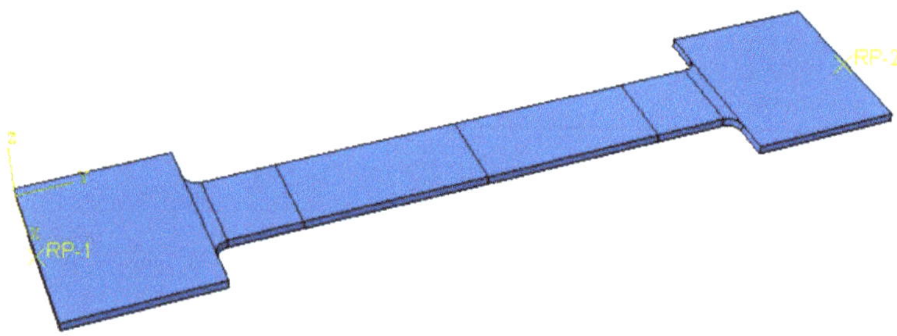

Figure 11. FEM part.

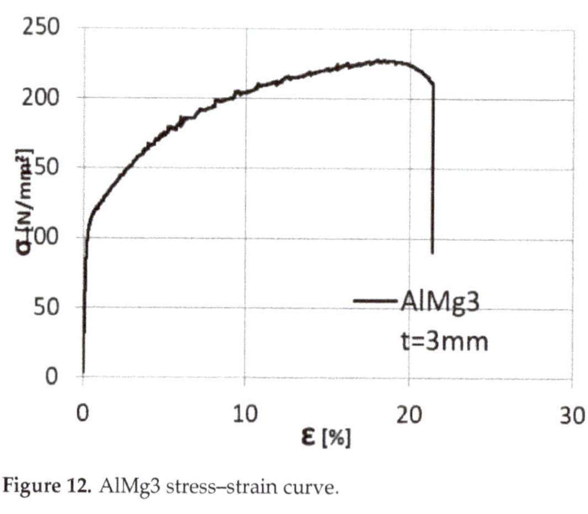

Figure 12. AlMg3 stress–strain curve.

The two parts were assembled using the *TIE* command, resulting in the specimen presented in Figure 13a, where the division lines in the parallel length represent the monitored distance of the tensile test extensometer. Rigid body constraints were defined at the ends of the specimen using two reference points. The reference points were used to define the boundary conditions. A fixed support was defined at one end, while a displacement of 30 mm was defined for the other end; Figure 13b.

The C3D8R (8-node linear brick, reduced integration, and hourglass control) finite element type was chosen for the analysis. It had an approximate size of 3 mm.

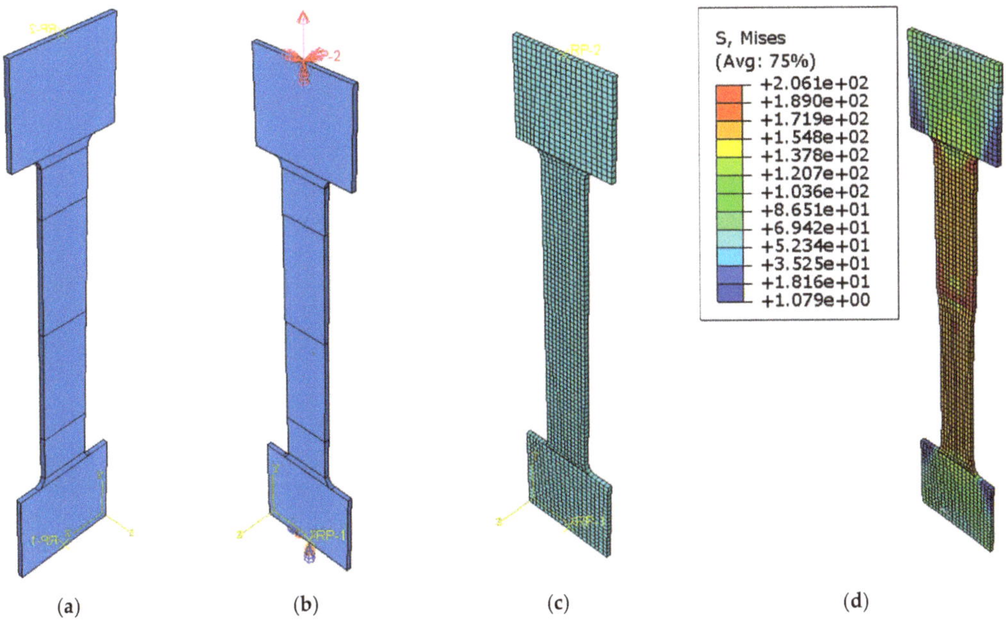

Figure 13. (**a**) Numerical model of the specimen, (**b**) boundary conditions, (**c**) finite element mesh, and (**d**) stress distribution in the specimen.

With the presented data above, a static analysis step was defined, resulting in the response of the specimen presented in Figure 14.

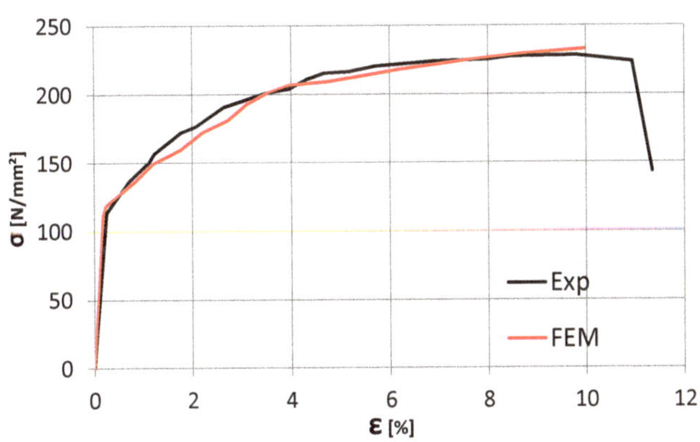

Figure 14. Comparison between the Experimental and Finite element analysis of the specimen. The two curves show a good similarity with a small discrepancy observed in the elastic range of the specimen behaviour.

5. Other Examples of Joints Obtained on Dissimilar Material on FSW Principle

There are references to various attempts to obtain dissimilar joints based on aluminium alloys using the friction stir welding process (FSW). For example, attempts have been made with good results to obtain a joint between the AlSi1MgMn alloy (EN AW 6082) and some pure copper plates using a conventional installation [12].

The pictures below present some tests performed on different materials and material combinations in the same welding lab and team.

A dissimilar joint was obtained by joining an alloy of Cu with AlMg3 (Figure 15), with the following technology.

The joining parameters were the following:

- The rotation speed of the tool—1000 rpm.
- The advance speed of the tool—0.28 m/min.
- The length of the welding pin—2.7 mm

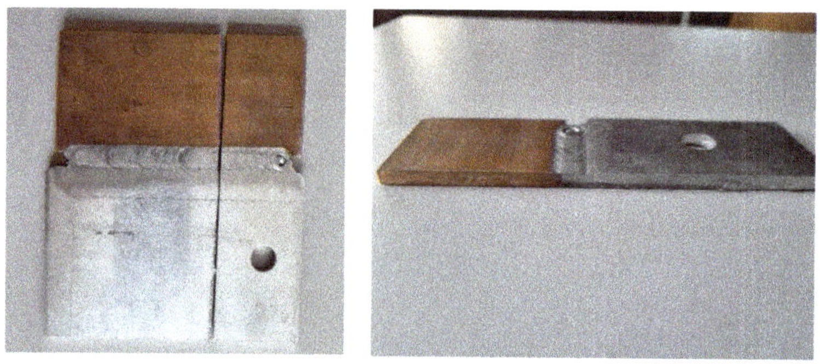

Figure 15. A joint of the AlMg3 + Cu alloy.

All of these examples are joints made with a modified milling machine on which a pin has been installed to obtain a process such as FSW in the welding lab in the Department of Materials and Manufacturing Engineering at the Faculty of Mechanics in Timisoara. All the experiments used the same length of welding pin (approximately 2.7 mm).

Microscopic structure of specimens:

In the case of the Al-Cu bonding in the microscopic images, the interaction between the two materials is better observed. From the images, the aluminium is the most engaged because the pin was on the aluminium side; moreover, in this joint, the aluminium only reached the plasticising phase. The darker-coloured aluminium zone is the zone where the aluminium has plasticised, i.e., the thermo-mechanically affected zone (TMAZ).

The nugget zone (NZ) can be observed in Figure 16b,c; in the last figure, from the base of the joint, some CU inclusions can be observed. In this case of the joint, the inclusions are more linear and follow the route of the pin.

In this joint, it is observed that there are no specific imperfections that could occur in these areas (insufficiently stirred root, voids, cracks, tunnel defects, surface lack of fill, seam underfill, blisters or surface galling, excessive surface flash formation, and irregular seam surface).

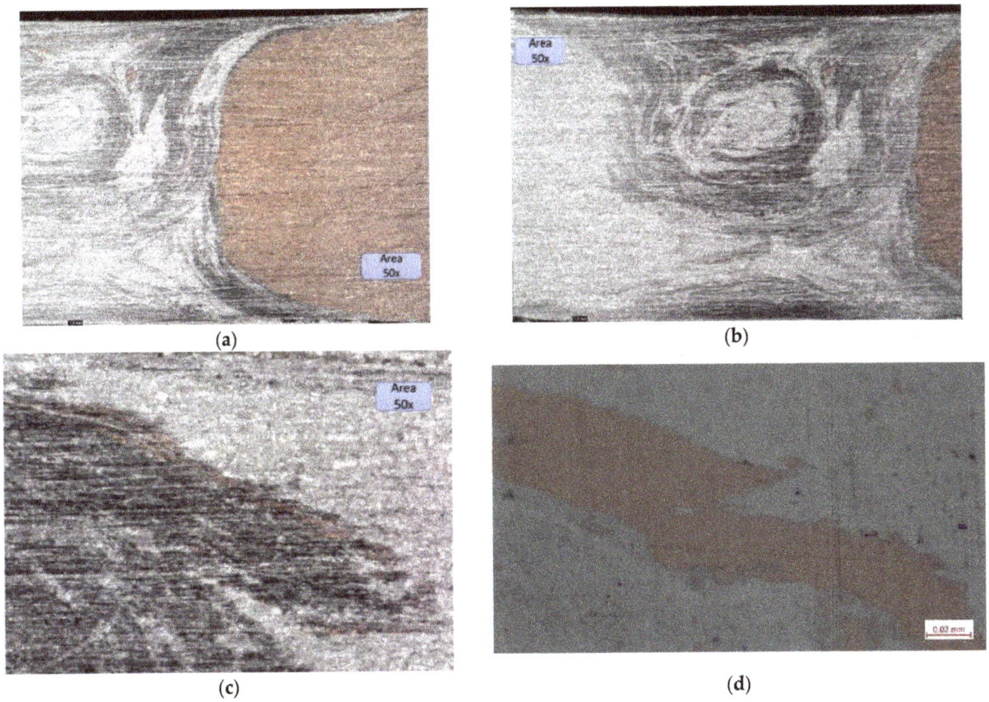

Figure 16. (**a**) the interface between the two alloys, (**b**) aluminium zone where there are few CU inclusions, (**c**) area at the base of the joint, and (**d**) microscopic image at 500×.

6. Conclusions

This study, carried out on a specific type of joint, shows that the joint is sustainable; only two out of six specimens failed in tensile tests. The two specimens failed because they were sampled from the end areas of the obtained specimen, i.e., the areas where the pin enters and exits (Figure 17).

The optimised procedure joined the specimens on both sides; in this way, the area with the lack of penetration disappeared. Before this study, the joining technology was tested and validated on two different materials, and on AlMg3 sheets only after optimising the welding technology and the infrastructure used on previous studies.

In the microstructure of the weld, in the thermally influenced zone, were observed some dark points; however, the conclusion was that the particles of stainless-steel X2CrNiMo17-12-2 alloys were incorporated into the aluminium structure. The X2CrNiMo17-12-2 particles were distributed on opposite sides by the stirring forces of the tool.

FSW demonstrated that it can be a productive and, at the same time, relatively inexpensive welding process that does not require the use of filler material, while saving the price of a professional plant.

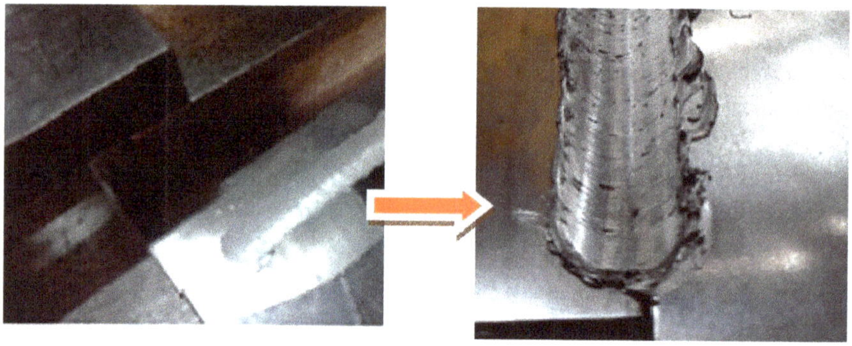

(**a**) (**b**)

Figure 17. End areas of the joint. (**a**) heterogeneous joining obtaoned with FSW process. (**b**) flashing at the end of the joining.

The much lower cost of the milling machine compared to a dedicated FSW machine, together with the possibility of using several types of pin geometries and obtaining high quality joints, are strong arguments in favour of the development of high performance-dedicated FSW-joining machines based on conventional milling machines. The research should continue to improve the flexibility of the milling machine by adding more motors, stiffening it, and developing force control solutions to prevent damage to the equipment. The research is also intended to be continued by modifying the length of the pin and the number of milling cutter revolutions, as both factors can generate a non-union zone in the joint.

This innovative solid-state method of joining opens a whole new range of welding possibilities—the low melting points of soft non-ferrous metals no longer pose a problem and, in the future, maybe such joints can be also used in civil engineering fields.

Even if FSW leads to low defect rates, the process must be controlled, and the joint must ensure that there are no defects that could compromise the integrity of the tool. One of the challenges of FSW is detecting defects in welded joints, as some of the defects associated with FSW are difficult to observe non-destructively.

The aesthetics of the weld are very satisfactory, and are without the need for further processing. Compared to other joining processes, the preparation of the parts is much easier, and in some cases, not even necessary.

Deformations resulting from the welding process are very small or almost non-existent.

As the number of specimens produced has been quite small, the research will continue in this direction because the joint can be obtained relatively quickly, without too many preparation operations and with the help of infrastructures that are quite low cost compared to machines designed exclusively for FSW.

Comparing them metallographically, both joints are obtained in the same laboratory, and it is observed that the microstructures provided are adequate and show a correct dissimilar joint without specific defects of the FSW process (wormhole, kiss bonding, etc).

The numerical model was defined to assess the response of the entire tensile test specimen. As the tensile test failure occurred with four specimens in the base material, the considered analysis is useful for the assessment of joint elements and not necessarily the joining.

Author Contributions: Conceptualisation, A.F. and I.B.; methodology, A.F.; modelling, A.F and I.B; experiments and validation, A.F.; resources, A.F.; data curation, all equally; writing—original draft preparation, E.P.; writing—review and editing, A.F.; supervision, A.F.; project administration, I.B. and A.F.; funding acquisition, A.F. All authors have read and agreed to the published version of the manuscript.

Funding: This research received no external funding.

Institutional Review Board Statement: Not applicable.

Informed Consent Statement: Not applicable.

Data Availability Statement: The data presented in this study are available on request from the corresponding author.

Acknowledgments: The collective would like to thank the technical colleagues who helped to produce the specimens for testing. The project leading to these results has received funding from the European Union's Horizon 2020 research and innovation program (H2020-WIDESPREAD-2018, SIRAMM) under grant agreement No. 857124.

Conflicts of Interest: The authors declare no conflict of interest. The funders had no role in the design of the study; in the collection, analyses, or interpretation of data; in the writing of the manuscript; or in the decision to publish the results.

References

1. Blaga, L.; dos Santos, J.F.; Bancila, R.; Amancio-Filho, S.T. Friction Riveting (FricRiveting) as a new joining technique in GFRP lightweight bridge construction. *Constr. Build. Mater.* **2015**, *80*, 167–179. [CrossRef]
2. Blaga, L.; Amancio-Filho, S.T.; dos Santos, J.F.; Bancila, R. Fricriveting of civil engineeting composite laminates for bridge construction. In Proceedings of the 70th Annual Technical Conference of the Society of Plastics Engineers, Orlando, FL, USA, 2–4 April 2012.
3. Feier, A.; Becheru, A.; Brindusoiu, M.; Blaga, L. Process Transferability of Friction Riveting of AA2024-T351/Polyetherimide (PEI) Joints Using Hand-Driven, Low-Cost Drilling Equipment. *Processes* **2021**, *9*, 1376. [CrossRef]
4. Modi, S.; Stevens, M.; Chess, M. *Mixed Material Joining Advancements and Challenges*; Center for Automotive Research: Ann Arbor, MI, USA, 2017; Available online: http://www.cargroup.org/wp-content/uploads/2017/05/Joining-Whitepaper-Final_May16.pdf (accessed on 1 May 2021).
5. Bola, R.; Feier, A.; Tavares, C.; Zifcak, M.P. Uran-Friction Stir Welding Handbook-European Friction Stir Welding Specialist and Engineer. Online Version. 2017.
6. Johnson, R.O.; Burlhis, H.S. Polyetherimide: A new high-performance thermoplastic resin. *J. Polym. Sci. Polym. Symp.* **2007**, *70*, 129–143. [CrossRef]
7. Cipriano, G.P.; Blaga, L.A.; Dos Santos, J.F.; Vilaça, P.; Amancio-Filho, S.T. Fundamentals of Force-Controlled Friction Riveting: Part I—Joint Formation and Heat Development. *Materials* **2018**, *11*, 2294. [CrossRef] [PubMed]
8. Cipriano, G.P.; Blaga, L.A.; Dos Santos, J.F.; Vilaça, P.; Amancio-Filho, S.T. Fundamentals of Force-Controlled Friction Riveting: Part II—Joint Global Mechanical Performance and Energy Efficiency. *Materials* **2018**, *11*, 2489. [CrossRef] [PubMed]
9. Borges, M.; Amancio-Filho, S.; dos Santos, J.; Strohaecker, T.; Mazzaferro, J. Development of computational models to predict the mechanical behavior of Friction Riveting joints. *Comput. Mater. Sci.* **2012**, *54*, 7–15. [CrossRef]
10. Panaitescu, Ș. *Sudare Prin Frecare cu Element Activ Rotator*; Editura Sudura: Timițoara, Romania, 2010.
11. Atanasiu, C.; Canta, T.R.; Caracostea, A.I.; Crudu, I.; Dragan, I.; Giriada, C.; Horbaniuc, D.; Lascu-Simion, N.; Leon, D.; Mocanu, D.R.; et al. *Încercarea Materialelor*; Editura Tehnică: București, Romania, 1982.
12. Feier, A. Timisoara 2018, Raport Proiect Disapora—PN-III-P11.1-MCT-2018-0032.
13. Hynes, N.R.J.; Vignesh, N.J.; Velu, P.S. Low-speed friction riveting: A new method for joining polymer/metal hybrid structures for aerospace applications. *J. Braz. Soc. Mech. Sci. Eng.* **2020**, *42*, 434. [CrossRef]
14. Feier, A.; Banciu, F. Research on the solid-state joining of AlMg3. *IOP Conf. Ser. Mater. Sci. Eng.* **2022**, *1235*, 012013. [CrossRef]
15. Cam, G.; Mistikoglu, S. Recent developments in friction stir welding of Al-alloys. *J. Mater. Eng. Perform.* **2014**, *23*, 1936–1953. [CrossRef]
16. Rodrigues, C.; Blaga, L.; dos Santos, J.; Canto, L.; Hage, E.; Amancio-Filho, S. FricRiveting of aluminum 2024-T351 and polycarbonate: Temperature evolution, microstructure and mechanical performance. *J. Mater. Process. Technol.* **2014**, *214*, 2029–2039. [CrossRef]
17. Friction Stir Welding, Technical Handbook, ESAB. Available online: https://assets.esab.com (accessed on 12 October 2022).
18. Filho, S.T.A.; Blaga, L.-A. *Joining of Polymer-Metal Hybrid Structures: Principles and Applications*; Wiley: Hoboken, NJ, USA, 2018.
19. Amancio-Filho, S.T.; Beyer, M.; Santos, J.F. Method of Connecting a Metallic Bolt to a Plastic Workpiece. U.S. Patent No. 7,575,149, 18 August 2009.
20. Radu, D.; Feier, A.; Petzek, E.; Băncilă, R. Refurbishment of existing steel structures—An actual problem. *Period. Polytehnica Civ. Eng.* **2017**, *61*, 621–631. [CrossRef]
21. De Proença, B.C.; Blaga, L.; Dos Santos, J.F.; Canto, L.B.; Amancio-Filho, S.T. Friction riveting ('FricRiveting') of 6056 T6 aluminium alloy and polyamide 6: Influence of rotational speed on the formation of the anchoring zone and on mechanical performance. *Weld. Int.* **2017**, *31*, 509–518. [CrossRef]

22. Feier, A.; Magda, A.; Banciu, F.; Petzek, E. Deflection and precambering study of VFT beams in the welded joints. In Proceedings of the Advanced Materials and Structures (AMS'20) Conference, Timișoara, Romania, 7–9 October 2020; Elsevier Ltd.: Amsterdam, The Netherlands. [CrossRef]
23. Pitkänen, J.; Haapalainen, J.; Lipponen, A.; Sarkimo, M. *NDT of Friction Stir Welds PLFW 1 to PLFW 5 (FSWL 98, FSWL 100, FSWL 101, FSWL 102, FSWL 103) NDT Data Report*; U.S. Department of Energy: Washington, DC, USA, 2014.

materials

MDPI

Article

Effects of Different Pre-Heating Welding Methods on the Temperature Field, Residual Stress and Deformation of a Q345C Steel Butt-Welded Joint

Jie Yuan [1], Hongchao Ji [1,2,*], Yingzhuo Zhong [3], Guofa Cui [4], Linglong Xu [3] and Xiuli Wang [2]

[1] College of Mechanical Engineering, North China University of Science and Technology, Tangshan 063210, China; yuanjie99530@163.com
[2] School of Materials Science and Engineering, Zhejiang University, Hangzhou 310030, China; wangxl@zju.edu.cn
[3] China MCC 22 Group Corporation Limited, Tangshan 063035, China; zhongyingzhuo@126.com (Y.Z.); 18000351356@163.com (L.X.)
[4] Tangshan Longquan Machinery Co., Ltd., Tangshan 063300, China; 13832829121@163.com
* Correspondence: jihongchao@ncst.edu.cn or jihongchao@zju.edu.cn; Tel.: +86-0315-8805-440

Abstract: Heavy plate welding has been widely used in the construction of large projects and structures, in which the residual stress and deformation caused by the welding process are the key problems to address to reduce the stability and safety of the whole structure. Strengthening before welding is an important method to reduce the temperature gradient, control the residual stress and reduce the deformation of welds. Based on the ABAQUS software, the thermal elastoplastic finite element method (FEM) was used to simulate the welding thermal cycle, residual stress and deformation of low-alloy, high-strength steel joints. Based on the finite element simulation, the influences of flame heating and ceramic heating on the temperature field, residual stress distribution and deformation of a Q345C steel butt-welded joint were studied. The results showed that the thermal cycle of the ceramic sheet before welding had little influence on the whole weldment, but had great influence on the residual stress of the weldment. The results show that the maximum temperature and residual stress of the welded parts are obviously weakened under the heating of ceramic pieces, and the residual stress of the selected feature points is reduced by 5.88%, and the maximum temperature of the thermal cycle curve is reduced by 22.67%. At the same time, it was concluded that the weld shapes of the two were basically the same, but the weld seams heated by ceramic pieces had a better weld quality and microstructures through comparing the macro- and micro-structures between the welded parts heated by ceramic pieces and the simulated weld. Heating before welding, therefore, is an effective method to obtain a high weld quality with less residual stress and deformation.

Keywords: ceramic heating; flame heating; butt welding; residual stress; finite element simulation

Citation: Yuan, J.; Ji, H.; Zhong, Y.; Cui, G.; Xu, L.; Wang, X. Effects of Different Pre-Heating Welding Methods on the Temperature Field, Residual Stress and Deformation of a Q345C Steel Butt-Welded Joint. *Materials* **2023**, *16*, 4782. https://doi.org/10.3390/ma16134782

Academic Editor: Cosmin Codrean

Received: 14 June 2023
Revised: 27 June 2023
Accepted: 30 June 2023
Published: 2 July 2023

1. Introduction

Q345-series steel accounts for the largest proportion in the output of thick plate factories and covers the largest range of varieties and specifications. Under the new equipment conditions, it is a technical problem that most thick plate enterprises are concerned with maximizing the potential performance [1].In the construction and manufacturing of complex large-scale projects and structures, welding is the main assembly and connection process [2].The strength of a welded joint determines the service quality and service life of the whole structure. The magnitude and distribution of residual stress in the component or structure affects the subsequent machining and the life prediction and evaluation of structural reliability [3,4]. Excessive welding deformation will reduce the dimensional accuracy of welding and increase additional stress [5]. It is an urgent task, therefore, to find a reasonable and effective way to reduce the residual stress and deformation.

Due to the complexity of the welding process, more and more scholars have made use of the possibility provided by modern software to conduct numerical simulations of the welding and heat treatment processes. A numerical simulation is an effective tool to study the welding process and to evaluate the mechanical properties of welded parts [6]. The finite element analysis welding method is a powerful and reliable technique to observe the temperature field and stress field in the welding process and to predict the residual stress and deformation in a welding structure, which can directly show the performance of a welded joint [7,8]. Yegaie et al. [9] used the finite element method to build a three-dimensional heat engine model in order to predict the temperature and residual stress distribution in the process of gas-shielded tungsten welding (GTAW) with a radiator. According to the experimental data, it was verified that the high temperature area was only near the heat source, and low levels of longitudinal and transverse residual stress were generated near the welding area. In order to find an effective optimization method for welding deformation, Islam et al. [10] developed a numerical optimization framework based on a coupled genetic algorithm (GA) and finite element analysis (FEA), and they performed a classical weakly-coupled thermo-mechanical analysis using thermos-elastoplastic assumptions to predict the deformation of numerical models. The results showed that the proposed framework can significantly improve the quality of final welding products. Ai et al. [11] built a 3D numerical simulation model to study the weld characteristics generated in fiber laser keyhole welding. Based on the numerical simulation results, the evolution of the weld contour features was demonstrated, and the results were used to estimate the shape and size of the generated weld. Kik et al. [12] presented the calibration of the numerical model for the multi-layer welding of X22CrMoV12-1 steel. The calibration of the heat source model based on the actual geometric size and single pass shape allowed the accurate determination of process parameters, and they obtained results that were highly consistent with the actual welding tests. However, many research studies have only focused on the simulation analysis of the single pass welding of a thin plate, while research on the temperature field, stress field and post-welding deformation of a thick plate in the multi-layer and multi-pass welding processes are relatively few.

Among the many factors affecting welding behavior, the structure of the pre-welding member [13], parameters of welding [14], selection of a heat source [15], welding methods and optimization [16] and the treatment method after welding [17,18], affect the quality of welding, the service life of a component and the safe operation of it. These factors are the guarantee of a high-performance weld and good mechanical properties [11,19,20]. In addition to focusing on the change of the heat input conditions, the study also obtained the microstructure through a pre-welding heat treatment, thereby obtaining a moderate hardness and good weld to enhance the service life and reduce cracks [21]. Peng et al. [22] studied the effect of a pre-welding heat treatment on the microstructure and mechanical properties of an electron beam-welding IN738LC superalloy. The results revealed that the pre-welding heat treatment reduced the non-equilibrium segregation of boron during the cooling process, and it inhibited the formation of a liquid film at the grain boundary and liquefied cracking, thus improving the microhardness of the weld zone (WZ), and affecting the microhardness of the base metal (BM) and the heat-affected zone (HAZ). Panov et al. [23] studied the influence of a preheating and post-welding heat treatment on the microstructure and mechanical properties of the laser-welded joints of a Ti alloy. An increase in the preheating temperature led to an increase in the width of the WZ and HAZ, and an increase in the porosity and gaseous element content (i.e., O and N), while the microhardness of the joint was lower than that of the base material. Luo et al. [24] studied the application of a preheating treatment in an aluminum alloy 5052 welding plan. The experimental results showed that the contact resistance at the W/W (workpiece-to-workpiece) faying interface after the preheating was very consistent and could be reduced by two orders of magnitude. The uncertain variation in the contact resistance of the W/W mapping surface was almost reduced or eliminated, and the quality of the spot welding in terms of the peak load and nugget diameter was checked and showed great improvement.

In practical engineering, different types of thick plate members are usually welded together to produce complex structures and efficient industrial requirements; therefore, it is very important to predict and control the distribution of the thermal cycle curve and residual stress in the welding process to improve the design rationality and safety of actual engineering structures. In order to meet this requirement, this paper presents a new heating method before welding—i.e., ceramic sheet heating. The ceramic electric heater uses the heating element as the heat source when a workpiece is heated. The preheating method is mainly composed of a temperature control box, a caterpillar heater and a temperature measuring instrument. It can be widely used in the local heat treatment of various alloy steel-welded structures. The ceramic heating device changes the defects of the traditional process such as an unstable quality, high energy consumption and poor working conditions. The secondary development of the MIG welding process was carried out by the ABAQUS software. Three-dimensional modeling and pre-welding strengthening measures were carried out on welds under low temperature conditions, and important parameters such as an appropriate welding voltage, the welding current and welding speed were selected as the fixed values within a certain range. The changes in the thermal cycle, residual stress and deformation under two pre-welding heating methods were compared and analyzed, so as to understand the detailed welding process, and obtain the best welding simulation quality of the pre-welding heating method. The results of the simulation and welding test were compared and analyzed to provide a reliable theoretical basis for further research.

2. Experiment

Q345 structural steel is a Chinese, standard, low-alloy, medium-tensile strength steel produced by a hot rolling process. Due to its comprehensive mechanical properties and welding properties, it has been widely used in railways, bridges, industrial plants, boilers, pressure vessels, steel fuel tanks, power stations and other manufacturing structural components. The main microstructures of the Q345C steel used in the test were ferrite and pearlite, which has strong mechanical and fatigue properties such as strength, toughness and weldability [25]. Its chemical composition is shown in Table 1. After a high-temperature repeated cooling test [26], it was found that the Q345C steel had good mechanical properties when it was heated to 200–500 °C and cooled to the ambient temperature; however, with an increase in the circulating heating and cooling times and heating temperature, its microstructure changed significantly and its mechanical properties decreased significantly. The thick plate Q345C steel was welded under a traditional flame and ceramic plate heating, respectively, to study the influence of different preheating methods on welding stress and deformation. The ceramic heating device is shown in Figure 1. According to the heat treatment workpiece, the required crawler ceramic electric heater was connected with stainless steel wire (where the distance should not exceed the wall thickness), and then covered on the heating workpiece, then the ceramic electric heater was tied with stainless steel wire or a stainless steel belt. When welding the outside, the heater was arranged inside the furnace body. When welding the inside, the heater was arranged outside.

Table 1. The chemical composition of Q345C steel.

	C	Mn	Si	S	P	Cr	Mo	V	Cu	Ni
Composition (%)	0.15	1.44	0.23	0.005	0.013	0.025	0.0077	0.031	0.016	0.0095

The ceramic heater has the following characteristics: (1) it has a high power density, can be rapidly heated, and its heating speed is much greater than for induction heating. (2) It has a small size, a simple and reasonable structure, and is light weight for light handling and dismantling labor. (3) The number of ceramic electric heaters can be determined according to the needs of a heat treatment work, and it is not bound by any conditions. (4) The ceramic heater is directly covered on a heat treatment workpiece, and the outer layer is covered with a layer of thermal insulation blanket (i.e., a needled blanket), which does not require any material with a large capacity, so that the heater's heat loss is small, and the effect on energy saving is significant.

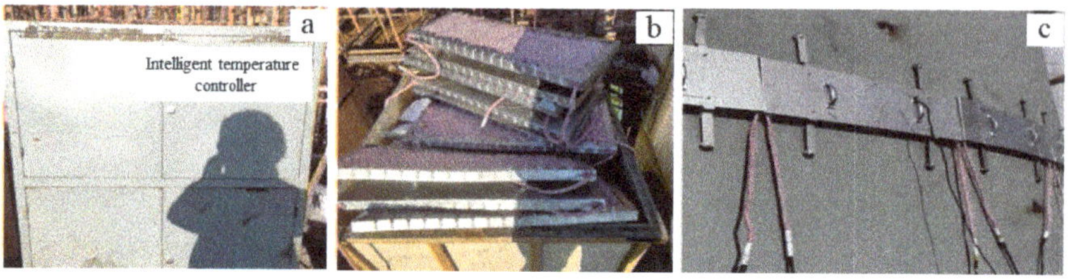

Figure 1. Ceramic heating-related device. (**a**) Temperature control box, (**b**) ceramic heating plate, and (**c**) ceramic heating construction.

The thick plates under both preheating methods had the same geometric dimensions, i.e., 300 mm × 300 mm × 25 mm. Under the same welding parameters, the MIG welding process was performed on the two plates. In order to prevent lamellar tearing of the thick plate welding, gas-shielded welding was used for the welding.

In order to make the finite element simulation correspond to the actual welding conditions of the butt-joint parts, the welded parts were prepared into tensile specimens to study the mechanical properties and microstructures of the welded joints. For the preparation of the tensile specimens, two tensile specimens obtained from the welded joints of thick plates under different heat treatment conditions were cut with a wire cutting machine. According to the relevant standards, the cross-sectional area of the sample was 10 mm × 2 mm. In addition, the sample was clamped to the SANS Electronic universal testing machine with a tensile rate of 1 mm/min, as shown in Figure 2.

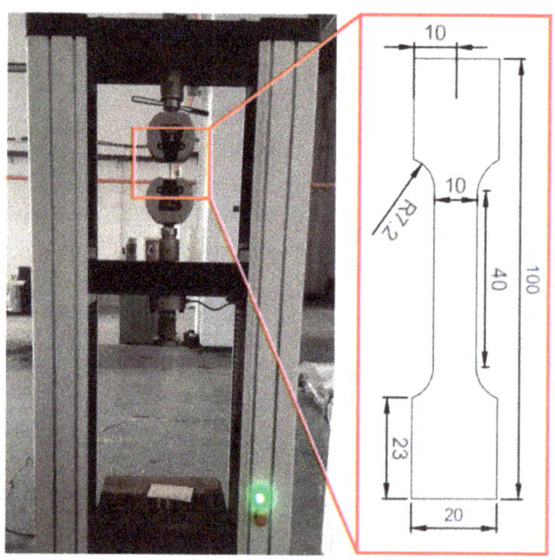

Figure 2. Tensile test through clamping specimens.

At the same time, in order to analyze the changes in the mechanical properties and microstructures of T-welded joints under different welding currents or thermal inputs, and to study the influence of welding defects on the mechanical properties in detail, some measurements and sample preparations were carried out. For example, a sample for metallographic observation was cut with a wire-cut machine from a joint perpendicular to the welding direction and cold-set after wiping the sample. After grinding and polishing, the sample was etched in a nitric acid reagent for 5–10 s. Then, in order to clearly see the distribution of the welding defects and the chemical composition of the materials, the morphology and microstructure of the obtained metallographic samples were analyzed by optical microscopy (OM) and scanning electron microscopy (SEM).

3. Numerical Simulation Framework

3.1. Characteristics of Materials

In the numerical simulation of the welding process, the temperature-dependent performance parameters of different welding zones are required, which are related to the chemical composition. According to the reference in [1] and the material property simulation software, JmatPro, the temperature-related performance parameters were given, as shown in Figure 3. For a welding temperature field analysis, the specific heat is important. In addition, it is necessary to determine the thermodynamic properties of the stress field simulation, such as via Poisson's ratio and Young's modulus.

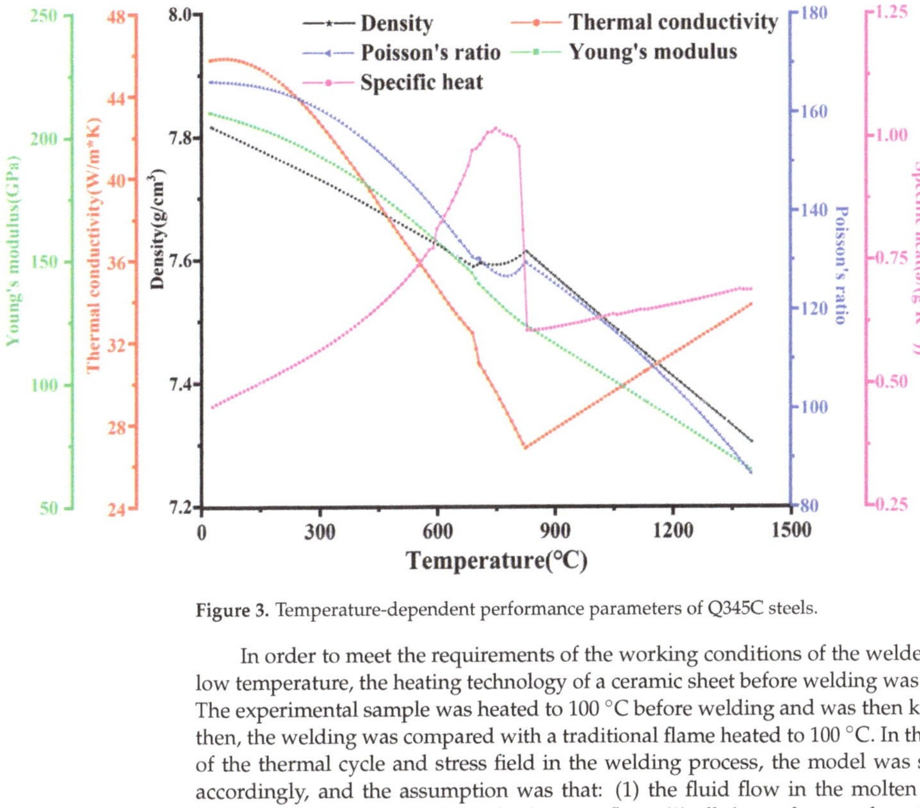

Figure 3. Temperature-dependent performance parameters of Q345C steels.

In order to meet the requirements of the working conditions of the welded sheet at low temperature, the heating technology of a ceramic sheet before welding was proposed. The experimental sample was heated to 100 °C before welding and was then kept warm, then, the welding was compared with a traditional flame heated to 100 °C. In the analysis of the thermal cycle and stress field in the welding process, the model was simplified accordingly, and the assumption was that: (1) the fluid flow in the molten pool was Newtonian, incompressible, and a laminar flow; (2) all the molten pools and droplets were symmetric with the weld center axis; and (3) the arc axis in the welding process was straight; while (4) changes in the physical properties of the weld metal caused by mixing in the weld pool were not considered.

3.2. Finite Element Model

The butt joint was an X-type groove created with CO_2 gas-shielded clean root welding, and then, the front and back with MIG filling welding. As the plate was of a medium thickness, it adopted three layers with three passes of welding. The welding process parameters of each weld are shown in Table 2. Firstly, the root was cleared. The upper groove opening angle was 60° for the double-layer double-pass welding, and the lower groove opening angle was 45° for the single-layer single-pass welding. A 3D welding model based on a finite element analysis was established. Figure 4 shows the meshing and boundary conditions of the welded finite element model, using C3D8R hexahedral elements and creating a fine mesh near the WZ for more accurate results. During the welding process, the welding parts and the heat source exchange heat with the surrounding medium through heat-transfer and heat-radiation. The heat radiation around the molten pool is dominant, and the convective heat-transfer away from the molten pool is dominant. The formula for calculating the heat flux loss is [11]:

$$q_s = h_c(T_s - T_0) + \varepsilon\sigma\left[T_s^4 - T_0^4\right] \tag{1}$$

where:

q_s—heat flux loss during welding;
h_c—coefficient of heat transfer;
T_s—the temperature of the welded part;
T_0—ambient temperature;
ε—surface emissivity, and for the high-strength steel Q345C, it was assumed to be 0.09;
σ—Stefan–Boltzmann constant.

Table 2. Welding process parameters.

Parameters	Unit	Welding		
		1	2	3
Voltage of welding	V	21	22	23
Current of welding	A	110	115	120
Power of welding	W	1733	1898	2070
Speed of welding	mm/s		0.1	

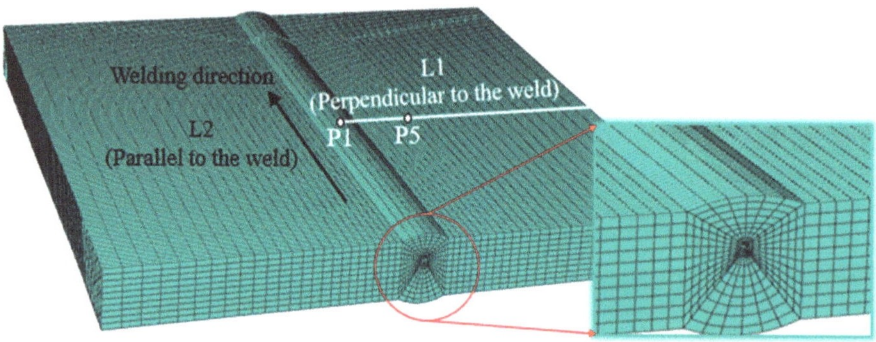

Figure 4. Finite element model meshing and simulated measurement position.

Based on a thermal simulation, the stress field can be calculated according to the thermoelastoplastic theory. The calculated temperature field is used as the thermal load to calculate the stress field. Appropriate constraints must be applied during a thermal simulation. The boundary conditions of the welding fixture were completely constrained, and the lower surface of the weldment was constrained along the thickness direction. Because the project was carried out in a temperature retaining shed, the default initial ambient temperature was 5 °C.

3.3. Heat Source Model

By modeling the heat source, it was easy to determine the location of the heat field. A double ellipsoidal heat source distribution function was used as the heat source load, and the heat source distribution model is shown in Figure 5. The double ellipsoidal heat source distribution model (i.e., Goldak heat source) is a heat source model improved by Goldak on the basis of Gauss. The heat flow not only acts on the surface, but also has heat flow at a certain depth, namely, the volume heat source. Moreover, the heat flux is the Gaussian distribution in the width, length and depth directions. The influence of the arc stiffness was taken into account in this model, in which the welding method with a large arc impact is more accurate. In addition, if the depth of the molten pool is large, the change of the welding heat source in the depth direction needs to be considered, and the Goldak model is usually used in this case. It is assumed that the heat flux distribution on the body conforms to the double ellipsoid distribution, which can be expressed by the following formula [27]:

$$q_1(x,y,z) = \frac{6\sqrt{3}(f_1Q)}{a_1bc\pi\sqrt{\pi}}e^{[-3(\frac{x^2}{a_1{}^2}+\frac{y^2}{b^2}+\frac{z^2}{c^2})]}, x \geq 0 \quad (2)$$

$$q_2(x,y,z) = \frac{6\sqrt{3}(f_2Q)}{a_2bc\pi\sqrt{\pi}}e^{[-3(\frac{x^2}{a_1{}^2}+\frac{y^2}{b^2}+\frac{z^2}{c^2})]}, x < 0 \quad (3)$$

$$f_1 + f_2 = 2.0 \quad (4)$$

where:

a_1, a_2, b, c—the heat source shape parameter model;
Q—the effective power of the heat source;
f_1, f_2—the ratio of the heat source between the front and the back.

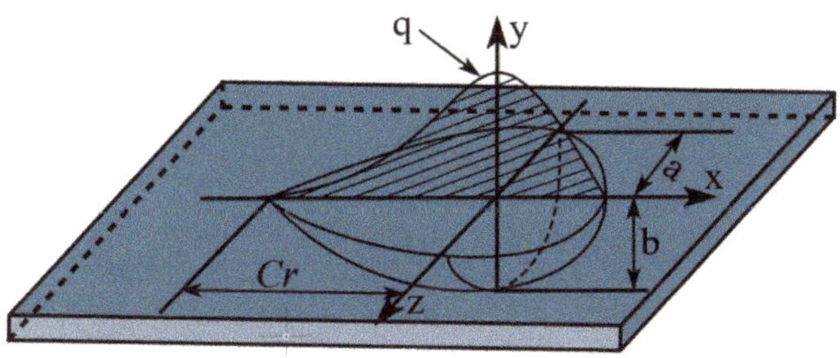

Figure 5. Double ellipsoid heat source distribution model.

4. Results and Analysis

4.1. Thermal Cycle Curve Distribution

Figure 6 shows a thick plate Q345C steel welded under a conventional flame and ceramic sheet heating. With the MIG welding process, both kinds of plates had a good weld with a full penetration and a similar welding width due to the same welding parameters. The macro-geometric and micro-structural characteristics of different welding zones are related to the welding thermal cycle. According to the microstructure, there are three different zones in the weld including the WZ, the HAZ and the BM [28,29]. There was no WZ before the welding began because the WZ is formed by the continuous melting of welding wire. With the movement of the welding heat source, the welding began to produce a WZ and the whole WZ was formed until the end of welding. The welding process not only changed the geometric characteristics of the welded joints, but also changed the metallurgical and mechanical properties of materials in the different welding zones. Figure 7 shows the cross section outline of the weld. The cross section of the welded joint was divided into two typical zones, namely, the WZ and the HAZ. Through measurements, the width of the top of the WZ obtained by the simulation in Figure 7a and the experimental results in Figure 7b were 1.830 mm and 1.655 mm, respectively. The comparison of the results in Figure 6 shows that the experimental weld morphology was basically consistent with the simulated weld morphology.

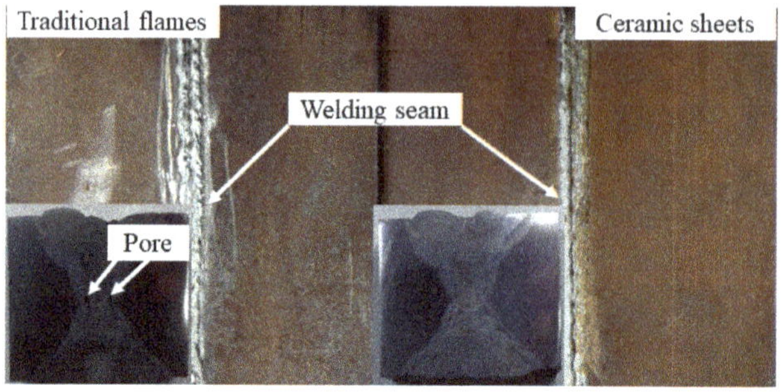

Figure 6. Welding results of Q345C steel under two heating methods.

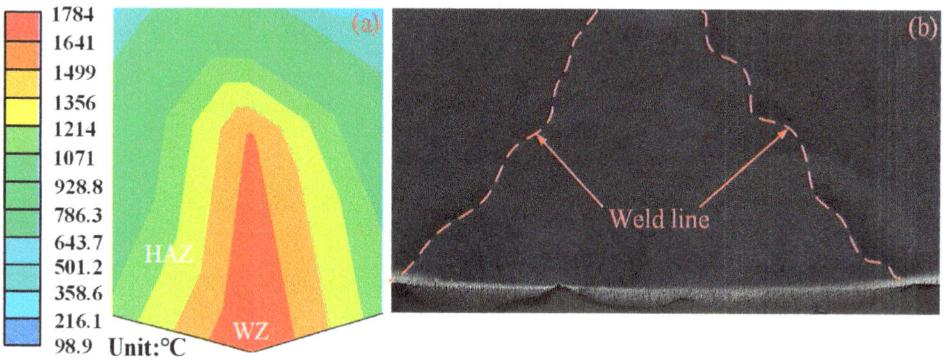

Figure 7. Temperature field of the welding seam of ceramic sheet under heating: (**a**) calculated weld geometry; (**b**) test weld geometry.

In order to study the influence of pre-welding heating of a ceramic sheet and flame on the welding process, the two simulated welding parameters were in the same operation, with the welding parameters shown in Table 2. Welding is a rapid heating and cooling process. In the heating stage, the metal material quickly melted, and after welding it quickly cooled. During this process, the material was first heated and expanded to create compressive stresses in the weld area. While during the cooling phase, the weld began to contract to create tensile stresses. The welding thermal cycle directly leads to the different mechanical properties of welds in different areas; therefore, the maximum temperature distribution helps to determine the geometric boundary of each welding area [30]. Figure 8 shows the 3D finite element numerical simulation results of the welding temperature under two working conditions under the same scale when the welding time was 468 s (i.e., the second weld). According to the peak temperature in the welding process in different areas, the cross section of the welded joint was divided into the three typical areas described above. Figure 8 shows the symmetrical geometric features of the WZ and HAZ on the upper surface of the workpiece along the welding direction. Figure 8a shows the simulated temperature field under traditional flame heating, and Figure 8b shows the simulated welding temperature field under ceramic plate heating. Where the liquid temperature of the WZ exceeded 1109 °C, the temperature of the HAZ was between the phase transition isotherm (1109 °C) and the solution temperature (660 °C). It could be seen that the size of the molten pool and the WZ obtained with the ceramic sheet heating welding was smaller and the area of the HAZ was reduced compared with the traditional flame heating welding method, which effectively improved the temperature gradient of the WZ.

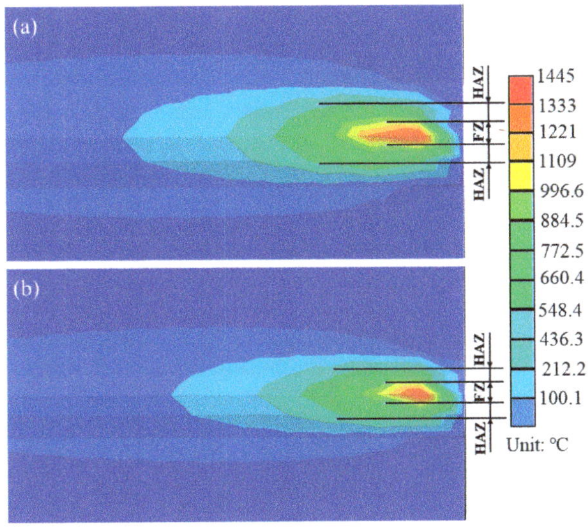

Figure 8. Numerical simulation results of the welding finite element temperature. (**a**) Conventional flame heating, and (**b**) ceramic sheet heating.

In order to clarify the complex thermal cycle curve and stress-strain evolution mechanism in the welding process, the historical temperature data points of five points on and near the weld were extracted between P0–P4 on L1, as shown in Figure 4. The coordinates of the five points were P0 (150, 25, 150), P1 (160, 25, 150), P2 (170, 25, 150), P3 (180, 25, 150), and P4 (200, 25, 150). The point of P0 was located on the weld, and points P1–P4 were at some points adjacent to the weld at a certain distance. Figure 9 shows the maximum temperature reached by the nodes perpendicular to the weld during the welding process. It was found that the maximum temperature significantly decreased in the welding process and the curve dropped more gently, with a smaller temperature gradient. It can be noted that the ceramic sheet heating method will affect the maximum temperature of the molten pool and the temperature field behind the heat source when comparing it with the traditional flame heating method.

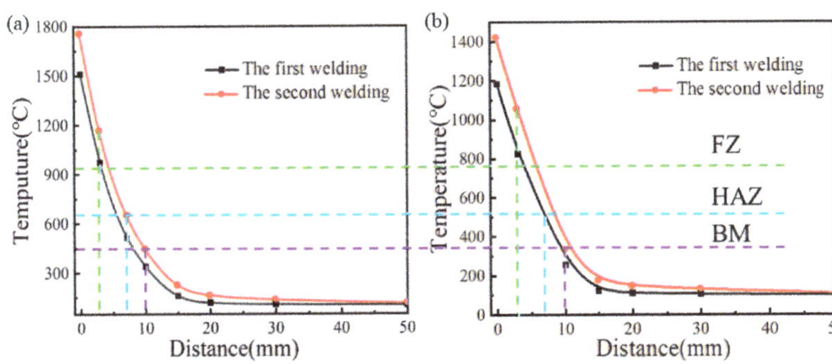

Figure 9. The highest temperature reached during welding. (**a**) Conventional flame heating, and (**b**) ceramic sheet heating.

The temperature thermal cycle of three coordinate points from P0 to P4 is shown in Figure 10. For the five measuring points perpendicular to the welding direction, the maximum temperature decreased with an increasing distance from the WZ. As the welding heat source approached, the temperature at P0 increased rapidly. Although the molten pool expanded due to the increase in temperature, compressive plastic strain appeared in the welding area due to the restriction of the surrounding base metal. After the heat source had passed through P0, the temperature at the point dropped rapidly, and the welding area solidified and shrunk. Due to the metal limitation around the molten pool, the tensile longitudinal plastic strain was generated after cooling, which cancelled out the compressive plastic strain during the heating process [31]; therefore, after the laser heat source had passed through P0, the longitudinal plastic strain decreased rapidly and reached a constant value. For P0, when the heat source passed by the point through the two heating methods of a traditional flame and ceramic sheet, there was an obvious difference between the maximum temperature of P0 in the process of the three-welds welding. When the first weld was welded, the maximum temperature under traditional flame heating was 1500 °C, and the maximum temperature of the ceramic sheet heating was 1180 °C, which decreased by 21.33%. In the process of the second welding, the maximum temperature of the heat source passing through the point under the two heating methods was 750 °C and 580 °C, respectively, which decreased by 22.67%. The temperature variation trend of the four points of P1 to P4 in the process of the three-pass welding was similar to P0.

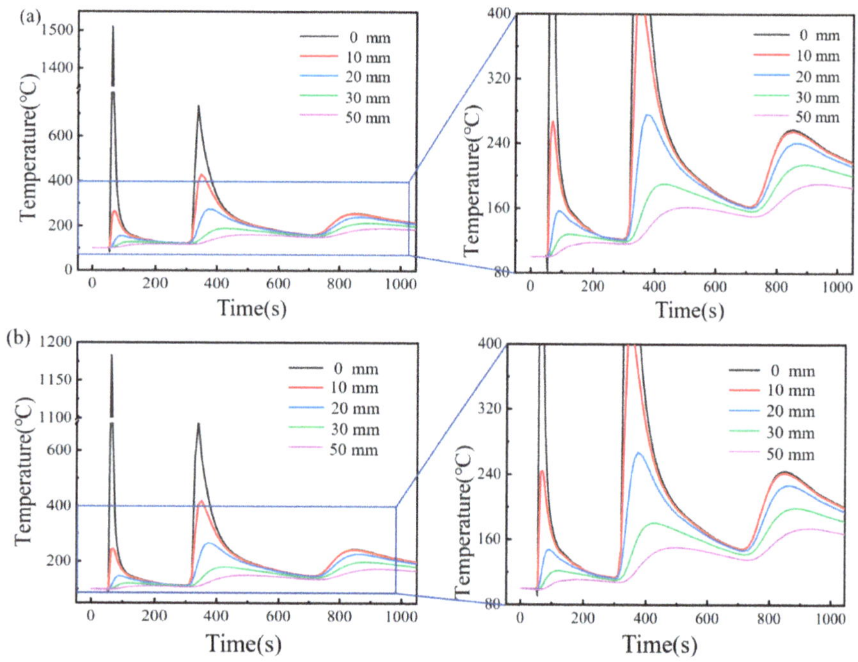

Figure 10. Heat cycle curve of characteristic points. (**a**) Conventional flame heating, and (**b**) ceramic sheet heating.

4.2. Residual Stress Distribution

Figure 11 shows the contours of the global, longitudinal and transverse residual stresses under the pre-welding heating of a conventional flame and ceramic sheet, respectively. By comparison, it can be found that the selection of the pre-welding heating method has great influence on the distribution of the overall residual stress, transverse and longitudinal residual stress. Although the distribution of residual stress was similar in the two cases, the magnitude of the residual stress was very different. It can be seen from Figure 11a that the equivalent residual stress near the weld of the weldment and the whole sheet under the traditional flame heating was larger, and the maximum equivalent residual stress was 464.8 MPa. The ceramic heating method changed this phenomenon and reduced the area of the equivalent residual stress region. The maximum residual stress of the welded specimens decreased significantly to 404.7 MPa, down by 12.93%. If the end effect of the weld was ignored, the equivalent residual stress in the middle of the weld was evenly distributed. Figure 11b shows the longitudinal residual stress distribution of the welded parts heated by s traditional flame and ceramic sheet, which indicated that the longitudinal stress of the weld and its vicinity was tensile stress. Because the filler wire and the base metal underwent rapid heating and cooling during the welding process, the tensile residual stress reached the yield strength of the base metal [32]. As the distance from the weld increased, the tensile stress gradually became compressive stress [33]. Figure 11c compares the transverse residual stress distribution of the two samples. The contour lines in the figure show that the transverse residual stress had a similar distribution in both cases. The transverse residual stress was high-tensile stress near the weld and WZ. With the increase in the distance from the weld, the tensile stress decreased gradually, and the transverse tensile stress weakened gradually. At the same time, the transverse residual stress near the weld under the heating of the ceramic sheet became the compression stress with a low value. It can be clearly compared from the above three figures that the residual stress was obviously improved by heating the ceramic sheet. Compared with the transverse residual stress, the longitudinal residual stress was more significantly affected by the two pre-welding heating methods.

Figure 12 shows the longitudinal and transverse residual stresses of the butt weld calculated by the finite element model along the direction perpendicular to the weld under the two pre-welding heating conditions. It is obvious that the residual stress perpendicular to the welding direction was symmetrically distributed along the weld center. For Figure 12a, the longitudinal residual stress near the WZ and HAZ was tensile stress. The peak stress under the conventional flame heating was about 270 MPa, while the peak stress under the ceramic plate heating was about 400 MPa. Although the heating stress of the ceramic sheet near the weld and HAZ was higher than that of the traditional flame, the longitudinal residual stress was significantly reduced and evenly distributed in the area outside until it gradually approached zero at the free surface of the end edge [34]. Figure 12b shows the transverse residual stress perpendicular to the weld, which was almost uniformly distributed along the X-axis, and the overall transverse residual stress was basically zero. It can be clearly seen that the transverse residual stress after welding was obviously reduced and distributed evenly with minimal fluctuation. In addition, the longitudinal residual stress was much greater than the transverse residual stress, namely, about ten times greater than the transverse residual stress, which may be the main cause of the plate buckling. The longitudinal stress was basically tensile stress, which was due to strict constraints. Due to strict constraints, the longitudinal stress had a high-tensile value near the weld and became compressed away from the weld, and the transverse residual stress distribution was almost constant relative to the longitudinal stress distribution. This made the longitudinal stress basically a tensile stress [35]. It can be concluded that most of the stress in the sample was mainly caused by the longitudinal stress and was mainly concentrated in the WZ.

The residual stress along the welding direction is shown in Figure 13. The longitudinal stress of the two pre-welding methods was negative at the beginning and end of the weld, indicating the existence of compressive residual stress, which gradually increased along the length of the weld and became tensile. Within the range of 50~250 mm in the direction of the weld, there was a stable high-tensile stress zone, in which the maximum longitudinal residual stress reached 450 MPa, as shown in Figure 13a. As a result, the welded joint was subjected to significant tensile stress and deformation along the length of the weld. As shown in Figure 13b, the transverse residual stress along the weld was basically zero and the stress value increased gradually away from the weld, indicating that a large tensile stress had been generated due to the constraint effect of the edge. Compared with the residual stress perpendicular to the weld direction, the residual stress along the weld direction was less affected by the pre-welding heating method, but it still had a slightly improved effect. It can be seen from the above analysis that, compared with the traditional flame, the heating method of the ceramic sheet had a significant weakening effect on the residual stress of the weldment. Its tensile residual stress was significantly reduced, and especially, the effect on the residual stress perpendicular to the weld direction was more significant. The lower tensile residual stress reduced the risk of failure under a cyclic or dynamic load [36].

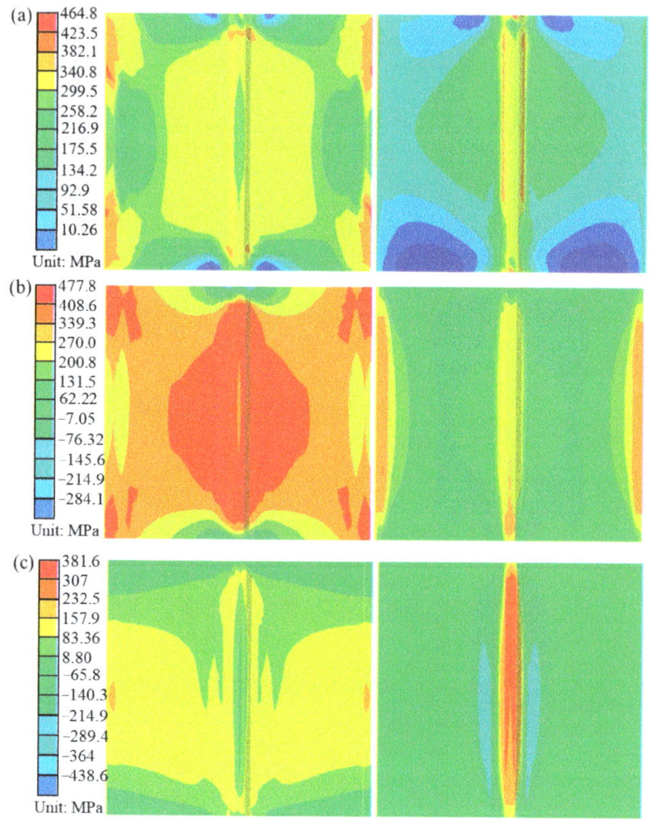

Figure 11. Residual stress distribution after welding. (**a**) Overall residual stress, (**b**) longitudinal residual stress, and (**c**) transverse residual stress.

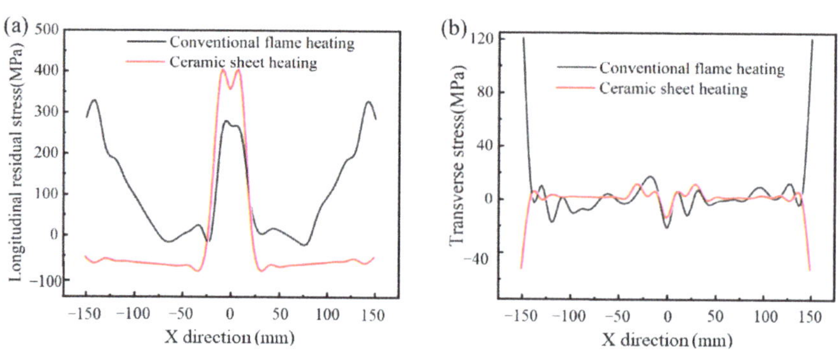

Figure 12. Residual stress distribution along the direction perpendicular to the weld. (**a**) Longitudinal residual stress, and (**b**) transverse residual stress.

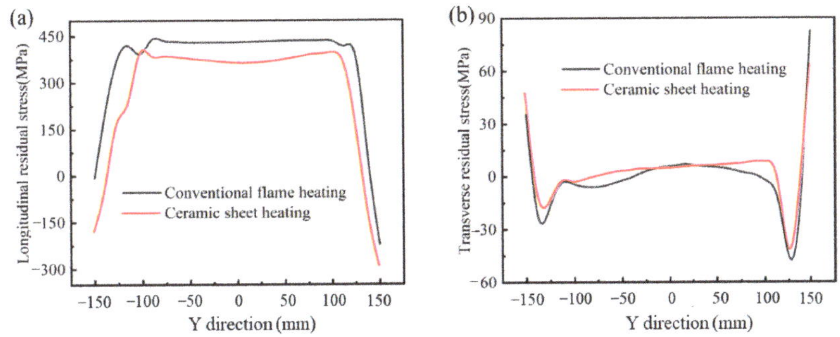

Figure 13. Residual stress distribution along the direction parallel to the weld. (**a**) Longitudinal residual stress, and (**b**) transverse residual stress.

4.3. Deformation

Similarly, in order to analyze the influence of traditional flame and ceramic sheet heating on welding deformation, the feature points as shown in Figure 4 were extracted in directions perpendicular to the weld and parallel to the weld, respectively. Figure 14 describes the out-of-plane deformation curves of L1 and L2 lines under different cold source distances (Case A is the traditional flame heating mode, and Case B is the ceramic sheet heating mode). As can be seen from Figure 14, under these two conditions, the maximum total post-welding deformation of the ceramic sheet heating was 1.782 mm, which increased by 9.06% compared with the maximum welding deformation of 1.634 mm under the traditional flame heating. Although the deformation on the weld increased in a concentrated way, the total deformation at both ends of the weld surface had an obvious decreasing trend and the regional distribution of the overall deformation also decreased obviously, while the deformation after welding was improved. Generally speaking, the smaller the binding force during welding, the smaller the welding displacement. Although the binding force is the main cause of welding deformation, the bending deformation caused by a temperature gradient in the thickness and width direction of a sample is a necessary factor to cause buckling and to influence welding deformation [37]. In order to clarify the cause of this phenomenon in detail, the displacement data perpendicular to and along the welding direction (L1) in Figure 4 were extracted. As shown in Figure 15a, the total deformation, and the *Y*-axis and *Z*-axis deformation curves perpendicular to the weld direction were symmetrically distributed with the weld as the axis of symmetry. The deformation near the weld was the largest, and the amount of deformation away from the weld decreased gradually, which may be attributed to the temperature gradient. Relative to the total deformation and the *Z*-axis deformation, the *Y*-axis deformation was almost unchanged. Similarly, the welding deformation along the weld (L2) direction is shown in Figure 15b, where the welding deformation varied in a certain proportion. The total deformation parallel to the weld (L2) was significantly reduced when the ceramic sheet was heated compared with the traditional flame welding, indicating that the pre-welding heating method of the ceramic sheet can effectively reduce or mitigate welding buckling in the welded joint.

4.4. Welding Results and Microstructure

According to the tensile results of the base metal and butt-welded joint in the experimental description above, the yield strength and tensile strength are shown in Table 3. It can be clearly seen from the ratio of the strength of the welded joint to the strength of BM that the yield strength and tensile strength were significantly lower than that of the base material. In order to better study the cause of this result and to analyze the difference of the microstructure of the weld under two heat treatment methods, the metallography and microhardness of the butt-welded joint were measured, and the changes in the microstructure and microhardness of the joint were analyzed.

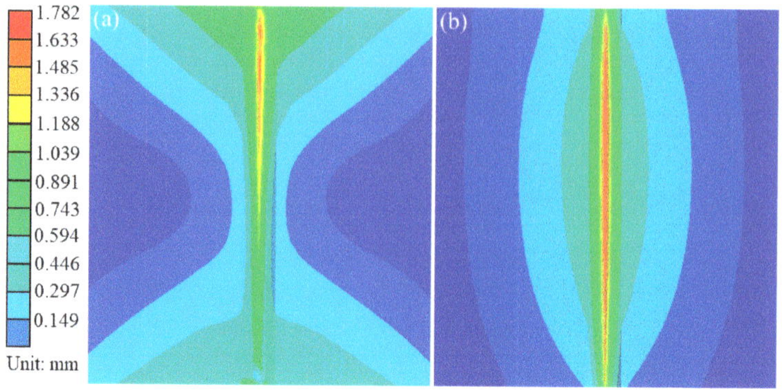

Figure 14. Distribution of deformation after welding. (**a**) Conventional flame heating, and (**b**) ceramic sheet heating.

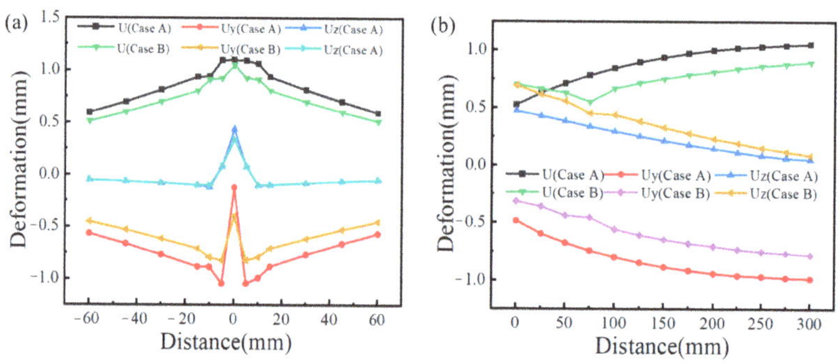

Figure 15. Welding deformation curves under different heating methods. (**a**) Perpendicular to the weld direction (L1), and (**b**) parallel to the weld direction (L2).

Table 3. Mechanical properties of the welded joint.

Heat Treatment Method	Yield Strength (MPa)	Tensile Strength (MPa)
Flame heating	503	523
Ceramic heating	542	569
BM	587	612

The weld microstructure and grain boundary morphology of the WZ and HAZ under the different heat treatments are shown in Figure 16. A metallographic observation was carried out using OM to obtain the microstructure characteristics of the BM to the WZ, as shown in Figure 17. Combined with the two figures, it is clear that there were some welding defects in the butt joint under the flame heat treatment mode, such as an incomplete fusion defect and a porosity defect of the WZ and HAZ. Welding defects have a great influence on the tensile strength, especially porosity defects, which are important factors affecting the tensile properties.

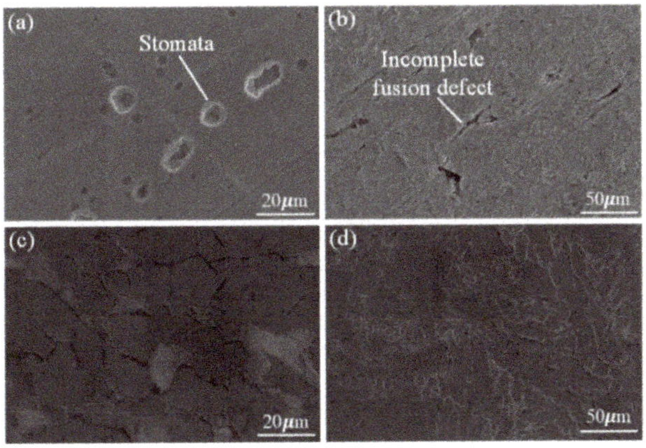

Figure 16. Weld of the WZ and HAZ under SEM; (**a**,**b**) flame heating mode; (**c**,**d**) ceramic sheet heating method.

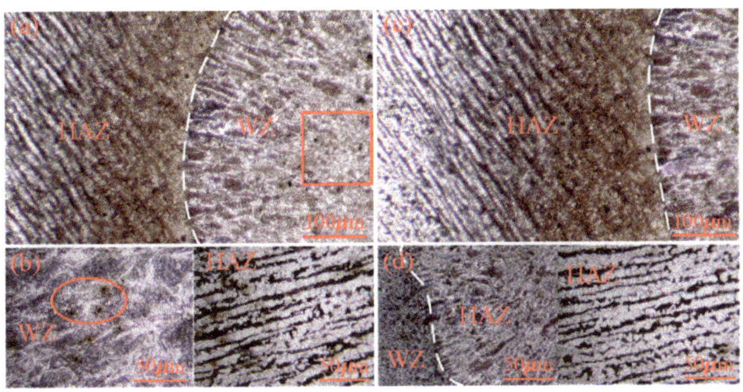

Figure 17. Microstructure features of the WZ and HAZ; (**a**,**b**) flame heating mode; (**c**,**d**) ceramic sheet heating method.

In Figure 17, due to the obvious porosity defect in the WZ under flame heating, one of the reasons for this is that the yield strength and tensile strength of the butt joint were somewhat lower than that of the base material; however, the weld strength after the heat treatment of the ceramic sheet was higher than that obtained by the flame heating, and no obvious porosity and cracks were observed in the FZ or the weld edge. Figure 17a,c represents the evolution of the microstructure of the welded joint from the HAZ to the weld center. It was composed of the BM, HAZ (i.e., a mixed zone of equiaxed dendrites and equiaxed grains with a coarse rolling-aged microstructure) and WZ. There were some differences in the microstructure characteristics of the WZ and HAZ. Due to the large temperature gradient near the weld, a clear columnar grain structure could be observed at the boundary of the WZ, which made the grain grow along the solidification front [38], resulting in different grain sizes. In the HAZ near the center of the weld, although the temperature gradient in the weld decreased, the higher temperature provided conditions for the growth of grain, which made the ferritic and pearlite of the base metal grow, while the distance narrowed.

5. Conclusions

In this paper, the influence of two kinds of pre-welding heating methods of, namely, a traditional flame and a ceramic sheet, on the thermal cycle, residual stress and deformation after welding of a Q345C-steel, thick plate was studied by the finite element theory, which can be used to improve the mechanical properties of the butt joint weld and improve the structural strength and service life of a whole component. The effects of the different heating methods on the welding evolution were studied by simulation and an experimental verification was carried out. The main conclusions of this study are as follows:

(1) A new heating method before welding was proposed. The method used a ceramic electric heater composed of heating elements as the heating source before welding. Compared with the traditional flame heating method, this welding method is controllable, relatively safe and of low cost. The weld shape was basically the same as the simulated weld shape and the quality was good.

(2) A three-layer and three-pass welding model of the thick plate weldment was established. The results show that the experimental weld morphology was basically consistent with the simulated weld morphology. It showed that although the pre-welding heating method had little influence on the welding temperature field, the maximum temperature of the node affected by the welding process was significantly reduced, and the maximum temperature difference between the two at the same characteristic point was 170 °C. The heating mode of the ceramic sheet affected the maximum temperature of the molten pool and the temperature field behind the heat source, and it effectively improved the thermal cycle curve.

(3) The influence of the two heating methods on the reduction or even elimination of the residual stress and deformation caused by welding was analyzed. The results show that the ceramic heating had a significant weakening effect on the residual stress of the welded parts, and especially, the effect on the residual stress perpendicular to the weld direction was more significant, which decreased by 5.88%.

(4) The welding deformation under the heating mode of the ceramic plate was significantly less than that under the traditional flame heating, and especially, the weakening effect of the deformation parallel to the weld direction was more obvious. The maximum total post-welding deformation of the ceramic sheet after heating was 1.782 mm, which was 9.06% higher than the maximum welding deformation of 1.634 mm under the traditional flame heating.

(5) The microcosmic experiment showed that the ceramic heating method produced a better strength and joint quality. In the flame heat treatment, the butt joint had some welding defects, such as incomplete fusion defects, and WZ and HAZ porosity defects. The weld strength of the ceramic sheet after the heat treatment was higher than that after the flame heating, while there were no obvious pores and cracks in the weld zone and weld edge, and it had a better tensile strength.

Author Contributions: J.Y.: Writing—original draft, review and editing, project administration and conceptualization; H.J.: Writing—original draft, project administration and conceptualization; Y.Z.: Review and editing and investigation; G.C.: Experimental conduct and review and editing; L.X.: Experimental conduct, and review and editing; X.W.: Conceptualization and project administration. All authors have read and agreed to the published version of the manuscript.

Funding: This work is supported by the National Natural Science Foundation of China (51905501), while this work is also supported by the Tangshan talent foundation innovation 514 team (A202202008) and is funded by the S&P Program of Hebei (Grant No. 22281802Z).

Institutional Review Board Statement: The authors claim that there are no ethical issues involved in this research.

Informed Consent Statement: All the authors consented to participate in this research and contributed to the research.

Data Availability Statement: No data or materials are available for this research.

Conflicts of Interest: The authors declare no conflict of interest.

References

1. Yang, N.; Su, C.; Wang, X.-F.; Bai, F. Research on damage evolution in thick steel plates. *J. Constr. Steel Res.* **2016**, *122*, 213–225. [CrossRef]
2. Zhang, X.; Wu, L.; Andrä, H.; Gan, W.; Hofmann, M.; Wang, D.; Ni, D.; Xiao, B.; Ma, Z. Effects of welding speed on the multiscale residual stresses in friction stir welded metal matrix composites. *J. Mater. Sci. Technol.* **2019**, *35*, 824–832. [CrossRef]
3. Lee, J.-H.; Jang, B.-S.; Kim, H.-J.; Shim, S.H.; Im, S.W. The effect of weld residual stress on fracture toughness at the intersection of two welding lines of offshore tubular structure. *Mar. Struct.* **2020**, *71*, 102708. [CrossRef]
4. James, M.N. Residual stress influences on structural reliability. *Eng. Fail. Anal.* **2011**, *18*, 1909–1920. [CrossRef]
5. Adamkowski, A.; Lewandowski, M. Analytical model of stress concentration for the welded joints with angular distortion of thin-walled pipelines. *Thin-Walled Struct.* **2015**, *97*, 101–113. [CrossRef]
6. Sadeghian, B.; Taherizadeh, A.; Atapour, M. Simulation of weld morphology during friction stir welding of aluminum- stainless steel joint. *J. Mater. Process. Technol.* **2018**, *259*, 96–108. [CrossRef]
7. Barsoum, Z.; Lundbäck, A. Simplified FE welding simulation of fillet welds–3D effects on the formation residual stresses. *Eng. Fail. Anal.* **2009**, *16*, 2281–2289. [CrossRef]
8. Zeng, P.; Gao, Y.; Lei, L.P. Welding process simulation under varying temperatures and constraints. *Mater. Sci. Eng. A* **2009**, *499*, 287–292. [CrossRef]
9. Yegaie, Y.S.; Kermanpur, A.; Shamanian, M. Numerical simulation and experimental investigation of temperature and residual stresses in GTAW with a heat sink process of Monel 400 plates. *J. Mater. Process. Technol.* **2010**, *210*, 1690–1701. [CrossRef]
10. Islam, M.; Buijk, A.; Rais-Rohani, M.; Motoyama, K. Simulation-based numerical optimization of arc welding process for reduced distortion in welded structures. *Finite Elem. Anal. Des.* **2014**, *84*, 54–64. [CrossRef]
11. Ai, Y.; Jiang, P.; Shao, X.; Li, P.; Wang, C.; Mi, G.; Geng, S.; Liu, Y.; Liu, W. The prediction of the whole weld in fiber laser keyhole welding based on numerical simulation. *Appl. Therm. Eng.* **2017**, *113*, 980–993. [CrossRef]
12. Kik, T.; Moravec, J.; Novakova, I. Numerical simulations of X22CrMoV12-1 steel multilayer welding. *Arch. Metall. Mater.* **2019**, *64*, 1441–1448.
13. Chen, G.; Shu, X.; Liu, J.; Zhang, B.; Zhang, B.; Feng, J. Investigation on microstructure of electron beam welded WC-Co/40Cr joints. *Vacuum* **2018**, *149*, 96–100. [CrossRef]
14. Bal, K.S.; Dutta Majumdar, J.; Roy Choudhury, A. Effect of electron beam accelerating voltage on the melt zone area, secondary-dendrite arm spacing and fusion line microstructure of bead-on-plate welded Hastelloy C-276 sheet. *Optik* **2019**, *183*, 355–366. [CrossRef]
15. Cho, D.-W.; Cho, W.-I.; Na, S.-J. Modeling and simulation of arc: Laser and hybrid welding process. *J. Manuf. Process.* **2014**, *16*, 26 55. [CrossRef]
16. He, G.-Q.; Yang, Z.-G.; Xi, B.; Zhang, Y.-P.; Wang, D.; Liu, B.; Li, S. Deformation control during the welding of AP1000 main pump casing and steam generator. *Nucl. Mater. Energy* **2021**, *29*, 101090. [CrossRef]
17. Zhao, M.S.; Chiew, S.P.; Lee, C.K. Post weld heat treatment for high strength steel welded connections. *J. Constr. Steel Res.* **2016**, *122*, 167–177. [CrossRef]
18. Li, S.; Xu, W.; Xiao, G.; Zhou, Z.; Su, F.; Feng, J. Effects of Sc on laser hot-wire welding performance of 7075 aluminum alloy. *Mater. Res. Express* **2020**, *7*, 106506. [CrossRef]
19. Heinze, C.; Schwenk, C.; Rethmeier, M. The effect of tack welding on numerically calculated welding-induced distortion. *J. Mater. Process. Technol.* **2012**, *212*, 308–314. [CrossRef]
20. Palanisamy, V.; Solberg, J.K.; Salberg, B.; Moe, P.T. Weld thermal simulation of API 5CT L80 grade steel. *Weld. World* **2021**, *65*, 1983–1995. [CrossRef]
21. Köse, C. Heat treatment and heat input effects on the dissimilar laser beam welded AISI 904L super austenitic stainless steel to AISI 317L austenitic stainless steel: Surface, texture, microstructure and mechanical properties. *Vacuum* **2022**, *205*, 111440. [CrossRef]
22. Ola, O.T.; Ojo, O.A.; Chaturvedi, M.C. On the development of a new pre-weld thermal treatment procedure for preventing heat-affected zone (HAZ) liquation cracking in nickel-base IN 738 superalloy. *Philos. Mag.* **2014**, *94*, 3295–3316. [CrossRef]
23. Panov, D.; Naumov, S.; Stepanov, N.; Sokolovsky, V.; Volokitina, E.; Kashaev, N.; Salishchev, G.; Ventzke, V.; Dinse, R.; Riekehr, S.; et al. Effect of pre-heating and post-weld heat treatment on structure and mechanical properties of laser beam-welded Ti2AlNb-based joints. *Intermetallics* **2022**, *143*, 107466. [CrossRef]
24. Luo, Z.; Ao, S.; Chao, Y.J.; Cui, X.; Li, Y.; Lin, Y. Application of Pre-heating to Improve the Consistency and Quality in AA5052 Resistance Spot Welding. *J. Mater. Eng. Perform.* **2015**, *24*, 3881–3891. [CrossRef]
25. Guo, L.; Zhang, X.-Z.; Feng, C.-X. Continuous bending and straightening technology of Q345c slab based on high-temperature creep deformation. *J. Iron Steel Res. Int.* **2017**, *24*, 595–600. [CrossRef]
26. Zhang, C.; Gong, M.; Zhu, L. Post-fire mechanical behavior of Q345 structural steel after repeated cooling from elevated temperatures with fire-extinguishing foam. *J. Constr. Steel Res.* **2022**, *191*, 107201. [CrossRef]
27. Dai, W.-H.; Song, Y.-T.; Xin, J.-J.; Fang, C.; Wei, J.; Wu, J.-F. Numerical simulation of the ITER BTCC prototype case enclosure welding. *Fusion Eng. Des.* **2020**, *154*, 111538. [CrossRef]
28. Lezaack, M.B.; Simar, A. Avoiding abnormal grain growth in thick 7XXX aluminium alloy friction stir welds during T6 post heat treatments. *Mater. Sci. Eng. A* **2021**, *807*, 140901. [CrossRef]

29. Kumar, U.; Gope, D.K.; Srivastava, J.P.; Chattopadhyaya, S.; Das, A.K.; Krolczyk, G. Experimental and Numerical Assessment of Temperature Field and Analysis of Microstructure and Mechanical Properties of Low Power Laser Annealed Welded Joints [J/OL]. *Materials* **2018**, *11*, 11091514. [CrossRef]
30. Wang, J.; Chen, X.; Yang, L.; Zhang, G. Sequentially combined thermo-mechanical and mechanical simulation of double-pulse MIG welding of 6061-T6 aluminum alloy sheets. *J. Manuf. Process.* **2022**, *77*, 616–631. [CrossRef]
31. Chen, L.; Zhang, Y.; Xue, X.; Wang, B.; Yang, J.; Zhang, Z.; Tyrer, N.; Barber, G.C. Investigation on shearing strength of resistance spot-welded joints of dissimilar steel plates with varying welding current and time. *J. Mater. Res. Technol.* **2022**, *16*, 1021–1028. [CrossRef]
32. Wu, T.; Ma, Y.; Xia, H.; Geng, P.; Niendorf, T.; Ma, N. Measurement and simulation of residual stresses in laser welded CFRP/steel lap joints. *Compos. Struct.* **2022**, *292*, 115687. [CrossRef]
33. Yang, B.; Lin, D.; Xia, H.; Li, H.; Wang, P.; Jiao, J.; Chen, X.; Tan, C.; Li, L.; Wang, Q.; et al. Welding characterization evolutions for dual spot laser welded-brazed Al/steel joint with various spot configurations. *J. Mater. Res. Technol.* **2022**, *19*, 697–708. [CrossRef]
34. Muaz, M.; Choudhury, S.K. Experimental investigations and multi-objective optimization of MQL-assisted milling process for finishing of AISI 4340 steel. *Measurement* **2019**, *138*, 557–569. [CrossRef]
35. Keränen, L.; Nousiainen, O.; Javaheri, V.; Kaijalainen, A.; Pokka, A.-P.; Keskitalo, M.; Niskanen, J.; Kurvinen, E. Mechanical properties of welded ultrahigh-strength S960 steel at low and elevated temperatures. *J. Constr. Steel Res.* **2022**, *198*, 107517. [CrossRef]
36. Lee, C.-H.; Chang, K.-H.; Van Do, V.N. Finite element modeling of residual stress relaxation in steel butt welds under cyclic loading. *Eng. Struct.* **2015**, *103*, 63–71. [CrossRef]
37. Chen, G.; Xue, W.; Jia, Y.; Shen, S.; Liu, G. Microstructure and mechanical property of WC-10Co/RM80 steel dissimilar resistance spot welding joint. *Mater. Sci. Eng. A* **2020**, *776*, 139008. [CrossRef]
38. Wei, H.L.; Elmer, J.W.; Debroy, T. Crystal growth during keyhole mode laser welding. *Acta Mater.* **2017**, *133*, 10–20. [CrossRef]

materials

MDPI

Article

Analysis of Kerf Quality Characteristics of Kevlar Fiber-Reinforced Polymers Cut by Abrasive Water Jet

Dinu-Valentin Gubencu, Carmen Opriș * and Adelina-Alina Han *

Mechanical Engineering Faculty, Politehnica University of Timisoara, 300222 Timisoara, Romania; dinu.gubencu@upt.ro
* Correspondence: carmen.opris@upt.ro (C.O.); adelina.han@upt.ro (A.-A.H.)

Abstract: Abrasive water jet machining has become an indispensable process for cutting Kevlar fiber-reinforced polymers used in applications such as ballistics protection, race cars, and protective gloves. The complex and diffuse action of a large number of input parameters leads to the need to evaluate the quality characteristics of the technological transformation as a result of the deployment of experimental studies adapted to the specific processing conditions. Thus, the paper focuses on identifying the influence of different factors and modeling their action on the characteristics that define the quality of the cut parts, such as the kerf taper angle and the *Ra* roughness parameter, by applying statistical methods of design and analysis of experiments.

Keywords: Kevlar fiber-reinforced polymers; abrasive water jet cutting; full factorial experiment; surface roughness; kerf taper angle

Citation: Gubencu, D.-V.; Opriș, C.; Han, A.-A. Analysis of Kerf Quality Characteristics of Kevlar Fiber-Reinforced Polymers Cut by Abrasive Water Jet. *Materials* **2023**, *16*, 2182. https://doi.org/10.3390/ma16062182

Academic Editor: Yuanqing Li

Received: 13 February 2023
Revised: 28 February 2023
Accepted: 7 March 2023
Published: 8 March 2023

1. Introduction

Kevlar® is an organic fiber belonging to the aromatic polyamides (aramids) family. It was developed in 1965 by the scientists from DuPont. Due to their exceptional characteristics, such as high specific strength, excellent fatigue resistance, good chemical behavior with respect to fuels, vibration damping, and thermal stability, these fibers became an adequate option as composite reinforcements for a large number of high-demanding industrial applications [1,2].

Currently, there are many variants of fibers in the Kevlar® family with different combinations of properties, which makes them suitable to meet various requirements. For instance, Kevlar® 29 (K-29) has a high toughness and low stiffness, and is mainly used in industrial applications such as cables, brake linings, boats hull reinforcement, asbestos replacement, or protective armors [2,3]. Regarding antiballistic applications, K-29 is used for both civil and military purposes and to manufacture panels for lightweight military vehicles, pilot seats, or body vests. These applications require the absorption of a bullet or projectile kinetic energy, avoiding penetration and back face deformation of the armor and, thus, any major injury to the person [4].

Some of the important limitations of Kevlar® fibers are their weakness in compression, poor resistance to UV rays, and hygroscopicity [2,5]. Among the disadvantages of Kevlar fiber-reinforced polymers (KFRP) are their low adhesion with impregnating resins and difficult machining by means of classical processes using hand-held tools. These processes have low reproducibility and accuracy and involve important tool wear. As a consequence, new technologies, such as abrasive water jet machining (AWJM), are widely used for cutting composite materials, including KFRP. However, compared with traditional cutting processes, the material treatment by AWJM is much more elaborate [6]. This cost-effective and flexible technique can achieve cutting of complicated shape parts with high precision and low residual stresses in the material structure. Although during the process no dust is generated in the ambient air, there are some environmental concerns related to loud background noise, unclean working areas, and high abrasive wear [7].

Basically, abrasive water- et cutting (AWJC) consists of using thin water jets under high pressure which pass through a convergent shape nozzle (Figure 1), determining the conversion of pressure energy into the kinetic energy of water [8]. Then, a high-speed water jet passes through a mixing chamber which is directly connected to the water nozzle. A stream of abrasive particles is added into the mixing chamber where high-energy water transfers onto the abrasive particles [9], accelerating them by the momentum exchange [10].

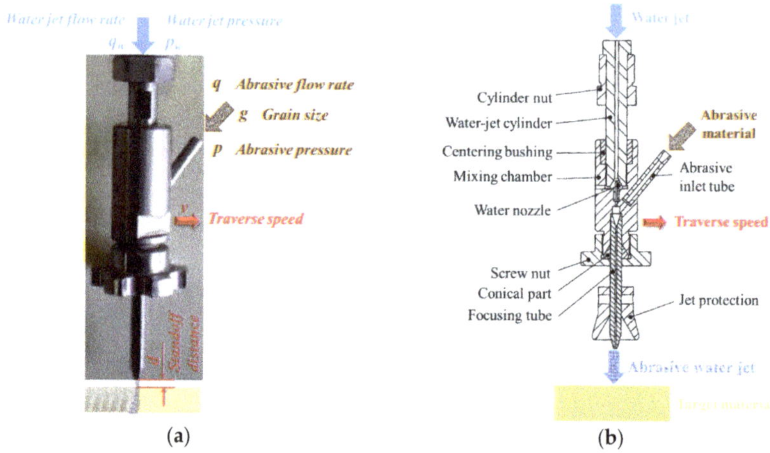

Figure 1. Principle of AWJC: (**a**) process parameters (**b**) components of cutting head.

Following this, a mixture of water and abrasive particles produces a high coherent jet that passes through a focusing tube nozzle, which is directed towards the working area to cut the target material by means of erosion. When the induced stress surpasses the ultimate shear stress of the material, small chips of the material are removed [8]. The cutting head is relatively moved to the workpiece depending on the process type operated in order to expose new material to be machined.

The quality characteristics of the kerf after AWJC of composite materials can be estimated by analyzing several parameters, such as surface roughness, surface waviness (Figure 2a), kerf top width, k_t, kerf bottom width, k_b, and kerf taper angle, k_a (Figure 2b).

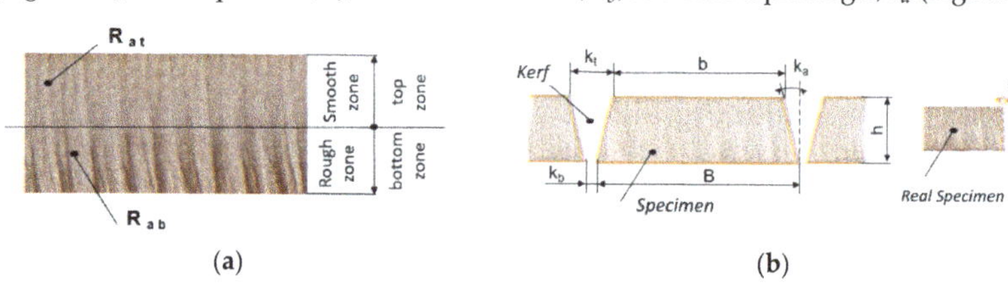

Figure 2. Quality characteristics of cut surfaces: (**a**) roughness zones on KFRP (**b**) dimensional parameters of the kerf. (b represents the specimen top width and B represents the specimen bottom width).

Several studies have highlighted that under an initial damage zone caused by jet expansion before impact and the radial variation of jet energy, the cut surface can be separated into two regions: the smooth (cutting wear) zone or top zone and the rough (deformation wear) zone or bottom zone (Figure 2a) [11–13].

The cutting process occurs at the top zone of the machined surface under conditions in which most of the particles have a higher kinetic energy than the required destruction energy of processed material, resulting a smooth surface [8]. The higher roughness of the bottom zone is explained by the loss of jet penetration energy, which provokes its deflection and, as a result, profile waviness [11].

The kerf taper angle is influenced by the cutting capacity of the jet, so a higher kinetic energy increased by water jet pressure and a small increase in the abrasive mass flow rate improve this output parameter [14].

As the first step of the conducted research, a fishbone diagram was constructed, as shown in Figure 3, in order to organize the different categories of causes that can influence the quality characteristics of kerf obtained by AWJC. Some of the factors listed in the cutting method category and in subcategories, such as hydraulic system, abrasive system, and mixing system, can be easily controlled.

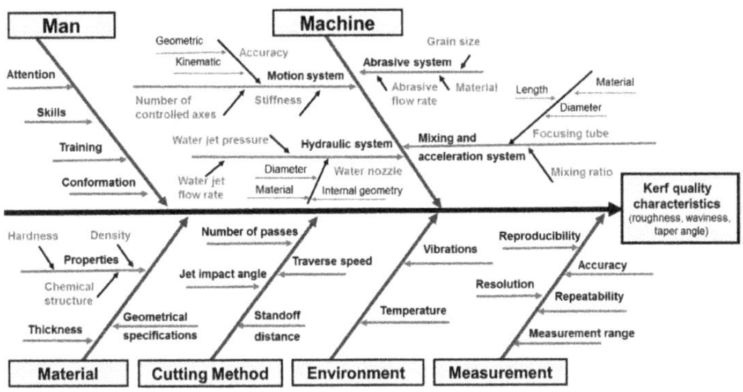

Figure 3. Cause–effect diagram for AWJC.

Therefore, numerous studies have focused on analyzing and quantifying the influence of these input parameters on surface roughness and taper angle of the kerf achieved by AWJC.

Siddiqui and Shukla carried out experimental research on AWJC of Kevlar epoxy composites, applying the Taguchi method and response surface methodology. They agreed, for the selected experimental range of each input parameter, that high water jet pressure, low abrasive flow rate, and low traverse speed ensure an optimum surface finish [11].

Sambruno et al. reported after performing an ANOVA analysis that when processing thermoplastic CFRP the slot walls become more vertical at a high water pressure and a low traverse rate [14].

Studying AWJM of carbon epoxy composites, Dhanawade and Kumar concluded that a medium value of the traverse speed, namely v = 100 mm/min in the experiment, ensures an acceptable surface roughness and high productivity [15]. They also observed the slight influence of the abrasive flow rate on the surface roughness and recommended a decrease in this parameter in order to reduce abrasive consumption.

Doreswamy et al. investigated composites with different reinforcements and established that kerf width is larger with the increase in water pressure and the stand-off distance and with the decrease in traverse speed. Abrasive flow rate had a marginal effect on kerf width [16]. Abidi et al. reported after analyzing AWJM of woven fabric CFRP that minimum surface roughness (Ra = 3.21 μm) was achieved for medium traverse speed (v = 2000 mm/min) and abrasive flow rate (q = 300 g/mm), associated with a minimum stand-off distance (h = 2 mm). On the other hand, kerf taper increased with the growth of both traverse speed and abrasive flow rate [17].

Azmir et al. [18] explored AWJC of Kevlar–phenolic composites using Taguchi's experimental design and reached the conclusion that both investigated output parameters, surface roughness Ra and kerf taper ratio, decreased when reducing the traverse speed and stand-off distance. They also determined that increasing the hydraulic pressure resulted in lower values of those parameters, while abrasive flow rate showed no effect.

After carrying out an experimental program with five input parameters by means of Taguchi's orthogonal array, Vikas and Srinivas [19] found that when processing glass epoxy composites a lower kerf taper angle can be achieved by increasing water jet pressure and decreasing grain size.

Shanmugam and Masood [20] conducted experimental research on AWJC of graphite epoxy and glass epoxy composites and decided that a combination of high water pressure, low traverse speed, and low standoff distance ensures the minimization of kerf taper angle.

Sathishkumar et al. [21] carried out an experimental study on AWJC of basalt–Kevlar–glass fiber-reinforced epoxy laminates, aiming to optimize its output parameters. The results showed that input parameters, such as traverse speed, abrasive mass flow rate, and standoff distance, generated an opposite effect on kerf quality characteristics. Thus, surface roughness decreased, and kerf taper angle increased when either the abrasive flow rate increased, or the traverse speed or standoff distance decreased.

Investigating AWJC of other hybrid composite laminates, consisting of basalt fiber and different loadings of fly ash particles as reinforcements for a vinyl ester polymer matrix, Ramraji et al. [22] concluded that high water jet pressure, low traverse speed, and standoff distance is recommended to reduce both kerf taper and surface roughness.

Many studies have been dedicated to the study of factors effects on the kerf surface finish when AWJC is conducted on other materials with high mechanical properties. Thus, in cutting Hardox 500 steel, Perec found an experimental model for the *Sq* roughness parameter [23], considering as input parameters traverse speed, abrasive flow rate, and water pressure, by means of a response surface methodology. While the pressure variation between 350 and 400 MPa had a statistically insignificant effect on *Sq*, the speed increase between 100 and 300 mm/min exerted the greatest influence on the roughness of the cut surface, causing its growth. Conversely, changing the flow rate to between 250 and 450 g/min resulted in an improvement in the *Sq* parameter.

The same three input parameters set at different levels were selected by Gupta et al. for analyzing kerf top width and kerf taper angle in the case of AWJC of marble [24]. Traverse speed exerted the most significant effect on top kerf width, followed by water pressure. For achieving minimum top kerf width, the settings of these two process parameters must be at the highest levels of 100 mm/min and 340 MPa, respectively. The kerf taper angle was significantly affected only by transverse speed and its minimum value was obtained at the lowest level of travel speed adopted in the experiment, i.e., 50 mm/min.

Major challenges for AWJC processes are to minimize both surface finish and kerf taper. The influence of factors and the results obtained in all previously presented studies are valid only under the particular conditions of the experiments performed, depending on the type and thickness of the material investigated or the values assigned to input variables. Thus, in order to choose appropriate combinations for the input parameters that are useful for achieving effective processes, a customized study on AWJC of KFRP composites must be conducted. Therefore, the objectives pursued in this work were to experimentally determine the mathematical link between the quality characteristics of AWJC of KFRP and the process parameters, adjusted between limits of technological interest for industrial applications.

2. Materials and Methods

The experimental approach focuses on modeling the process based on measured data, starting from an input–output model similar to a black box. That is, the in-depth analysis of the physical phenomena occurring in the workspace is deliberately neglected, aiming only to establish the link between the investigated objective functions and the influence factors using mathematical statistics.

The experimental program was designed using a full factorial experiment applied using 4 relevant influence factors with the purpose of finding the regression models.

2.1. Material

The specimens to be analyzed in this work (Figure 4) were AWJ cut into a rectangular shape of 60 mm × 30 mm, from a vacuum-bagged and autoclave-cured Kevlar epoxy laminate. This material was processed by Duqueine Composites for ballistic applications, resulting in a 9.5 mm thickness laminate formed by 18 layers of prepregs with 40% resin content. For prepreg reinforcement, woven fabric was used with the specifications presented in Table 1 [25]. Fabrics, which comprise of at least 2 woven-together threads (the warp and the weft) are available in several styles, selected according to necessary crimping and drapeability [26]. The tested material used a plain weave style, with each warp strand floating over and then under one fill strand, suitable for flat surface applications because the lay-up process over complex shaped molds is quite difficult [27].

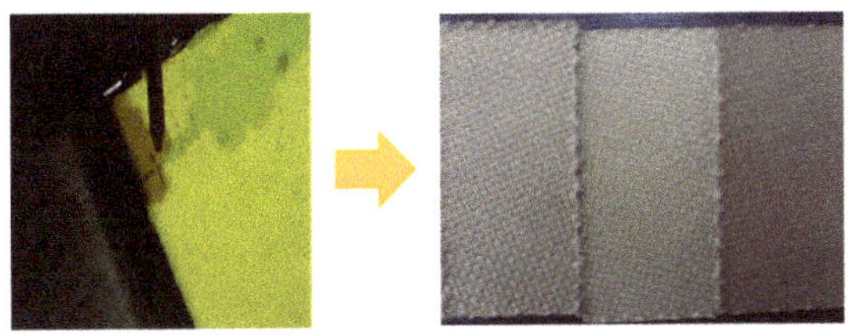

Figure 4. KFRP specimens cut by AWJC.

Table 1. Fabric specifications.

Characteristic	Specification
Composition	100% para-aramid Kevlar 29
Warp	3300 dtex
Weft	3300 dtex
Thread count	$6.65 \times 6.65 \pm 0.20$ ends/cm
Density	440 ± 10 g/m^2
Tensile strength	2863 MPa
Tensile modulus	67 GPa
Elongation at break	3.7%
Weave	Plain
Treatment	Washed

2.2. *Machining Equipment*

The process was conducted on a JEDO CNC water jet abrasive machine (Figure 5). The equipment includes the following component units [8–12]:

- Water supply system
- High-pressure water generator
- Abrasive delivery system
- Cutting head
- CNC router
- Catcher and drain system.

The water supply system is basically a high-pressure water pump system including 2 different storage tanks: the cutting tank and the cooling water tank.

The high-pressure generating system includes 2 main units, the intensifier and the accumulator. The intensifier, acting as an amplifier, provides ultra-high pressurized water at approximately 200–400 MPa. The accumulator temporarily stores the energy of high-pressure water until is required. In order to prevent the fragmentation of the water jet, long-chain polymers are added to the water as stabilizers.

The abrasive delivery system consists of an abrasive hopper and a pneumatically operated valve to control the abrasive mass flow rate. The abrasive material typically used in most applications is garnet, which has a hardness of 8 on the Mohs scale [8]. The GMA garnet originated from Australia and was used for the investigations. This product presents some advantages over other abrasives, as it is totally natural, chemically inert, and free of any toxic metals or crystalline silica. The GMA garnet characteristics and composition [28] are presented in Tables 2 and 3, respectively.

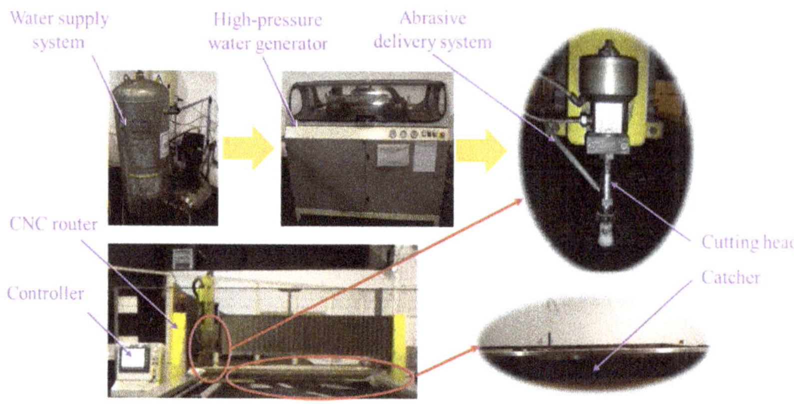

Figure 5. JEDO AWJC equipment setup.

Table 2. Physical characteristics of GMA garnet (typical).

Property	Value
Bulk Density	2.3 T/m^3
Specific Gravity	4.1
Hardness (Mohs)	7.5–8.0
Melting Point	1250 °C
Shape of natural grains	Sub-angular

Table 3. Composition of GMA garnet (typical).

Mineral and Metal Composition		Average Chemical Composition	
Garnet (Almandite)	97–98%	SiO_2	36%
Ilmenite	1–2%	Al_2O_3	20%
Zircon	0.2%	FeO	30%
Quartz (free silica)	<0.5%	Fe_2O_3	2%
Others	0.25%	TiO_2	1%
Ferrite (free iron)	<0.01%	MnO	1%
Lead	<0.002%	CaO	2%
Copper	<0.005%	MnO	6%
Other Heavy Metals	<0.01%		

The cutting head, with the structure shown in Figure 1, is the equipment component where the mixing of high-pressure water and abrasive material occurs, thus achieving material removal as explained previously. The motion of the cutting head is controlled by a CNC router to desired coordinates of the working area, ensuring flexibility, productivity, and high dimensional precision, as well as accuracy of the machining process.

A catcher tank supports the rigid frame worktable and collects the pressurized water after cutting, thus dissipating energy and reducing noise.

2.3. Objective Functions and Influence Factors

The characterizations of the processed parts were assessed by determining 3 objective functions:

- arithmetic average roughness, Ra_t, at the top zone
- arithmetic average roughness, Ra_b, at the bottom zone
- kerf taper angle, k_a, measured for the longer sides of the specimens.

The *Ra* roughness parameter was chosen because it is the most widely used as a global evaluation of the roughness amplitude on a profile and, also, is defined similarly in all standards, as the arithmetic mean of the absolute ordinate values Z(x), within the sampling length, l_r, numerically equal to the characteristic wavelength of the profile filter λ_c, which suppress the longwave component [29]:

$$Ra = \frac{1}{l_r} \int_0^{l_r} |Z(x)|\, dx \tag{1}$$

where Z(x) represents the heights of the assessed profile at any position x.

The kerf taper angle, k_a, was calculated based on the difference between the measured top and bottom sides of the processed specimens, b and B, respectively, and the measured thickness of the specimen, h (Figure 3), using the standard formula:

$$k_a = \tan^{-1}\frac{B - b}{2\,h} \times \frac{180^\circ}{\pi} \tag{2}$$

Input parameters were selected to test their influence not only on the quality characteristics of the cut, but also on the efficiency of KFRP processing process, the analysis of which was the objective of a separate study.

Thus, 4 influence parameters were chosen for carrying out the experimental programme:

- traverse speed, v (mm/min)
- focusing tube diameter, D (mm)
- abrasive flow rate, q (g/min)
- abrasive grain size, g (mesh #)

The experimental range for each factor (Table 4) was selected taking into account not only previous conducted studies by several researchers and industrial expertise, but our own preliminary experiments. At the same time, some limitations regarding the equipment, which is supplied with 2 focusing tubes each with a focusing tube diameter of 0.76 mm or 1.00 mm, determined the levels for this input parameter. The abrasive flow rate levels were selected in correlation with the other process parameters to avoid saturation problems and difficulty in the transfer of momentum to the abrasive particles.

Table 4. Experimental levels and ranges of variation for the influence factors.

Factors	Symbol	Central Point 0	Range of Variation Δ_j	Lower Level −1	Higher Level +1
Traverse speed, v (mm/min)	x_1	300	200	100	500
Focusing tube diameter, D (mm)	x_2	–	0.12	0.76	1.00
Abrasive flow rate, q (g/min)	x_3	180	20	160	200
Abrasive grain size, g (mesh #)	x_4	150	50	100	200

The initial parameter settings, which were kept unchanged during all experimental trials, are presented in Table 5. Their values were selected considering the presumed positive effect on the objective functions.

Table 5. AWJC process experimental settings.

Parameter	Specification
Water pressure, pw (MPa)	300
Water nozzle diameter, d0 (mm)	0.33
Standoff distance, d (mm)	3
Jet angle of attack, γ (deg.)	90
No of passes (–)	1
Abrasive material	GMA garnet

3. Results and Discussions

For modeling the action of a number of p influence factors selected, x_j, on a performance characteristic, y, by means of a 2^p full factorial experiment, the experimental results are used to find the constants, b_0, b_j, b_{jk}, of the polynomial [30]:

$$y = b_0 + \sum_{j=1}^{p} b_j\, x_j + \sum_{j,k=1,\ j\neq k}^{p} b_{jk}\, x_j\, x_k \tag{3}$$

This experimental design ensures that a high precision model is obtained with a minimum number of trials, thus reducing the investigation costs.

3.1. Experimental Design and Measured Results

The experimental matrix (Table 6) was constructed according to the principles of designing full factorial experiments, containing all the possible combinations of the factors levels [30]. The method selected for estimating the experimental error was to replicate each run, i, of the experiment. Table 6 presents results for all three objective functions investigated.

Table 6. Matrix and results of the 2^4 full factorial experiments.

Run No	A: v		B: D		C: q		D: g		Ra_t (μm)		Ra_b (μm)		k_a (deg)	
i	Coded	(mm/min)	Coded	(mm)	Coded	(g/min)	Coded	(mesh #)	Ra_{t1}	Ra_{t2}	Ra_{b1}	Ra_{b2}	k_{a1}	k_{a2}
1	−1	100	−1	0.76	−1	160	−1	100	10.7	10.1	14.8	16.4	1.15	1.34
2	+1	500	−1	0.76	−1	160	−1	100	12.2	11.6	18.6	16.2	1.32	1.53
3	−1	100	+1	1.00	−1	160	−1	100	9.7	10.6	15.9	15.1	1.19	1.26
4	+1	500	+1	1.00	−1	160	−1	100	15.2	14.4	21.2	22.4	1.96	1.79
5	−1	100	−1	0.76	+1	200	−1	100	6.8	7.2	10.2	12.4	0.74	0.95
6	+1	500	−1	0.76	+1	200	−1	100	10.8	9.6	14.7	15.8	1.16	1.23
7	−1	100	+1	1.00	+1	200	−1	100	9.4	8.6	12.6	13.8	0.94	1.12
8	+1	500	+1	1.00	+1	200	−1	100	12.1	11.2	19.2	18.2	1.49	1.32
9	−1	100	−1	0.76	−1	160	+1	200	6.7	7.8	9.8	11.4	0.97	0.76
10	+1	500	−1	0.76	−1	160	+1	200	9.8	8.6	12.8	15.2	1.34	1.17
11	−1	100	+1	1.00	−1	160	+1	200	11.5	9.9	15.2	16.3	1.26	1.31
12	+1	500	+1	1.00	−1	160	+1	200	14.0	12.8	20.8	19.2	1.58	1.64
13	−1	100	−1	0.76	+1	200	+1	200	6.1	7.2	10.1	11.1	0.68	0.77
14	+1	500	−1	0.76	+1	200	+1	200	9.8	8.6	13.2	12.1	1.18	0.95
15	−1	100	+1	1.00	+1	200	+1	200	7.8	7.2	10.6	12.5	0.87	1.06
16	+1	500	+1	1.00	+1	200	+1	200	10.9	12.1	15.4	17.6	1.26	1.42

For processing, analyzing, and plotting the experimental data, the statistical dedicated software Minitab® 17 was utilized.

3.2. Analysis of the Surface Roughness

Surface roughness, namely the *Ra* parameter, was measured using a handheld Taylor Hobson Surtronic 25 stylus tester (Figure 6). This parameter was established separately for the smooth and the rough zone of the kerf, Ra_t and Ra_b, respectively (Figure 2a), adopting a unitary procedure for both investigated areas by keeping the most unfavorable value among those determined after performing two measurements on each longer side cut surface of the samples. For all experimental trials, stylus traverse movements perpendicular to the lay direction were measured and the Gaussian filter was applied.

The cutoff length values were selected according to standard recommendations [31] as follows: λ_c = 2.5 mm, if Ra < 10 μm or λ_c = 8 mm, if Ra > 10 μm. On the other hand, there were differences between the initial settings for the two kerf zones regarding the number of sampling lengths considered to establish the evaluation lengths. Thus, this number for the kerf top zone was N_{top} = 5, but for the bottom zone, characterized by insufficient space on the sample surface, the agreed value was N_{bottom} = 2.

The effects of selected input parameters on the top zone roughness, Ra_t, are shown in Figure 7a and on the bottom zone roughness, Ra_b, in Figure 7b. From both figures, it can be observed that increasing the abrasive flow rate, q, and mesh #, g, ensures a lower surface roughness; meanwhile, increasing the other two factors, traverse speed, v, and focusing tube diameter, D, have an opposite effect.

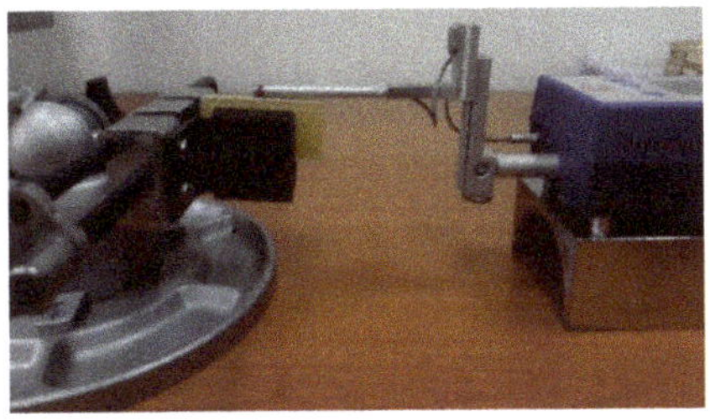

Figure 6. Experimental measuring of the surface roughness.

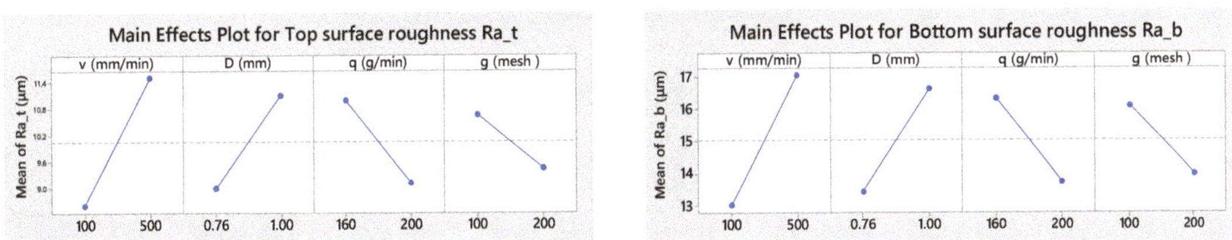

(**a**) (**b**)

Figure 7. Main effects plot: (**a**) for top surface roughness, Ra_t (**b**) for bottom surface roughness, Ra_b.

The positive effect of q growth can be clarified by the proximity of the trajectories of the abrasive particles that exert the erosive effect on the surface of the cut. The higher the number of particles involved, the higher the number of initial geometrically superposed craters formed by the particles at the impacted surface. These craters will initiate closer micro-cracks resulting in a better surface finish. One of the micro-mechanisms involved in material separation, plastic deformation, results in the appearance of traces on the kerf surface and the sizes are dependent on the abrasive grain size, thus explaining the decrease in roughness with a higher mesh # of the sand.

On the other hand, a lower D value leads to a reduction in the impact surface of the jet on the workpiece and more close trajectories of the abrasive particles are generated, causing a decrease in roughness parameter Ra in both zones. A similar effect on particle paths is caused by the decreases of the traverse speed, v, combined with a smaller jet deflection, explaining the improvement of surface finish.

An analysis of variance (ANOVA) was carried out to distinguish the statistically significant effects on top surface roughness, Ra_t (Table 7), and on bottom surface roughness, Ra_b (Table 8).

Since the p-value is less than 0.05 for all selected influence factors, these have a statistically significant effect both on Ra_t and Ra_b at the 95% confidence level. Contrarily, since the p-value is higher than 0.05 (with one exception) the interaction effects are insignificant; as a result, they can be eliminated from the future model.

Consequently, the regression equations of the fitted models developed for coded values of input variables are:

$$Ra_t = 10.031 + 1.450\, v + 1.056\, D - 0.944\, q - 0.606\, g \quad (4)$$

$$Ra_b = 15.025 + 2.013\, v + 1.600\, D - 1.306\, q - 1.069\, g + 0.612\, v\, D \quad (5)$$

By examining the regression coefficients values it can be seen that all the input parameters have a higher effect on Ra_b than on Ra_t. Furthermore, the hierarchy of factors, in order of their influence on surface roughness, is the same for the top and bottom zones (Figure 8).

Table 7. ANOVA for top surface roughness, Ra_t.

Source	DF	Adj SS	Adj MS	F-Value	*p*-Value
Model	10	152.707	15.2707	20.20	0.000
Linear	4	143.244	35.8109	47.47	0.000
V	1	67.280	67.2800	89.19	0.000
D	1	35.701	35.7012	47.33	0.000
q	1	28.501	28.5013	37.78	0.000
g	1	11.761	11.7613	15.59	0.001
Two-way interaction	6	9.464	1.5773	2.09	0.098
v × D	1	2.880	2.8800	3.82	0.064
v × q	1	0.320	0.3200	0.42	0.522
v × g	1	0.080	0.0800	0.11	0.748
D × q	1	1.711	1.7113	2.27	0.147
D × g	1	2.761	2.7613	3.66	0.069
q × g	1	1.711	1.7112	2.27	0.147
Error	21	15.841	0.7543		
Lack-of-Fit	5	7.681	1.5363	3.01	0.042
Pure error	16	8.160	0.5100		
Total	31	168.549			

Table 8. ANOVA for bottom surface roughness, Ra_b.

Source	DF	Adj SS	Adj MS	F-Value	*p*-Value
Model	10	325.248	32.525	21.31	0.000
Linear	4	302.678	75.669	49.58	0.000
V	1	129.605	129.605	84.91	0.000
D	1	81.920	81.920	53.67	0.000
q	1	54.601	54.601	35.77	0.000
g	1	36.551	36.551	23.95	0.000
Two-way interaction	6	22.570	3.762	2.46	0.058
v × D	1	12.005	12.005	7.87	0.011
v × q	1	0.061	0.061	0.04	0.843
v × g	1	1.051	1.051	0.69	0.416
D × q	1	3.511	3.511	2.30	0.144
D × g	1	4.961	4.961	3.25	0.086
q × g	1	0.980	0.980	0.64	0.432
Error	21	32.052	1.526		
Lack-of-Fit	5	11.232	2.246	1.73	0.186
Pure error	16	20.820	1.301		
Total	31	357.300			

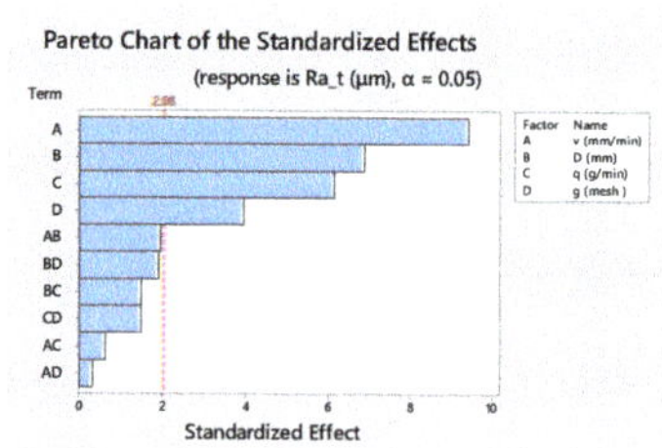

(a)

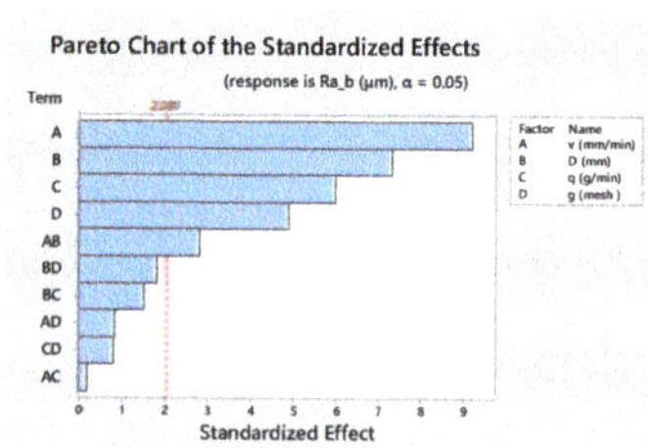

(b)

Figure 8. Hierarchy of factors effects: (**a**) for top surface roughness, Ra_t (**b**) for bottom surface roughness, Ra_b.

The normal probability plot (Figure 9) is a scatter plot with the theoretical percentiles of the normal distribution on the *x*-axis and the sample percentiles of the residuals on the *y*-axis, used to test the normal distribution of residuals. The residuals, represented on the plot with blue dots, are placed on a straight line for both response functions, showing that the relationships between the theoretical percentiles and the sample percentiles are approximately linear and proving that errors are indeed normally distributed.

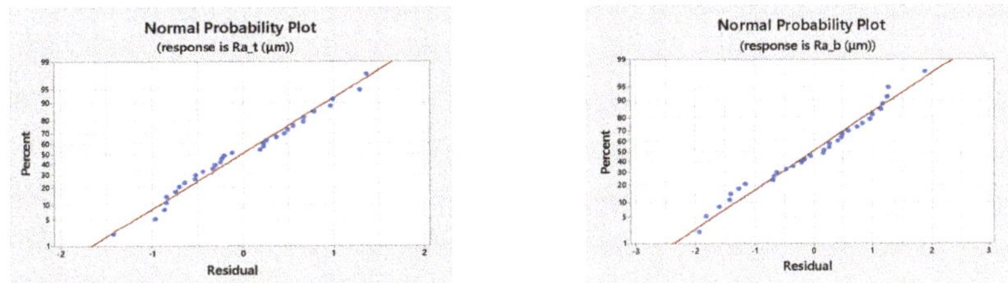

(**a**) (**b**)

Figure 9. Normal probability plot: (**a**) for top surface roughness, Ra_t (**b**) for bottom surface roughness, Ra_b.

Figure 10 illustrates the agreement between the experimental data (represented in the graph with red squares) and the predicted values, confirming the consistency of data for both investigated roughness zones.

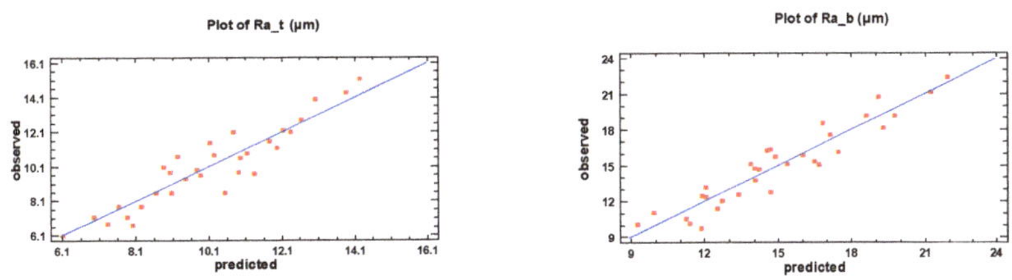

(**a**) (**b**)

Figure 10. Comparison of experimental and model predicted surface roughness: (**a**) for top surface roughness, Ra_t (**b**) for bottom surface roughness, Ra_b.

The empirical models can be graphically represented considering various combinations of two factors as independent variables and setting the other factors at different constant values. As an example, three-dimensional graphs of both fitted models with natural values of the factors are presented in Figure 11. Figure 12 shows the contour lines of constant responses Ra_t and Ra_b in the selected two factors plane.

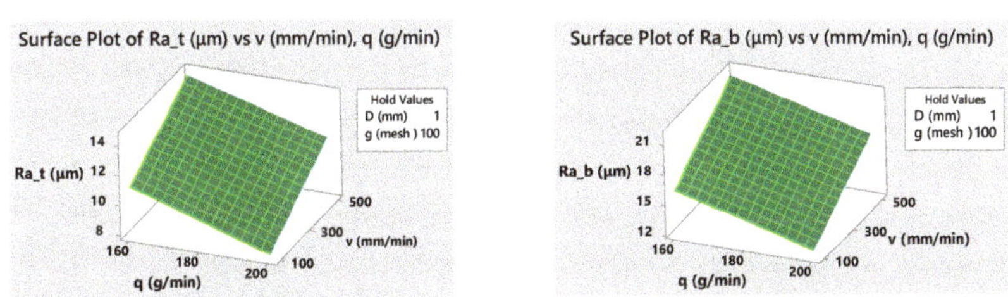

(**a**) (**b**)

Figure 11. Response surfaces of roughness versus traverse speed and abrasive flow rate at D = 1 mm, g = 100: (**a**) for top surface roughness, Ra_t (**b**) for bottom surface roughness, Ra_b.

3.3. Analysis of the Kerf Taper Angle

The kerf taper angle, k_a, was evaluated by carrying out two measurements for each specimen with a FARO-Edge arm and retaining for the analysis the higher value (Table 6). The main effects of the influence factors were plotted in Figure 13a, showing that the increase in traverse speed, v, and the focusing tube diameter, D, leads to a higher kerf taper angle, k_a. Contrarily, the rise in abrasive flow rate, q, and in mesh number, g, generates the decrease in k_a. The magnitude of factor influence allowed their ranking in Figure 13b.

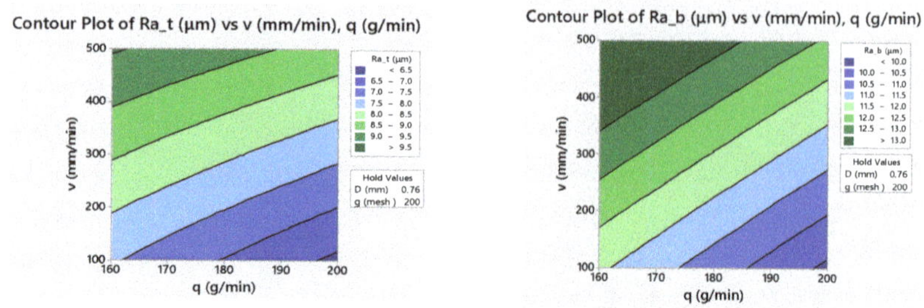

(**a**) (**b**)

Figure 12. Contour plots of surface roughness versus traverse speed and abrasive flow rate at $D = 1$ mm, $g = 100$: (**a**) for top surface roughness, Ra_t (**b**) for bottom surface roughness, Ra_b.

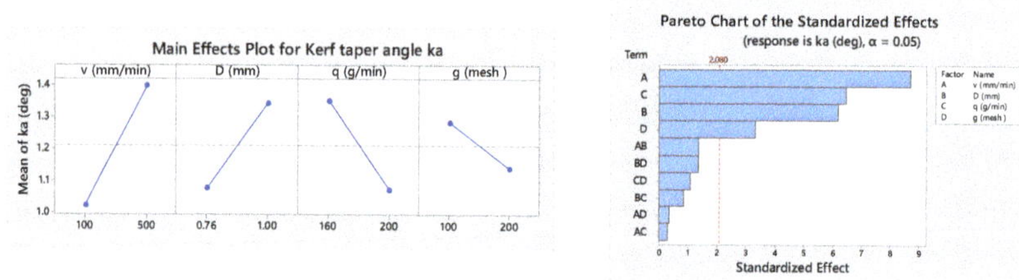

(**a**) (**b**)

Figure 13. Effects on kerf taper angle, k_a: (**a**) main effects plot (**b**) effects hierarchy.

The unfavorable effect on k_a of traverse speed increasing can be explained by the reduction in the overlap of abrasive grain impacts and, as a consequence, of its cutting capacity. On the other hand, increasing q in the selected experimental range avoids collisions between abrasive particles and its edge rounding and determines an improvement of k_a due to the greater cutting capacity of the jet.

A common, simplified interpretation of the effects of the input parameters on the investigated quality characteristics can be formulated after the analysis of the physical phenomena that occur in the working space. Material removal mechanisms involved in AWJC of fiber-reinforced polymers consist of both erosive processes of the matrix and shearing and pulling-out processes of the fibers [32]. Basically, these processes depend on the amount of kinetic energy in the solid particles at impingement, following the rule that an increase in the energy level leads to an improvement in the quality characteristics of the kerf, i.e., to lower roughness and taper angle.

Thus, an increase in the traverse speed reduces the kinetic energy per unit of time and a higher focusing tube diameter diminishes the kinetic energy density of the jet upon impact on the processed material. As a result, a higher surface roughness and kerf taper angle are obtained. Contrarily, when the abrasive flow rate and abrasive mesh # are increased, more particles strike the target area. Therefore, the jet will gain a higher kinetic energy that enhances its penetration capacity of the workpiece. As a consequence, smoother cut surfaces will be obtained and the top and bottom widths of the kerf will have close values, determining a smaller taper angle.

An ANOVA was performed to analyze the statistical significance of input parameter effects and of their interactions with the kerf taper angle (Table 9).

Table 9. ANOVA for kerf taper angle, k_a.

Source	DF	Adj SS	Adj MS	F-Value	*p*-Value
Model	10	2.533	0.25331	17.32	0.000
Linear	4	2.447	0.61181	41.82	0.000
V	1	1.113	1.1137	76.13	0.000
D	1	0.559	0.5591	38.22	0.000
q	1	0.613	0.6132	41.92	0.000
g	1	0.161	0.1612	11.01	0.003
Two-way interaction	10	2.533	0.25331	17.32	0.464
v × D	6	0.085	0.0114	0.98	0.188
v × q	1	0.027	0.0270	1.85	0.762
v × g	1	0.001	0.0013	0.09	0.718
D × q	1	0.002	0.0019	0.13	0.398
D × g	1	0.011	0.0108	0.74	0.188
q × g	1	0.027	0.0270	1.85	0.285
Error	21	0.307	0.0146		
Lack-of-Fit	5	0.094	0.0188	1.41	0.272
Pure error	16	0.213	0.0133		
Total	31	2.8403			

Analyzing the *p*-value column proves that all factor effects are significant at the 95% confidence level, but the interaction effects are not, so they will be neglected in the regression model:

$$k_a = 1.2097 + 0.1866\,v + 0.1322\,D - 0.1384\,q - 0.0709\,g \tag{6}$$

Figure 14a reveals that the residuals, represented in the graph by blue dots, are positioned quite close to a straight line, attesting the normal distribution of errors. Figure 14b demonstrates good concordance between measured kerf taper angles, represented in the graph by red squares, and predicted values by the fitted model.

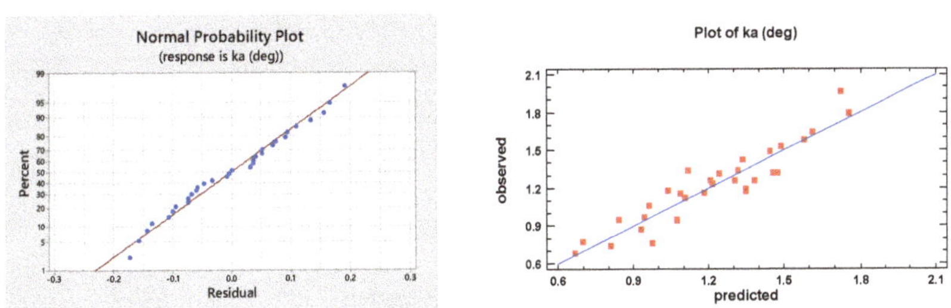

(a) (b)

Figure 14. Diagnostic plots for kerf taper angle, k_a: (**a**) normal probability plot (**b**) observed versus predicted values.

The empirical models can be graphically represented considering various combination of two factors as independent variables and setting the other factors at different constant values. An example of a response surface and contour plot for kerf taper angle, k_a, is presented in Figure 15. Contour plots in particular are of practical importance, allowing the selection of appropriate combinations of factor levels that ensure the fulfillment of specifications related to the objective function, in this case, kerf taper angle.

3.4. Investigations on Delamination

Visual and macroscopic examinations of the samples were conducted in order to emphasize the kerf aspect. An Olympus stereo microscope type SZX7 with Analysis 5.0 software was used for sample imaging.

Almost all samples showed a cut surface without delamination damage, unlike the following cases, where some problems occurred. For instance, sample 11 presented delamination and broad and accentuated kerf (Figure 16).

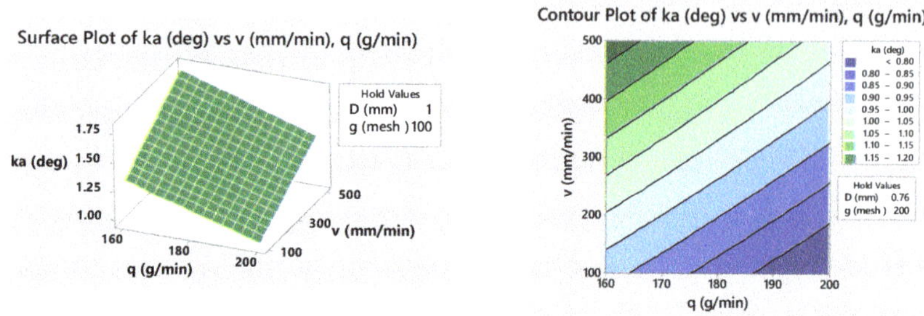

(**a**) (**b**)

Figure 15. Response plots for kerf taper angle, k_a, versus traverse speed and abrasive flow rate at $D = 1$ mm, $g = 100$: (**a**) 3D surface plot (**b**) contour plot.

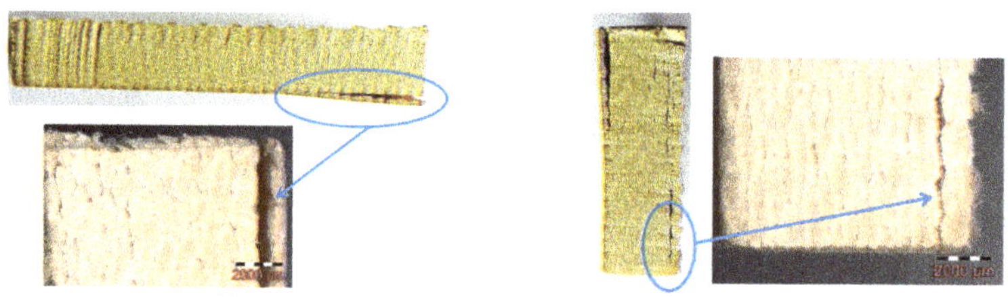

(**a**) (**b**)

Figure 16. Sample 11: (**a**) length surface showing delamination and broad kerf (**b**) width surface showing delamination.

The analysis revealed for sample 18 a discreet kerf with a minor defect at the start point (see detail in Figure 17a).

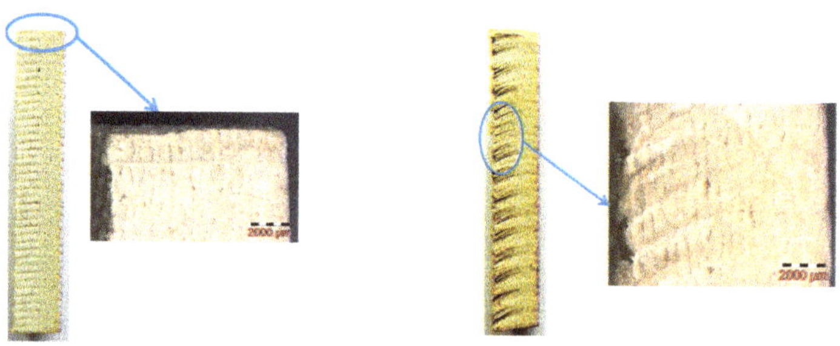

(**a**) (**b**)

Figure 17. Kerf aspect: (**a**) sample 18 (**b**) sample 13.

On the other hand, many samples, as seen in Figure 17b for sample 13, exhibited minor fiber pullouts at the exit of the abrasive water jet.

As a result of this analysis, a causal relationship could not be identified between the levels assigned to the factors in the experiment and the occurrence of these isolated cases of sample delamination.

4. Conclusions

Following the experiment carried out, the most important conclusions regarding AWJC of KFRP that can be drawn are:

- All the input parameters adjusted in the experiment proved to have a significant influence on the three analyzed quality characteristics.
- The study revealed that traverse speed had the greatest influence on surface roughness, both on the top cutting zone, Ra_t, and on the bottom cutting zone, Ra_b, followed by focusing tube diameter, abrasive flow rate, and abrasive grain size, but the amplitude of the effects produced by these factors is magnified at the bottom zone.
- The traverse speed had the most important influence on the kerf taper angle, and the hierarchy was continued by abrasive flow rate, focusing tube diameter, and abrasive grain size.
- Experimental models with coefficients of determination at approximately 90% were found to describe the action of selected factors on surface roughness Ra_t (R-sq = 90.60%), Ra_b (R-sq = 91.03%), and kerf taper angle k_a (R-sq = 89.18%).
- The increase in abrasive flow rate and in mesh number had a positive effect, determining the decrease in all three investigated quality characteristics, while the growth in traverse speed and focusing tube diameter generated a negative effect.
- Therefore, the optimum combination of factor levels that assures minimum surface roughness and kerf taper angle under the conditions of the experiment performed is: v = 100 mm/min, D = 0.76 mm, q = 200 g/min, g = 200 mesh #.

Author Contributions: Conceptualization, D.-V.G.; methodology, D.-V.G. and A.-A.H.; software, A.-A.H.; validation, D.-V.G., C.O. and A.-A.H.; formal analysis, C.O.; investigation, D.-V.G., C.O. and A.-A.H.; writing—original draft preparation, D.-V.G.; writing—review and editing, C.O. and A.-A.H.; visualization, C.O. and A.-A.H.; supervision, D.-V.G. All authors have read and agreed to the published version of the manuscript.

Funding: This research received no external funding.

Institutional Review Board Statement: Not applicable.

Informed Consent Statement: Not applicable.

Data Availability Statement: Not applicable.

Acknowledgments: The authors gratefully acknowledge the support of Duqueine Composites, Timisoara, for providing the investigated material and extending their AWJC facilities for this experimental work.

Conflicts of Interest: The authors declare no conflict of interest.

References

1. Kevlar®Aramid Fiber Technical Guide. Available online: www.dupont.com/content/dam/dupont/amer/us/en/safety/public/documents/en/Kevlar_Technical_Guide_0319.pdf (accessed on 30 September 2022).
2. Perrin Aqua Découpe. Available online: https://www.perrinaquadecoupe.fr/materiauxcarbonekevelar.html (accessed on 30 September 2022).
3. Materials Today. Available online: https://materials-today.com/kevlar-uses-properties-and-processing/ (accessed on 30 September 2022).
4. Çolakoğlu, M.; Soykasap, Ö. Ballistic Properties of Kevlar Composites. In Proceedings of the NATO workshop "Textile Composites", Kiev, Ukraine, 18–21 May 2009.
5. Hasan, Z. *Tooling for Composite Aerospace Structures: Manufacturing and Applications*; Butterworth-Heinemann: Oxford, UK, 2020; pp. 21–48.
6. Perec, A.; Radomska-Zalas, A.; Fajdek-Bieda, A. Experimental Research into Marble Cutting by Abrasive Water Jet. *Facta Univ. Ser. Mech. Eng.* **2022**, *20*, 145–156. [CrossRef]
7. Ma, C.; Deam, R.T. A correlation for predicting the kerf profile from abrasive water jet cutting. *Exp. Therm. Fluid Sci.* **2006**, *30*, 337–343. [CrossRef]
8. Saravanan, S.; Vijayan, V.; Jaya Suthahar, S.T.; Balan, A.V.; Sankar, S.; Ravichandran, M. A review on recent progresses in machining methods based on abrasive water jet machining. *Mater. Today Proc.* **2020**, *21*, 116–122. [CrossRef]

9. Alsoufi, M.S. State-of-the-Art in Abrasive Water Jet Cutting Technology and the Promise for Micro- and Nano-Machining. *Int. J. Mech. Eng. Appl.* **2017**, *5*, 1–14. [CrossRef]
10. Marušić, V.; Baralić, J.; Nedić, B.; Rosandić, Ž. Effect Of Machining Parameters On Jet Lagging In Abrasive Water Jet Cutting. *Teh. Vjesn.* **2013**, *20*, 677–682.
11. Siddiqui, T.U.; Shukla, M. Optimisation of surface finish in abrasive water jet cutting of Kevlar composites using hybrid Taguchi and response surface method. *Int. J. Mach. Mach. Mater.* **2008**, *3*, 382–402. [CrossRef]
12. Shukla, M. Predictive modelling of surface roughness and kerf widths in abrasive water jet cutting of Kevlar composites using neural network. *Int. J. Mach. Mach. Mater.* **2010**, *8*, 226–246. [CrossRef]
13. Natarajan, Y.; Murugesan, P.K.; Mohan, M.; Khan, S.A.L.A. Abrasive Water Jet Machining process: A state of art of review. *J. Manuf. Process.* **2020**, *49*, 271–322. [CrossRef]
14. Sambruno, A.; Bañon, F.; Salguero, J.; Simonet, B.; Batista, M. Kerf Taper Defect Minimization Based on Abrasive Waterjet Machining of Low Thickness Thermoplastic Carbon Fiber Composites C/TPU. *Materials* **2019**, *12*, 4192. [CrossRef] [PubMed]
15. Dhanawade, A.; Kumar, S. Experimental study of delamination and kerf geometry of carbon epoxy composite machined by abrasive water jet. *J. Compos. Mater.* **2017**, *51*, 3373–3390. [CrossRef]
16. Doreswamy, D.; Shivamurthy, B.; Anjaiah, D.; Sharma, Y. An Investigation of Abrasive Water Jet Machining on Graphite/Glass/Epoxy Composite. *Int. J. Manuf. Eng.* **2015**, 627218. [CrossRef]
17. Abidi, A.; Salem, S.B.; Yallese, M.A. Experimental and Analysis in Abrasive Water jet cutting of carbon fiber reinforced plastics. In Proceedings of the 24ème Congrès Français de Mécanique, Brest, France, 26–30 August 2019.
18. Azmir, M.A.; Ahsan, A.K.; Rahmah, A. Effect of abrasive water jet machining parameters on aramid fibre reinforced plastics composite. *Int. J. Mater. Form* **2009**, *2*, 37–44. [CrossRef]
19. Vikas, B.G.; Srinivas, S. Kerf analysis and delamination studies on glass-epoxy composites cut by abrasive water jet. *Mater. Today Proc.* **2021**, *46*, 4475–4481. [CrossRef]
20. Shanmugam, D.K.; Masood, S.H. An investigation on kerf characteristics in abrasive waterjet cutting of layered composites. *J. Mater. Process. Technol.* **2009**, *209*, 3887–3893. [CrossRef]
21. Sathishkumar, N.; Selvam, R.; Kumar, K.M.; Abishini, A.H.; Khaleelur Rahman, T.; Mohanaranga, S. Influence of garnet abrasive in drilling of Basalt–Kevlar–Glass fiber reinforced polymer cross ply laminate by Abrasive Water Jet Machining process. *Mater. Today Proc.* **2022**, *62*, 1361–1368. [CrossRef]
22. Ramraji, K.; Rajkumar, K.; Dhananchezian, M.; Sabarinathan, P. Key Experimental Investigations of cutting dimensionality by Abrasive Water Jet Machining on Basalt Fiber/Fly ash Reinforced Polymer Composite. *Mater. Today Proc.* **2020**, *22*, 1351–1359. [CrossRef]
23. Perec, A. Multiple Response Optimization of Abrasive Water Jet Cutting Process using Response Surface Methodology (RSM). *Procedia Comput. Sci.* **2021**, *192*, 931–940. [CrossRef]
24. Gupta, V.; Pandey, P.M.; Garg, M.P.; Khanna, R.; Batra, N.K. Minimization of kerf taper angle and kerf width using Taguchi's method in abrasive water jet machining of marble. *Procedia Mater. Sci.* **2014**, *6*, 140–149. [CrossRef]
25. NCV Industries. Protections Pare-Balles et Pare-Eclats (Casques). Balistique F 04007 1320 DSO 000. 2005. Available online: https://www.porcher-ind.com/en/solutions/textiles (accessed on 12 September 2022).
26. HexPly®Prepreg Technology. Publication No. FGU 017c. Hexcel Corporation. January 2013. Available online: https://kevra.fi/wp-content/uploads/Prepreg_Technology_pdf (accessed on 30 September 2022).
27. Composite Fabric Weave Styles. Available online: https://compositeenvisions.com/document/fabric-weave-styles/ (accessed on 30 September 2022).
28. GMA GARNET®. The Most Popular Garnet Abrasive, Worldwide—Near You! Available online: https://www.irangarnet.com/wp-content/uploads/2017/05/Waterjet-Brochure.pdf (accessed on 30 September 2022).
29. *ISO 4287:1997*; Geometrical Product Specification (GPS)—Surface Texture: Profile Method—Terms, Definitions, and Surface Texture Parameters. International Organization for Standardization: London, UK, 1997.
30. Montgomery, D.C. *Design and Analysis of Experiments*, 5th ed.; John Wiley and Sons: New York, NY, USA, 2001; pp. 218–276.
31. *ISO 4288:1996*; Geometrical Product Specifications (GPS)—Surface Texture: Profile Method—Rules and Procedures for the Assessment of Surface Texture. International Organization for Standardization: London, UK, 1996.
32. Wang, J.; Guo, D.M. A predictive depth of penetration model for abrasive waterjet cutting of polymer matrix composites. *J. Mater. Process. Technol.* **2002**, *121*, 390–394. [CrossRef]

materials

MDPI

Article

Fabrication of Flexible Supercapacitor Electrode Materials by Chemical Oxidation of Iron-Based Amorphous Ribbons

Mircea Nicolaescu [1,2], Melinda Vajda [2,3], Carmen Lazau [2], Corina Orha [2], Cornelia Bandas [2], Viorel-Aurel Serban [1] and Cosmin Codrean [1,*]

1 Department of Materials and Manufacturing Engineering, Faculty of Mechanical Engineering, Politehnica University Timisoara, Mihai Viteazu 1, 300222 Timisoara, Romania; mircea.nicolaescu@student.upt.ro (M.N.); viorel.serban@upt.ro (V.-A.S.)

2 National Institute for Research and Development in Electrochemistry and Condensed Matter, Dr. A. P. Podeanu 144, 300569 Timisoara, Romania; melinda.vajda@student.upt.ro (M.V.); carmenlazau@yahoo.com (C.L.); orha.corina@gmail.com (C.O.); cornelia.bandas@gmail.com (C.B.)

3 Department of Applied Chemistry and Engineering of Inorganic Compounds and Environment, Faculty of Industrial Chemistry and Environmental Engineering, Politehnica University Timisoara, Piata Victoriei 2, 300006 Timisoara, Romania

* Correspondence: cosmin.codrean@upt.ro

Abstract: A flexible electrode constructed from Fe-based amorphous ribbons decorated with nanostructured iron oxides, representing the novelty of this research, was successfully achieved in one-step via a chemical oxidation method, using a low concentration of NaOH solution. The growth of metal oxides on a conductive substrate, which forms some metal/oxide structure, has been demonstrated to be an efficient method for increasing the charge transfer efficiency. Through the control and variation of synthetic parameters, different structures and morphologies of iron oxide were obtained, including hexagonal structures with a hollow ball shape and rhombohedral structures with rhombus-like shapes. Structural and morphological characterization methods such as X-ray diffraction and SEM morphology were used on the as-synthesized composite materials. The supercapacitor properties of the as-developed amorphous ribbons decorated with Fe_2O_3 nanoparticles were investigated by cyclic voltammetry, galvanostatic charge discharge, and electrochemical impedance spectroscopy. The flexible supercapacitor negative electrode demonstrates a specific capacitance of 5.96 F g^{-1} for the 0.2 M NaOH treated sample and 8.94 Fg^{-1} for the 0.4 M NaOH treated sample. The 0.2 M treated negative electrodes deliver 0.48 Wh/kg at a power density of 20.11 W/kg, and the 0.4 M treated electrode delivers 0.61 Wh/kg at a power density of 20.85 W/kg. The above results show that these flexible electrodes are adequate for integration in supercapacitor devices, for example, as negative electrodes.

Keywords: amorphous ribbons; oxidation; supercapacitors; nanocrystalline Fe_2O_3

Citation: Nicolaescu, M.; Vajda, M.; Lazau, C.; Orha, C.; Bandas, C.; Serban, V.-A.; Codrean, C. Fabrication of Flexible Supercapacitor Electrode Materials by Chemical Oxidation of Iron-Based Amorphous Ribbons. *Materials* **2023**, *16*, 2820. https://doi.org/10.3390/ma16072820

Academic Editor: Shinichi Tashiro

Received: 8 March 2023
Revised: 23 March 2023
Accepted: 30 March 2023
Published: 1 April 2023

1. Introduction

Over the past decade, due to the increase in energy demands, electrochemical supercapacitors have gradually become one of the most promising energy storage devices because of their low cost, long life, and high power density and capacitance characteristics. Supercapacitors have many practical uses that demand high power outputs and fast charge/discharge rates, such as electric vehicles, renewable energy systems, and consumer electronics. In addition, they are considered more environmentally safe compared to traditional batteries because they lack toxic chemicals and heavy metals [1]. The performance of supercapacitors in general depends on the electrode materials; thus, new electrode materials have been designed to increase the performance of supercapacitors [2]. However, the main problem of supercapacitors is currently the issue of low energy density. This obstacle can be overcome by employing materials possessing a high theoretical capacitance.

Electrode materials can be classified into three groups: carbon-based materials [3], conducting polymers [4], and transition metal oxides [5]. One of the most promising and most used materials for fabricating electrodes in supercapacitors are porous-carbon-based materials. They have a large surface area, a high porosity, and a good electrical conductivity, which makes them a promising electrode material for storing electric charge [3]. Many researchers have approached these carbon-based materials with the aim of enhancing energy storage properties through various methods. Yifan Wang et al. doped porous carbon electrodes with nitrogen using a one-step activation carbonization method. By achieving a high N doping amount of 3.35 at.%, they obtained supercapacitor electrodes with a specific capacitance of 250 F g^{-1} at 50 A g^{-1} [6]. Gaigai Duan et al. obtained a nitrogen-doped carbon/ZnO composite electrode material by impregnating flax fibers with $Zn(NO_3)_2$ and EDTA Na_2Zn activators, followed by thermal treatment. The prepared electrode exhibited a specific capacitance of 292 F g^{-1} at a current density of 0.5 A g^{-1} [7]. Another class of electrode materials is represented by conducting polymers, which are a promising material for pseudo-capacitor devices due to their high electrical conductivity and redox properties. Among the conducting polymers, polypyrrole, polyaniline, and derivatives of polythiophene are the most extensively studied materials for use in pseudocapacitor devices. These conducting polymers have demonstrated promising results due to their ability to store charge via redox reactions and their high electrical conductivity, which is promising for use in electrodes for storing electric charge [4]. Jehan El Nady et al. developed a nanocomposite electrode for supercapacitor applications by using a one-step electrodeposition technique to deposit polypyrrole on a NiO substrate, forming a polypyrrole/NiO nanocomposite electrode. Compared to the pristine Ppy electrode, the supercapacitor performance of the Ppy/NiO nanocomposite electrode was significantly enhanced. The study revealed that the Ppy/NiO electrode deposited at 4 A/cm^{-2} demonstrated the highest specific capacitance of 679 F g^{-1} at 1 Ag^{-1} [8]. Azza Shokry et al. synthesized a supercapacitor electrode by combining polythiophene and single-walled carbon nanotubes (SWCNTs) in varying ratios to form nanocomposites. The maximum specific capacitance was achieved when the nanocomposite material contained 50% SWCNTs and had a specific capacitance of 245.8 F g^{-1} at a current density of 0.5 Ag^{-1} [9].

Nanostructures based on transition metal oxides and hydroxides [10], such as Fe_2O_3, Fe_3O_4, Co_3O_4, NiO, CuO, and FeOOH, are promising electrode materials for supercapacitors due to their high specific capacitance [11]. Suprimkumar D. Dhas et al. synthesized porous NiO nanoparticles for use in supercapacitor applications using a simple hydrothermal method. The supercapacitor metal oxide electrode demonstrated a specific capacitance of 116 F g^{-1} at 10 mAg^{-1} in KOH electrolyte and a maximum capacitance of 74 F g^{-1} at 10 mAg^{-1} in Na_2SO_4 electrolyte [12]. M. M. Momeni et al. conducted a study where they used chemical corrosion to produce CuO nanostructures on a pure copper plate. Corrosion was carried out in a mixture of 20 mL NaOH and 35 mM $(NH_4)_2S_2O_8$ aqueous solution at room temperature for varying durations. The specific capacitance of the supercapacitor electrode material was determined to be 158 F g^{-1} using cyclic voltammetry at a scan rate of 10 m s^{-1} [13]. Fe_2O_3 is considered a very promising candidate for electrode materials for supercapacitors and batteries because of its high theoretical specific capacitance, non-toxicity, and low cost [14]. Various methods have been used to synthesize Fe_2O_3 nanostructures, such as sol-gel, hydrothermal, electrodeposition, and vapor deposition methods [15]. Similar to other transition metal oxides, Fe_2O_3 usually suffers from poor conductivity and particle agglomeration, which reduces its performance, mainly at high rates. The formation of metal/oxide structures through the growth of metal oxides on conductive substrates has been shown t−o be an effective way to address the issue of conductibility by enhancing the efficiency of charge transfer [5]. Dealloying and corrosion processes permit the growth of uniform metallic oxide on the porous metallic core, with a better interface between the metal and oxide, and improve the uniformity distribution of the active materials by avoiding particle agglomeration [16,17]. Iron oxide, with a high theoretical capacity, was obtained through a dealloying/corrosion process by using materials both in powder form and in the form of a three-dimensional layer grown on the ribbon surface in the pure phase or in a mix of oxides [18]. The chemical or electrochemical oxidation synthesis of Fe_2O_3, compared to the other methods, is the simplest approach to obtain metal oxides on the surface of metal ribbons with a low production cost and high reproducibility. Baolong Sun et al. conducted a study on the synthesis of iron oxides by oxidizing low carbon steel ribbons in a 1 M NaOH aqueous solution for 6 h at 60 °C. The authors compared the areal capacitance of the obtained iron oxide decorated ribbons with that of the same ribbons coated with a layer of polypyrrole for use in supercapacitors. The results showed that the areal capacitance of the rust coated with polypyrrole was significantly higher than that of the uncovered rust. The coated rust electrode exhibited a capacitance of 2.202 F cm^{-2} compared to 0.878 F cm^{-2} for the uncovered electrode at 1 mA cm^{-2} [19]. A.A. Yadav et al. synthesized iron oxide negative electrodes using a chemical method for use in asymmetric supercapacitor devices. The α-Fe_2O_3 thin films obtained using a simple chemical bath deposition (CBD) method exhibited a nanosphere-like morphology. At a current density of 4 Ag^{-1}, the specific capacitance value of the α-Fe_2O_3 electrode was found to be 2125 F g^{-1} [20]. Jianhui Zhu et al. employed an iron rust byproduct from the steel industry, which was previously deemed useless and unwanted, and subjected it to a straightforward hydrothermal treatment in an aqueous solution containing HNO_3. The resulting product formed a sphere-shaped structure that was integrated into a negative electrode for a storage device. The iron rust electrode obtained through this process exhibited a maximum specific capacity of 269 mA h g^{-1} at 0.3 A g^{-1} [21].

An amorphous alloy was chosen for the fabrication of uniform nano/microstructures because of its disordered atomic scale structure, absence of defects, grain boundaries, secondary phases, element segregation, and heterogeneous structure [22]. These properties, combined with its super-elasticity, make it an ideal candidate for obtaining electrodes for supercapacitors [23]. Additionally, by controlling the morphology and microstructure of the grown Fe_2O_3 layer, the electrochemical performance can be improved. With different morphologies and microstructures, major changes in the electrochemical and storage properties of these devices have been reported. In this study, a chemical oxidation process on iron-based amorphous ribbons in an alkaline solution was used to grow different oxide nanocrystalline structures and Fe_2O_3 morphologies, such as hexagonal with a hollow ball structure and rhombohedral with a rhombus-like structure. Furthermore, the novelty of the research consists of the use of $Fe_{75}Si_{12}B_{10}Nb_3$ amorphous ribbons as a substrate for flexible electrodes constructed on Fe-based amorphous ribbons decorated with nanostructured iron oxides fabricated in a one-step chemical oxidation synthesis for supercapacitor applications.

2. Materials and Methods

2.1. Fabrication of Amorphous Ribbons

Quaternary alloys with a nominal composition of $Fe_{75}Si_{12}B_{10}Nb_3$ were synthesized by the arc melting process. To ensure accurate atomic percentages and the homogeneity of the raw materials, the melting process was repeated four times, using a mixture of pure elemental Fe, Nb metal, and Fe–B and Fe–Si ferroalloys under an argon atmosphere. The raw materials were prepared to match the nominal atomic percent of the $Fe_75Si_{12}B1_0Nb_3$ alloy. The melt spinning method [24] with a copper roll speed of 37.7 m/s was used to prepare amorphous Fe–Si–B–Nb ribbons with a 25 μm thickness and a 15 mm width. We used optical emission spectroscopy to determine the atomic percentage of the prepared sample, and ensured that the alloy compositions represented the nominal atomic percent.

2.2. Fabrication of Flexible Electrodes Decorated with Iron Oxide Nanoparticles

The corrosion process was carried out in a NaOH aqueous solution (10 mL) of 0.2 M and 0.4 M concentrations in free immersion conditions for 7 days [25]. The as-oxidized ribbons were removed and washed thoroughly with distilled water to clean any residual impurities. These steps are presented in the schematic diagram in Figure 1.

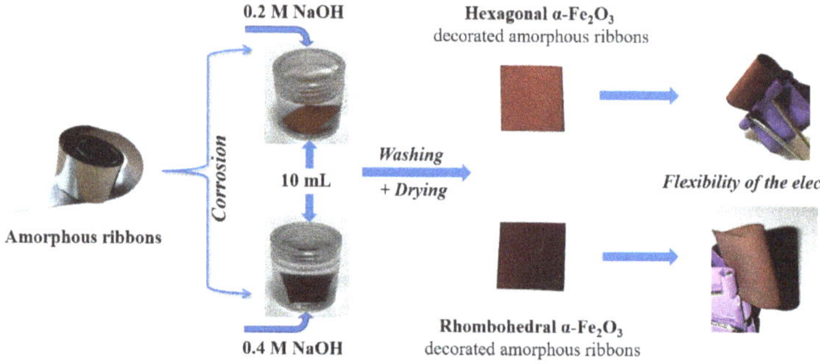

Figure 1. Schematic illustration of the one-step synthesis process of the flexible electrode.

The use of a NaOH solution can facilitate the transformation of Fe to FeOOH and Fe_2O_3 by creating an alkaline environment that can promote the formation of dehydrated iron oxide [25,26]. The concentration, the volume of the container, the amount of solution, and the temperature of the NaOH solution can directly affect the kinetics and thermodynamics of these reactions. These reactions need a long period of time and a stable environment to obtain the desired structure and morphology of iron oxide.

The chemical reactions that occur between the iron-based amorphous ribbons and the NaOH solution take place according to the following reactions [27]:

$$Fe + 2NaOH + 0.25O_2 + 0.5H_2O \rightarrow FeOOH + 2Na^+ + 2OH^- \quad (1)$$

$$FeOOH\ (goethite) \rightarrow \gamma\text{-}FeOOH\ (lepidocrocite) + H_2O \quad (2)$$

$$\gamma\text{-}FeOOH\ (lepidocrocite) \rightarrow Fe_2O_3\ (hematite) + H_2O \quad (3)$$

Hexagonal Fe_2O_3 structures in the form of hollow spheres were obtained by immersing the iron-based amorphous ribbons in a solution of NaOH with a concentration of 0.2 M at a temperature of 20 °C. The immersion process was carried out for a period of 7 days to allow the formation of a nanocrystalline structure.

To obtain orthorhombic and rhombohedral Fe_2O_3 structures, the material was immersed in a NaOH solution with a concentration of 0.4 M at a temperature of 25 °C for 7 days. During this process, the NaOH solution reacted with the Fe_2O_3 material, leading to the formation of the desired structures with a rhombus-like morphological shape.

2.3. Materials and Characterization

To determine the crystalline structure of the decorated oxides, X-ray diffraction (XRD) was used. XRD was performed using a PANalytical X'Pert PRO MPD diffractometer equipped with a monochromator used to filter out the fluorescent radiation, with Cu-Kα radiation (CuKa1: 1.540598 Å and CuKa2: 1.544426 Å) in the range of 2theta from 20 to 80°. The morphology of surface oxides was examined by scanning electron microscopy (SEM). An FEI Inspect S model coupled with an energy dispersive X-ray analysis detector (EDX) was used to inspect the surface morphologies. The electrochemical performances of the electrodes were determined on an electrochemical workstation (Voltalab Potentiostat model PGZ 402) using a standard system consisting of three electrodes, a flexible electrode (size 0.5 × 0.5 mm), a Ag/AgCl electrode (sat. KCl), and a Pt wire served as the working electrode, reference electrode, and counter electrode, in a 0.5 M Na_2SO_4 solution. Electrochemical measurements included cyclic voltammetry (CV), galvanostatic charge–discharge (GCD), and electrochemical impedance spectroscopy (EIS).

3. Results and Discussion

3.1. Structural and Morphological Properties

Structural investigations of the as-prepared flexible electrode before and after being decorated with Fe_2O_3 nanoparticles were carried out using X-ray diffraction (XRD) analyses (Figure 2). In the XRD patterns of the as-spun ribbons, a broad intensity peak was observed at a 2θ angle within the range of 37 to 50°. The shape of the diffraction peak was in line with other research studying iron-based amorphous ribbons [28,29]. This peak is indicative of the amorphous state of the iron-based alloy, as it is also present in the spectra of metallic ribbons decorated with iron oxide. This furthermore evidenced the presence of an amorphous core that gave the flexible application of the supercapacitor electrode. In the case of the ribbon immersed in a 0.2 M NaOH solution, α-Fe_2O_3 with diffraction peaks at 7.8° (002), 25.2° (004), 30° (106), and 47.3° (027) associated with a hexagonal structure (JCPDS No 01-076-1821) is present. Additionally, a rhombohedral structure (JCPDS No 01-073-0603) of α-Fe_2O_3, with diffraction peaks at 21.2° (012), 35.5° (110), 40.4° (113), and 53.7° (211) is also present in the sample maintained in a 0.4 M alkaline solution.

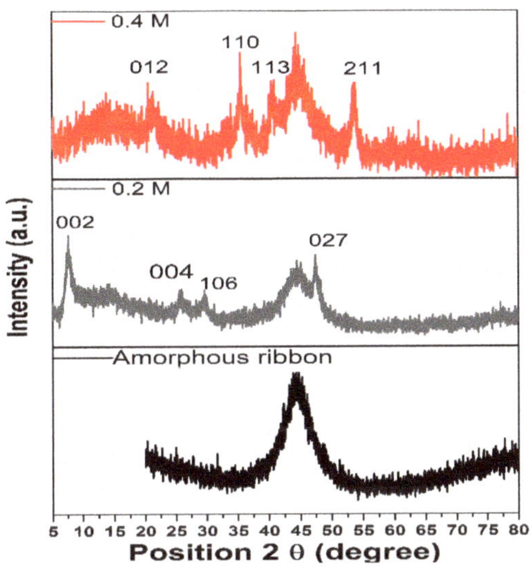

Figure 2. XRD pattern of amorphous ribbons and amorphous ribbons decorated with iron oxide nanoparticles.

Scanning electron microscopy (SEM) was used to study the morphology and uniformity of the surface oxide layer. In Figure 3a, the surface morphology of flexible ribbons decorated with oxide nanoparticles at an immersion time of 7 days and a concentration of NaOH solution of 0.2 M is presented. As a result of the high holding time of the ribbons in the alkaline solution, a porous hollow-ball-like oxide [30] is present on the surface of the flexible amorphous ribbons. The mesoporous structure of the oxide particles can increase the specific surface area of the film. The surface morphology of the oxide-decorated flexible ribbon treated in a solution of 0.4 M NaOH for 7 days is presented in Figure 3b, and it can be observed that uniform structures composed from rhomboidal formations are obtained. Figure 3c,d presents the cross-section images of the as-obtained electrodes, and it is observed that the oxide layer thickness is directly proportional to the concentration of the NaOH solution. Typically, hematite forms in a hexagonal crystal system with a trigonal structure. However, under certain conditions, hematite can exhibit a rhombohedral crystal structure, which gives it a rhombus-like appearance [31,32]. The optimization of the process parameters lead to the formation of a uniform oxide layer distributed on the surface of the samples.

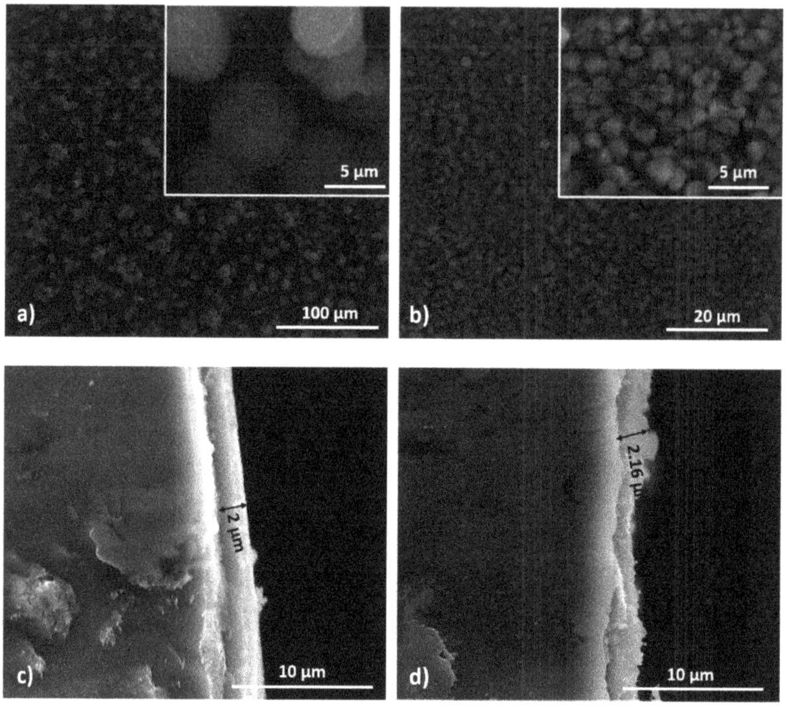

Figure 3. SEM images of the supercapacitor electrode surface obtained in a NaOH solution with a concentration of (**a**) 0.2 M and (**b**) 0.4 M. Cross-section images of electrodes obtained at (**c**) 0.2 M and (**d**) 0.4 M.

3.2. Electrochemical Performance of the Flexible Electrodes

Cyclic voltammetry (CV) studies of the hexagonal and rhomboidal Fe_2O_3-decorated electrodes were undertaken to assess the electrochemical behavior of the electrodes, and are presented in Figure 4a,b.

The measurements were performed at a potential window range of −0.7 to 0 V, using scan rates of 0.005, 0.01, 0.02, 0.05, and 0.1 V s^{-1}, and both materials presented negative electrode behavior [15,20].

The recorded current increased with the increasing scan rate for both flexible electrodes decorated with Fe_2O_3 nanoparticles. This observation suggests that the electrodes exhibit excellent supercapacitive behavior. The shape of the CV profile provides valuable information about the mechanism of charge storage and the performance of the electrode. The non-rectangular shape of the cyclic voltammetry (CV) profile is linked to the occurrence of faradaic reactions, namely reduction and oxidation, on the electrodes [33].

The capacitance (C_P) derived from the CV analysis was calculated according to Equation (4) [34]:

$$C_P = \frac{A}{km\Delta V} \tag{4}$$

where C_P is the capacitance, A is the area under the curve, k is the scan rate, m is the total mass of the sample, and ΔV is the potential window.

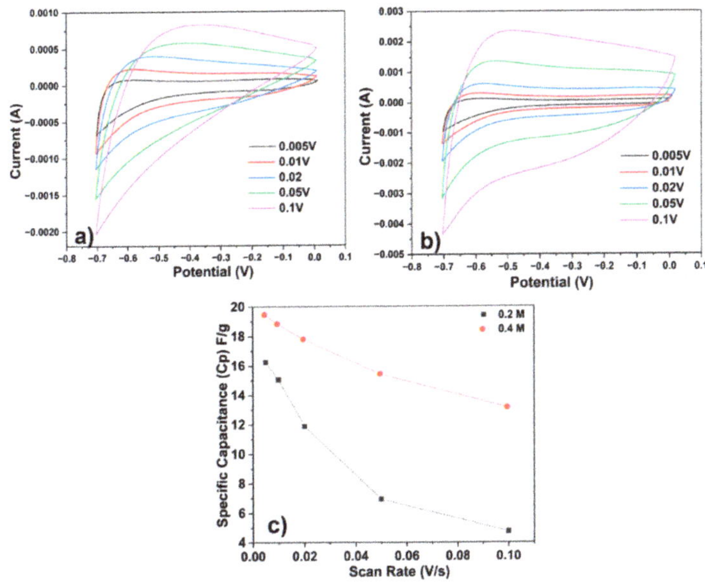

Figure 4. Cyclic voltammograms of (**a**) 0.2 M sample and (**b**) 0.4 M sample. (**c**) Plots of specific capacitance versus potential scan rate of samples.

The calculated capacitance versus the scan rate of the CV analysis for both negative electrodes is plotted in Figure 4c. At a 0.005V scan rate, the capacitance values are 16.25 F g^{-1} and 19.5 F g^{-1} for the amorphous ribbon treated with 0.2 M NaOH solution and 0.4 M alkaline solution, respectively. Figure 4c shows a clear decrease in the charge storage capacity from 16.25 to 4.74 F g^{-1} for the 0.2 M electrode and from 19.5 to 13.3 F g^{-1} for the 0.4 M electrode with increasing scan rate. This decrease is attributed to the ion exchange mechanism and is more pronounced for the 0.2 M electrode than for the 0.4 M electrode, probably because the highly porous hollow structures of the electrode need more time for the intercalation–deintercalation process during charge and discharge [35,36]. Consequently, only a small fraction of the electrode material can be utilized, whereas most of the material remains unutilized at high potentials.

Figure 5a,b illustrates the galvanostatic charge–discharge (GCD) curves as a function of time at different current densities (0.1, 0.2, 0.3, 0.4, and 0.5 A g^{-1}) for the negative electrodes (0.2 M and 0.4 M) to assess the electrode's performance.

As previously reported in the literature for various electroactive materials, an increase in the current density results in a decrease in the discharging time [37,38]. The GCD profiles for the negative electrode present non-linear behavior (more visible at a low current density), indicating that the capacitance of the supercapacitor is not constant over the entire charging/discharging process [35,39]. Furthermore, this indicates that the Faradic pseudocapacitive nature of the as-produced flexible electrodes agrees with the results obtained from the cyclic voltammetry measurements.The GCD curves of the 0.2 M tested electrode displays two main variation regions (Figure 5a): an initial rapid voltage drop due to the internal resistance followed by a subsequent linear region that can be attributed to the capacitive behavior [40]. The GCD curve of the 0.4 M-treated electrode exhibits a third region, which is likely caused by Faradaic processes, such as redox reactions or electrochemical adsorption/desorption, occurring at the interface between the electrode and the electrolyte [35].

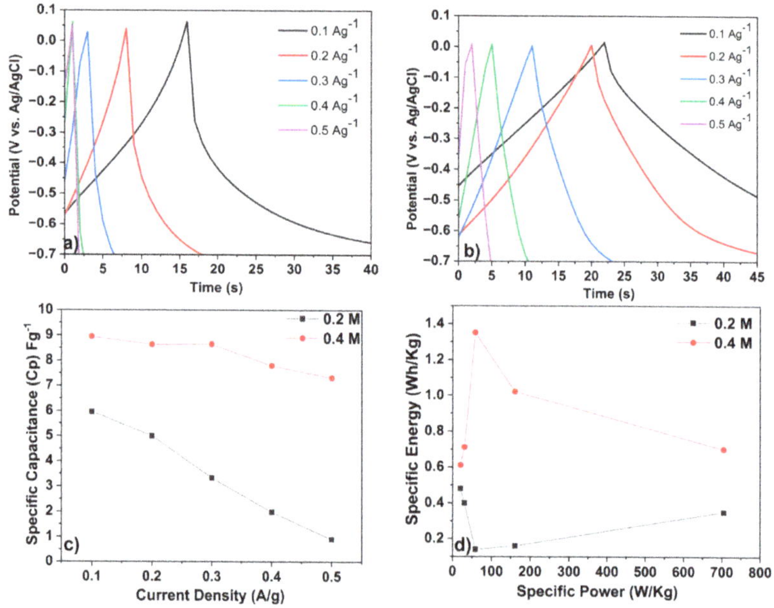

Figure 5. GCD curves of electrodes tested at different current densities for the (**a**) 0.2 M sample and (**b**) 0.4 M sample. (**c**) Specific capacitance of both electrodes at different current densities. (**d**) Ragone plot of both electrodes.

The specific capacitance (C_{SP}), energy density (E), and power density (P) were calculated from the GCD analysis according to the following formulas [37]:

$$C_{SP} = \frac{I\Delta t}{\Delta Vm} \tag{5}$$

$$E = \left(\frac{C_{SP}\Delta V^2}{2}\right) \tag{6}$$

$$P = \frac{\Delta VI}{2m} \tag{7}$$

where I represents the applied current (in A), m represents the mass of the active material (in g), and Δt and ΔV represent the discharging time (in s) and the discharge voltage, respectively.

In Figure 5c, the specific capacitance, calculated from the GDC analysis, is plotted against the current density. After increasing the power density from 0.1 A g^{-1} to 0.5 A g^{-1}, the specific capacitance of the 0.2 M electrode decreased from 5.96 F g^{-1} to 0.9 F g^{-1} and showed a more pronounced decrease compared to the 0.4 M sample. The 0.4 M sample exhibited a higher specific capacitance compared to the 0.2 M sample, and the capacitance was also found to be more stable, with only a slight decrease from 8.94 F g^{-1} to 7.41 F g^{-1} when the power density increased from 0.1 A g^{-1} to 0.5 A g^{-1}. Figure 5d displays the specific energy vs. specific power calculated from the GDC analysis at different current densities.

The behavior of the two tested electrodes differs in terms of energy by the amount of stored energy and the rate at which that energy can be delivered, as shown in the graph. However, the specific energy to specific power ratio remains constant for both electrodes, decreasing with the increasing current density. Both electrodes deliver the highest energy density at an applied current density of 0.1 A g^{-1}. The 0.2 M electrode delivers 0.48 Wh/kg at a power density of 20.11 W/kg and the 0.4 M electrode delivers 0.61 Wh/kg at a power density of 20.85 W/kg.

Electrochemical impedance spectroscopy (EIS) was performed over a frequency range of 0.1 Hz to 10,000 Hz with an amplitude of 0.01 V. The EIS spectra for 0.2 M and 0.4 M electrodes exhibit a semicircle and a line, as depicted in Figure 6a,b. The intersection of the EIS plots and the real axis (an impedance of zero) represents the ohmic resistance, which encompasses the ionic resistance of the electrolyte, the resistances of the iron oxide and iron base amorphous ribbons substrate, and the contact resistance, R_S. The semicircle observed in the EIS analysis is an indication of the charge transfer resistance (R_P) between the amorphous ribbons decorated with iron oxide and the electrolyte [41]. This resistance corresponds to the electrochemical activity of the active material in the system (the impact of both Faradaic and non-Faradaic reactions) [35]. The slope of the straight line observed in the low frequency region is attributed to the Warburg resistance and capacitive behavior of the electrode [42]. Both electrodes have a good capacitive behavior, demonstrated by an inclination greater than 45° with the decrease in frequency [35]. The 0.4 M electrode displays a higher capacitive behavior than the 0.2 M electrode, as evidenced by a steeper inclination in the frequency range near 90°. The inset of Figure 6a,b illustrates the equivalent circuit used for curve fitting. Based on the extracted parameters, it was observed that the 0.4 M negative electrode exhibited a lower ionic and electronic resistance, R_S = 5.44 Ω, compared to the 0.2 M negative electrode, which had R_S = 10.1 Ω. This suggests that the 0.4 M negative electrode has better conductivity. Moreover, the 0.4 M negative electrode has a lower interfacial charge transfer resistance, R_P = 3.30 Ω, compared to the 0.2 M negative electrode at R_P = 55 Ω. This can explain the superior performance of the 0.4 M negative electrode.

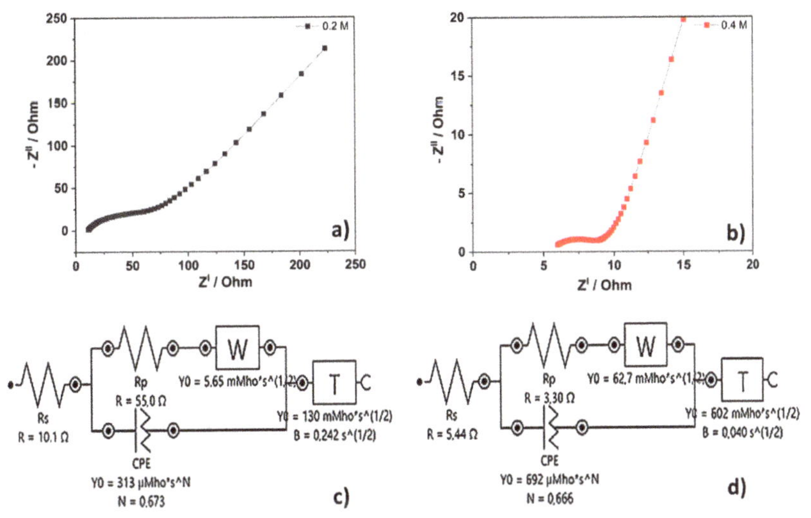

Figure 6. Nyquist plots of (**a**) 0.2 M sample and (**b**) 0.4 M sample. Equivalent circuit fitting of the samples: (**c**) 0.2 M and (**d**) 0.4 M.

4. Conclusions

In this work, a flexible negative electrode for supercapacitors was successfully prepared via a one-step chemical oxidation process by decorating amorphous ribbons with Fe_2O_3 nanoparticles. This process presumes the immersion at a high holding time of the iron-based amorphous ribbons in a low molarity NaOH solution. Uniform mesoporous hollow-ball-like oxide (0.2 M) and rhombus-like oxide (0.4 M) formations are obtained on the alloy surface. Varying the process parameters, such as the concentration, temperature, or holding time, during the oxidation process led to different morphologies or crystal structures. These different structures can be further optimized for other specific applications. From the XRD data, it was found that the hollow ball morphology samples have a hexagonal structure, and the rhombus-like morphology samples have a rhomboidal structure. The CV curves show that both samples work like negative electrodes and show a charge storage capacities of 16.25 F g^{-1} for the 0.2 M sample and 19.5 F g^{-1} for the 0.4 M sample at a 0.05 V/s scan rate. From the GCD analysis, the maximum specific capacitance was obtained, with values of 5.96 F g^{-1} for the 0.2 M sample and 8.94 Fg^{-1} for the 0.4 M sample at a power density of 0.1 A g^{-1}. Additionally, both electrodes deliver the highest energy density at an applied current density of 0.1 A g^{-1}. The 0.2 M electrode delivers 0.48 Wh/kg at a power density of 20.11 W/kg and the 0.4 M electrode delivers 0.61 Wh/kg at a power density of 20.85 W/kg. In conclusion, this one-step synthesis process is a simple, low-cost, effective, and efficient method for large-scale production of flexible negative electrode materials. According to the obtained data, the amorphous ribbons decorated with Fe_2O_3 nanoparticles have high potential in supercapacitor applications. The relatively small values of current density can be explained by the weight of the electrode, the total weight of the amorphous metal alloy electrode, and the iron oxide being taken as the electrode mass.

Author Contributions: M.N. was involved in conceptualization, methodology, validation, investigation, and writing—original draft; M.V. was involved in methodology, investigation, and writing—original draft; C.L. was involved in methodology, investigation, and writing—original draft; C.O. was involved in methodology, investigation, and writing—original draft; C.B. was involved in validation, investigation, and writing—original draft; V.-A.S. was involved in validation, investigation, and writing—original draft; C.C. was involved in methodology, investigation, writing—original draft, and supervision. All authors have read and agreed to the published version of the manuscript.

Funding: This work was supported by a grant from the Ministry of Research, Innovation and Digitization, CNCS-UEFISCDI, project number PN-III-P1-1.1-TE-2021-0963, within PNCDI III, with contract number TE13/2022 (DD-CyT) and project code PN 23 27 01 02 INOMAT, 23-27 29N/2023; by the Project "Network of excellence in applied research and innovation for doctoral and postdoctoral programs/InoHubDoc", project co-funded by the European Social Fund financing agreement no. POCU/993/6/13/153437.

Institutional Review Board Statement: Not applicable.

Informed Consent Statement: Not applicable.

Data Availability Statement: Not applicable.

Conflicts of Interest: The authors declare no conflict of interest.

References

1. Yaseen, M.; Khattak, M.A.K.; Humayun, M.; Usman, M.; Shah, S.S.; Bibi, S.; Hasnain, B.S.U.; Ahmad, S.M.; Khan, A.; Shah, N.; et al. A Review of Supercapacitors: Materials Design, Modification, and Applications. *Energies* **2021**, *14*, 7779. [CrossRef]
2. Miller, E.E.; Hua, Y.; Tezel, F.H. Materials for energy storage: Review of electrode materials and methods of increasing capacitance for supercapacitors. *J. Energy Storage* **2018**, *20*, 30–40. [CrossRef]
3. Wang, Y.; Zhang, L.; Hou, H.; Xu, W.; Duan, G.; He, S.; Liu, K.; Jiang, S. Recent progress in carbon-based materials for supercapacitor electrodes: A review. *J. Mater. Sci.* **2020**, *56*, 173–200. [CrossRef]
4. Li, L.; Meng, J.; Zhang, M.; Liu, T.; Zhang, C. Recent advances in conductive polymer hydrogel composites and nanocomposites for flexible electrochemical supercapacitors. *Chem. Commun.* **2022**, *58*, 185–207. [CrossRef]
5. Liang, R.; Du, Y.; Xiao, P.; Cheng, J.; Yuan, S.; Chen, Y.; Yuan, J.; Chen, J. Transition Metal Oxide Electrode Materials for Supercapacitors: A Review of Recent Developments. *Nanomaterials* **2021**, *11*, 1248. [CrossRef]
6. Wang, Y.; Li, H.; Yang, W.; Jian, S.; Zhang, C.; Duan, G. One step activation by ammonium chloride toward N-doped porous carbon from camellia oleifera for supercapacitor with high specific capacitance and rate capability. *Diam. Relat. Mater.* **2022**, *130*, 109526. [CrossRef]
7. Duan, G.; Zhao, L.; Zhang, C.; Chen, L.; Zhang, Q.; Liu, K.; Wang, F. Pyrolysis of zinc salt-treated flax fiber: Hierarchically porous carbon electrode for supercapacitor. *Diam. Relat. Mater.* **2022**, *129*, 109339. [CrossRef]
8. El Nady, J.; Shokry, A.; Khalil, M.; Ebrahim, S.; Elshaer, A.M.; Anas, M. One-step electrodeposition of a polypyrrole/NiO nanocomposite as a supercapacitor electrode. *Sci. Rep.* **2022**, *12*, 3611. [CrossRef]

9. Shokry, A.; Karim, M.; Khalil, M.; Ebrahim, S.; El Nady, J. Supercapacitor based on polymeric binary composite of polythiophene and single-walled carbon nanotubes. *Sci. Rep.* **2022**, *12*, 11278. [CrossRef]
10. Shi, F.; Li, L.; Wang, X.-l.; Gu, C.-d.; Tu, J.-p. Metal oxide/hydroxide-based materials for supercapacitors. *RSC Adv.* **2014**, *4*, 41910–41921. [CrossRef]
11. Owusu, K.A.; Qu, L.; Li, J.; Wang, Z.; Zhao, K.; Yang, C.; Hercule, K.M.; Lin, C.; Shi, C.; Wei, Q.; et al. Low-crystalline iron oxide hydroxide nanoparticle anode for high-performance supercapacitors. *Nat. Commun.* **2017**, *8*, 14264. [CrossRef] [PubMed]
12. Dhas, S.D.; Maldar, P.S.; Patil, M.D.; Nagare, A.B.; Waikar, M.R.; Sonkawade, R.G.; Moholkar, A.V. Synthesis of NiO nanoparticles for supercapacitor application as an efficient electrode material. *Vacuum* **2020**, *181*, 109646. [CrossRef]
13. Momeni, M.M.; Nazari, Z.; Kazempour, A.; Hakimiyan, M.; Mirhoseini, S.M. Preparation of CuO nanostructures coating on copper as supercapacitor materials. *Surf. Eng.* **2014**, *30*, 775–778. [CrossRef]
14. Li, T.; Yu, H.; Zhi, L.; Zhang, W.; Dang, L.; Liu, Z.; Lei, Z. Facile Electrochemical Fabrication of Porous Fe_2O_3 Nanosheets for Flexible Asymmetric Supercapacitors. *J. Phys. Chem. C* **2017**, *121*, 18982–18991. [CrossRef]
15. Nithya, V.D.; Arul, N.S. Review on α-Fe_2O_3 based negative electrode for high performance supercapacitors. *J. Power Sources* **2016**, *327*, 297–318. [CrossRef]
16. McCue, I.; Benn, E.; Gaskey, B.; Erlebacher, J. Dealloying and Dealloyed Materials. *Annu. Rev. Mater. Res.* **2016**, *46*, 263–286. [CrossRef]
17. Li, Z.; Gadipelli, S.; Yang, Y.; He, G.; Guo, J.; Li, J.; Lu, Y.; Howard, C.A.; Brett, D.J.L.; Parkin, I.P.; et al. Exceptional supercapacitor performance from optimized oxidation of graphene-oxide. *Energy Storage Mater.* **2019**, *17*, 12–21. [CrossRef]
18. Liu, H.; Wang, X.; Wang, J.; Xu, H.; Yu, W.; Dong, X.; Zhang, H.; Wang, L. High electrochemical performance of nanoporous Fe3O4/CuO/Cu composites synthesized by dealloying Al-Cu-Fe quasicrystal. *J. Alloys Compd.* **2017**, *729*, 360–369. [CrossRef]
19. Sun, B.; Yao, M.; Chen, Y.; Tang, X.; Hu, W.; Pillai, S.C. Facile fabrication of flower-like γ-Fe_2O_3 @PPy from iron rust for high-performing asymmetric supercapacitors. *J. Alloys Compd.* **2022**, *922*, 166055. [CrossRef]
20. Yadav, A.A.; Hunge, Y.M.; Ko, S.; Kang, S.W. Chemically Synthesized Iron-Oxide-Based Pure Negative Electrode for Solid-State Asymmetric Supercapacitor Devices. *Materials* **2022**, *15*, 6133. [CrossRef]
21. Zhu, J.; Li, L.; Xiong, Z.; Hu, Y.; Jiang, J. Evolution of Useless Iron Rust into Uniform α-Fe_2O_3 Nanospheres: A Smart Way to Make Sustainable Anodes for Hybrid Ni–Fe Cell Devices. *ACS Sustain. Chem. Eng.* **2016**, *5*, 269–276. [CrossRef]
22. Dan, Z.; Qin, F.; Yamaura, S.-i.; Sugawara, Y.; Muto, I.; Hara, N. Dealloying behavior of amorphous binary Ti–Cu alloys in hydrofluoric acid solutions at various temperatures. *J. Alloys Compd.* **2013**, *581*, 567–572. [CrossRef]
23. Barandiarán, J.M.; Gutiérrez, J.; García-Arribas, A. Magneto-elasticity in amorphous ferromagnets: Basic principles and applications. *Phys. Status Solidi (A)* **2011**, *208*, 2258–2264. [CrossRef]
24. Sowjanya, M.; Kishen Kumar Reddy, T. Cooling wheel features and amorphous ribbon formation during planar flow melt spinning process. *J. Mater. Process. Technol.* **2014**, *214*, 1861–1870. [CrossRef]
25. Nilsson, G. Behaviour of Activated Iron in Sodium Hydroxide Solutions. *Nature* **1946**, *157*, 586–587. [CrossRef]
26. Hang, B.T.; Anh, T.T. Controlled synthesis of various Fe_2O_3 morphologies as energy storage materials. *Sci. Rep.* **2021**, *11*, 5185. [CrossRef]
27. Cudennec, Y.; Lecerf, A. Topotactic transformations of goethite and lepidocrocite into hematite and maghemite. *Solid State Sci.* **2005**, *7*, 520–529. [CrossRef]
28. Li, Y.; Shen, N.; Zhang, S.; Wu, Y.; Chen, L.; Lv, K.; He, Z.; Li, F.; Hui, X. Crystallization behavior and soft magnetic properties of Fe–B–P–C–Cu ribbons with amorphous/α-Fe hierarchic structure. *Intermetallics* **2021**, *131*, 107100. [CrossRef]
29. Wu, X.; Li, X.; Li, S. Crystallization kinetics and soft magnetic properties of Fe71Si16B9Cu1Nb3 amorphous alloys. *Mater. Res. Express* **2020**, *7*, 016118. [CrossRef]
30. Zhu, S.-W.C.a.Y.-J. Hierarchically Nanostructured α-Fe_2O_3 Hollow Spheres: Preparation, Growth Mechanism, Photocatalytic Property, and Application in Water Treatment. *J. Phys. Chem. C* **2008**, *112*, 6253–6257. [CrossRef]
31. Guo, W.; Sun, W.; Lv, L.P.; Kong, S.; Wang, Y. Microwave-Assisted Morphology Evolution of Fe-Based Metal-Organic Frameworks and Their Derived Fe_2O_3 Nanostructures for Li-Ion Storage. *ACS Nano* **2017**, *11*, 4198–4205. [CrossRef] [PubMed]
32. Cornell, R.M.; Schwertmann, U. *The Iron Oxides: Structure, Properties, Reactions, Occurences and Uses*; WILEY-VCH: Weinheim, Germany, 2003. [CrossRef]
33. Azman, N.H.N.; Mamat Mat Nazir, M.S.; Ngee, L.H.; Sulaiman, Y. Graphene-based ternary composites for supercapacitors. *Int. J. Energy Res.* **2018**, *42*, 2104–2116. [CrossRef]
34. Ghosh, K.; Srivastava, S.K. Enhanced Supercapacitor Performance and Electromagnetic Interference Shielding Effectiveness of CuS Quantum Dots Grown on Reduced Graphene Oxide Sheets. *ACS Omega* **2021**, *6*, 4582–4596. [CrossRef] [PubMed]
35. Gund, G.S.; Dubal, D.P.; Chodankar, N.R.; Cho, J.Y.; Gomez-Romero, P.; Park, C.; Lokhande, C.D. Low-cost flexible supercapacitors with high-energy density based on nanostructured MnO_2 and Fe_2O_3 thin films directly fabricated onto stainless steel. *Sci. Rep.* **2015**, *5*, 12454. [CrossRef]
36. Bograchev, D.A.; Volfkovich, Y.M.; Sosenkin, V.E.; Podgornova, O.A.; Kosova, N.V. The Influence of Porous Structure on the Electrochemical Properties of $LiFe_{0.5}Mn_{0.5}PO_4$ Cathode Material Prepared by Mechanochemically Assisted Solid-State Synthesis. *Energies* **2020**, *13*, 542. [CrossRef]
37. Dong, Y.; Xing, L.; Chen, K.; Wu, X. Porous alpha-Fe_2O_3@C Nanowire Arrays as Flexible Supercapacitors Electrode Materials with Excellent Electrochemical Performances. *Nanomaterials* **2018**, *8*, 487. [CrossRef]

38. Yan, Y.; Tang, H.; Wu, F.; Wang, R.; Pan, M. One-Step Self-Assembly Synthesis α-Fe_2O_3 with Carbon-Coated Nanoparticles for Stabilized and Enhanced Supercapacitors Electrode. *Energies* **2017**, *10*, 1296. [CrossRef]
39. Du, X.; Xia, C.; Li, Q.; Wang, X.; Yang, T.; Yin, F. Facile fabrication of Cu O composite nanoarray on nanoporous copper as supercapacitor electrode. *Mater. Lett.* **2018**, *233*, 170–173. [CrossRef]
40. Gund, G.S.; Dubal, D.P.; Patil, B.H.; Shinde, S.S.; Lokhande, C.D. Enhanced activity of chemically synthesized hybrid graphene oxide/Mn_3O_4 composite for high performance supercapacitors. *Electrochim. Acta* **2013**, *92*, 205–215. [CrossRef]
41. Wang, J.; Zhang, L.; Liu, X.; Zhang, X.; Tian, Y.; Liu, X.; Zhao, J.; Li, Y. Assembly of flexible $CoMoO_4$@$NiMoO_4.xH_2O$ and Fe_2O_3 electrodes for solid-state asymmetric supercapacitors. *Sci. Rep.* **2017**, *7*, 41088. [CrossRef]
42. Basri, N.; Awitdrus, A.; Suleman, M.; Syahirah, N.; Nor, M.; Nurdiana, B.; Dolah, M.; Sahri, M.; Shamsudin, S.A. Energy and Power of Supercapacitor Using Carbon Electrode Deposited with Nanoparticles Nickel Oxide. *Int. J. Electrochem. Sci.* **2016**, *11*, 95–110.

materials

Article

Characteristics of Hydroxyapatite-Modified Coatings Based on TiO_2 Obtained by Plasma Electrolytic Oxidation and Electrophoretic Deposition

Roxana Muntean [1,*], Mihai Brîndușoiu [1], Dragoș Buzdugan [1], Nicoleta Sorina Nemeș [2], Andrea Kellenberger [3] and Ion Dragoș Uțu [1]

1 Department of Materials and Manufacturing Engineering, Politehnica University Timișoara, Piața Victoriei 2, 300006 Timișoara, Romania
2 Research Institute for Renewable Energy—ICER, Politehnica University Timișoara, Piața Victoriei 2, 300006 Timișoara, Romania
3 Faculty of Industrial Chemistry and Environmental Engineering, Politehnica University Timișoara, Piața Victoriei No. 2, 300006 Timișoara, Romania
* Correspondence: roxana.muntean@upt.ro; Tel.: +40-256-403795

Abstract: In order to modify the surface of light metals and alloys, plasma electrolytic oxidation (PEO) is a useful electrochemical technique. During the oxidation process, by applying a positive high voltage greater than the dielectric breakdown value of the oxide layer, the formation of a ceramic film onto the substrate material is enabled. The resulting surface presents hardness, chemical stability, biocompatibility, and increased corrosion wear resistance. The current study aims to investigate the corrosion resistance and tribological properties of PEO-modified coatings on titanium substrates produced by applying either direct or pulsed current in a silicate-alkaline electrolyte. In this way, a uniform TiO_2 layer is formed, and subsequently, electrophoretic deposition of hydroxyapatite particles (HAP) is performed. The morpho-structural characteristics and chemical composition of the resulting coatings are investigated using scanning electron microscopy combined with energy dispersive spectroscopy analysis and X-ray diffraction. Dry sliding wear testing of the TiO_2 and HAP-modified TiO_2 coatings were carried out using a ball-on-disc configuration, while the corrosion resistance was electrochemically evaluated at 37 °C in a Ringer's solution. The corrosion rates of the investigated samples decreased significantly, up to two orders of magnitude, when the PEO treatment was applied, while the wear rate was 50% lower compared to the untreated titanium substrate.

Keywords: plasma electrolytic oxidation; titanium; hydroxyapatite; corrosion and wear resistance

Citation: Muntean, R.; Brîndușoiu, M.; Buzdugan, D.; Nemeș, N.S.; Kellenberger, A.; Uțu, I.D. Characteristics of Hydroxyapatite-Modified Coatings Based on TiO_2 Obtained by Plasma Electrolytic Oxidation and Electrophoretic Deposition. *Materials* **2023**, *16*, 1410. https://doi.org/10.3390/ma16041410

Academic Editor: Juan José Santana-Rodríguez

Received: 12 January 2023
Revised: 1 February 2023
Accepted: 6 February 2023
Published: 8 February 2023

1. Introduction

The outstanding properties of titanium and its alloys (high strength-to-weight ratio, superior corrosion resistance, or biocompatibility) make them suitable for various advanced engineering applications, such as aerospace, marine, photocatalysis, or biomedical [1–3]. However, there are still various problems associated with the use of such materials in terms of surface features to ensure good stability, wear resistance, or bonding strength [2]. Although titanium performs well in most corrosive environments, it can be susceptible to corrosion under highly aggressive conditions. Commonly, titanium is very reactive in its metallic state and naturally forms a thin stable oxide layer (passivation layer typically between 1.5–5 nm) when exposed to normal conditions. Still, once high contact loads are applied combined with reducing or low oxygen environments, the native oxide film may be rapidly damaged, promoting galvanic or crevice corrosion. Consequently, several superficial treatment processes are frequently applied to enhance surface characteristics [3,4], including physical vapor deposition, thermal oxidation, plasma spraying, anodizing, or plasma electrolytic oxidation (PEO) [5]. Moreover, the tribological behavior of Ti-based materials can be significantly improved by applying an appropriate surface modification strategy [6]. The PEO is a widely used superficial treatment process to shield the surfaces of light metals and alloys by inducing the growth of an inert oxide layer. Since its first application in the early 1990s [3], scientists have constantly attempted to understand and improve this technology by varying the main process parameters and searching for new possible applications. The PEO process consists of a luminescent phenomenon of electrical discharge on the metallic working part (immersed in an electrolyte) that acts similar to an anode, accompanied by many crystallizing and melting events [7]. The oxidation occurs when a positive voltage greater than the dielectric breakdown value of the oxide film is reached between the sample and the counter electrode [8], and in this way, a highly adherent ceramic-like porous layer is formed. The coating consists of two different layers, the outer one, which presents numerous structural defects (pores and cracks), and the inner layer, which is denser and more compact [8]. Due to the thermochemical interactions and sparking discharges between the electrolyte and the metallic substrate, a localized high pressure (10^2–10^3 MPa) and temperature (10^3–10^4 K) are achieved [9]. Adjusting the current density and mode, voltage, or power provided to the cell, a wide range of polarization conditions, including direct current (DC) [10], alternating current (AC) [11], or pulsed current (PC) [12], are accessible for the production of such coatings. The PEO process has several advantages compared to other techniques, such as cost-effectiveness, eco-friendly, rapid and easy conversion of the metal surface [13], facile control of coating thickness and porosity, and finely tailored coatings, considering their chemical composition and microstructure [14]. Furthermore, the electrolyte used during the PEO process can strongly define the properties of the treated surfaces, including the microstructure or crystallinity [8,9,15]. PEO-developed coatings on titanium substrates exhibit great importance, primarily in medical applications, due to their ductility, biocompatibility, and corrosion resistance. In this case, to improve osteointegration, the incorporation of elements such as calcium, phosphorous, or silicon is frequently reported [16,17]. Biofunctionalization consists mainly of two different approaches: either the incorporation of inorganic components and their further conversion during the plasma treatment or the integration of organic substances directly into the oxide pores [18]. Consequently, electrolytes containing calcium acetate and glycerophosphate [2,8,14,16], sodium silicate and hexametaphosphate [4,13,15], phosphoric acid and copper nitrate [10,11], calcium citrate and potassium titanium oxalate [17], sodium aluminate and potassium dihydrogen phosphate [19] or molten salts mixtures (KNO_3 and $NaNO_3$) [12] are commonly used. It has been shown that PEO-modified titanium from a silicate-containing solution presents better osteointegration compared to untreated titanium [20]. Several studies in the literature confirm the significant effect of the electrolyte composition on the corrosion behavior of the PEO-developed coatings [9,13]. Shokouhfar et al. [13] concluded that changing the electrolyte composition without altering the rest of the process parameters has led to an increase in pores size generated on the surface of the samples, and furthermore, the corrosion resistance was considerably affected. Molaei et al. [9] investigated the effect of different sodium-based additives in the working electrolyte on the corrosion resistance of the PEO-modified pure titanium. Their study revealed that the additives have a clear influence on the microstructure, phase composition, and subsequently on the corrosion resistance of the PEO-modified samples. Another approach frequently reported in the literature is the production of composite PEO coatings by adding different particles, either directly in the working electrolyte during the PEO treatment or as an additional step once the oxidation process is completed [8]. In this way, several characteristics can be acquired, such as an antifouling effect, better adhesion, hardness, wettability, or better biological response.

The aim of the current study is to assess the corrosion resistance and wear behavior of hydroxyapatite-modified TiO_2 coatings produced by a two-step fabrication method. Firstly, a PEO treatment is proposed based on two different polarization conditions (either direct current or pulsed current) from a sodium silicate and potassium hydroxide alkaline solution, and subsequently, uniform incorporation of hydroxyapatite particles (HAP) is achieved by electrophoresis from an additive-free HAP suspension in isopropanol.

2. Materials and Methods

Commercially available pure titanium sheets (grade 1) of 3.5 mm thickness were cut in the form of rectangular samples with the dimensions of 50 mm × 20 mm and used further as substrate material. Prior to the PEO, the samples were ground with SiC paper (up to 4000 grit) and washed in distilled water and ethanol for 5 min in an ultrasonic bath. The surface roughness of the substrate was approximately 0.15 ± 0.02 μm. The PEO process was carried out in a single-compartment two-electrode cell containing 500 mL alkaline electrolyte prepared by mixing 20 g L^{-1} Na_2SiO_3 with 4 g L^{-1} KOH (for increasing the electrical conductivity), in distilled water, at a pH of 12. The solution was continuously stirred during the oxidation process using a magnetic stirrer to keep the temperature as low as possible (under 40 °C) and to avoid concentration gradients. The working electrodes (WE) in the form of Ti samples (employed as anodes) were immersed in the electrolyte using a threaded rod that assured electrical contact, and for the counter electrode (CE) (cathode), a rectangular stainless-steel plate, 70 mm × 100 mm (type 304) was employed. Two different polarization conditions were applied with the aid of an electrical power supply (max. 250 V/20 A, RFT Sparstelltrafo SST 250/20), one using direct current (DC) and the second one with pulsed current (PC). The process was carried out in galvanostatic mode at a total current value of 200 mA for all the samples, while the maximum reached voltage was 240 V. In the case of PC treatment, a unipolar current was applied with the aid of an electrical relay, setting a duty cycle (*d.c*) of 20%, at a frequency of 0.2 Hz. Subsequently, the PEO-treated samples were washed with distilled water and dried under warm air. From the prepared samples, the Ti PEO PC samples were selected and further subjected to the electrophoretic deposition of HAP from an additive-free suspension (2.5% wt.) prepared using only hydroxyapatite particles (commercially available p.a. ≥ 90%, $Ca_3(PO_4)_2$, Fluka Analytical) and isopropanol (99.9%). The commonly used chemicals reported in the literature for the preparation of HAP suspensions encompass the HAP particles, a solvent (n-butanol, isopropanol), a dispersant (triethanolamine, iodine), and sometimes other additives, such as PVP (polyvinyl pyrrolidone), PEG (polyethylene glycol), PEI (polyethylene amine). However, such additives are not desirable since, after the drying step, depending on their nature, they may leave residues on the surface of the modified sample. In this case, a graphite plate (70 mm × 100 mm × 5 mm) was used as an anode (CE), and the working electrode consisted of previously modified Ti-PEO samples. The schematic representation of the complete PEO process is shown in Figure 1, and the PEO treatment conditions are presented in Table 1.

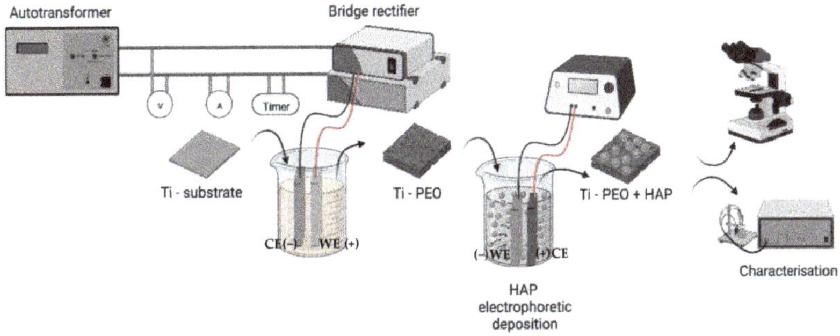

Figure 1. Schematic representation of the synthesis process (Created with BioRender.com–created on 12 January 2023).

Table 1. PEO treatment conditions for obtaining the Ti-modified samples.

Sample	Treatment Conditions				
	Current Density, *i* [mA cm^{-2}]	Voltage, [V]	Total Time, [s]	*d.c.*, [%]	HAP Deposition
Ti PEO DC	10	max 240	600	-	-
Ti PEO PC			1250	20	-
Ti PEO PC + HAP			1250	20	600 s at 150 V

The surface morphology of the resulting samples was investigated using an FEI Quanta™ FEG 250 (FEI Quanta™, Hillsboro, OR, USA) scanning electron microscope (SEM) equipped with an energy-dispersive spectroscopy analyzer (EDS) for chemical composition identification. For structure and phase evaluation, an X-ray Diffractometer (XRD) PANalytical X'Pert Pro Powder (Malvern Panalytical, Malvern, UK) was employed, with Cu-K_α radiation (λ = 1.54), and the measurements were carried out between 20° and 100° 2θ-angles, with a step size of 0.033, at 45 kV and 30 mA. The roughness of the samples before and after the PEO treatment was evaluated with a Mitutoyo SJ-201 portable surface tester (Mitutoyo, Aurora, IL, USA). The thicknesses of the coatings were checked with an eddy current thickness gauge (Namicon, Otopeni, Romania).

The corrosion resistance of PEO-treated samples and HAP-modified coatings was investigated via open circuit potential (OCP), potentiodynamic polarization, and electrochemical impedance spectroscopy (EIS) measurements and compared to the Ti substrate. All electrochemical measurements were carried out in a water jacket three-electrode configuration corrosion cell connected to an Autolab PGSTAT 302 N potentiostat/galvanostat (Metrohm, Herisau, Switzerland). The working electrode consisted of a titanium sample (1 cm^2 active surface) modified by PEO as previously described, the reference was an Ag/AgCl (3M NaCl) electrode positioned near the working electrode via a Luggin capillary, and the counter electrode was a Pt gauze. Potentiodynamic polarization curves were recorded in a Ringer's solution at 37 °C with a scan rate of 0.16 mV s^{-1}. The Ringer's solution was prepared by mixing 9 g L^{-1} NaCl, 0.42 g L^{-1} KCl, 0.2 g L^{-1} $NaHCO_3$, and 0.24 g L^{-1} $CaCl_2$ with distilled water. The main corrosion parameters were evaluated from the polarisation curves using the Tafel extrapolation method in the linear region of the anodic and cathodic branches. EIS measurements were carried out in the same conditions and electrolyte, at the corrosion potential value, by collecting 60 points for each spectrum, in the frequency range from 10^5 to 10^{-2} Hz and AC voltage amplitude of 10 mV rms. An electrical equivalent circuit was proposed to model the experimental EIS data by a complex non-linear least squares procedure using the ZView 3.0 software (Scribner Associates, Inc., Southern Pines, NC, USA). The dry sliding wear behavior of the Ti substrate and PEO-modified samples was evaluated using the ball-on-disc method, according to the ASTM standard G99. The tests were performed with the aid of a TR-20 Tribometer, Ducom-Materials Characterisation Systems, under dry sliding conditions and ambient temperature, with a normal load of 5 N, sliding rate of 300 rot min^{-1}, test time t = 100 min, and a total sliding distance around 1000 m. The static partner selected for the tests was a 100Cr6 steel ball of 6 mm diameter. During testing, the variation of the coefficient of friction (COF) with the distance was automatically recorded. The profile of the wear track was subsequently determined using a 3D optical profilometer, (Ducom-Materials Characterization Systems, Groningen, The Netherlands), and the loss of material and wear rate were estimated. For each sample, three measurements were performed in order to ensure the reproducibility of the results.

3. Results and Discussion

3.1. Coating Morphology and Composition

The PEO treatment changed the surface of the titanium samples from bright metallic to a porous dark grey ceramic coating, typical for a PEO-modified surface. Macroscopically, the samples have the same appearance, but the microscopic analysis revealed some significant differences in the microstructure of the samples due to the different treatment conditions. The SEM micrographs from Figure 2 show the complex porous surface of the oxidized titanium samples in different stages. According to the PEO process mechanism described in the literature [5], initially, at low voltage values, a uniform, dense, and thin (nano-scaled) oxide layer is formed on the surface of the samples. As soon as the applied voltage reaches the breakdown potential of the previously formed film, various sparking phenomena appear, leading to the formation of discharge channels, where temperature and pressure achieve higher values and different processes, such as melting and oxidation, may occur. During the PEO process, molten oxides may migrate through the discharge channels and then are rapidly cooled, generating an uneven microstructure with well-separated micropores, similar to a volcano-crater shape [9]. When DC is applied, the oxidized titanium surface presents numerous irregularities and open pores, which may be attributed to the intense gas evolution that takes place during the oxidation process. It is considered that oxygen is generated at the limit of the crystals [4], and the gas is released when the oxide film breaks, producing well-defined pores. Moreover, during the DC discharges, the heating process is more intense and may melt the oxides and eject them from the coating/substrate interface to the surface of the coating, promoting a higher roughness [21]. In these conditions, the surface roughness increases from the initial titanium substrate value $Ra = 0.15$ μm to $Ra = 0.4$ μm, as shown in Figure 3. In contrast, the pulsed current treatment generates a smoother surface, with smaller pores and a denser structure, the roughness reaching, in this case, almost 0.3 μm, confirming the microstructure of the PEO-modified surfaces is strongly influenced by the polarization conditions [22]. This aspect may be attributed to the intermittent (pulsed) current applied during the oxidation ($t_{on} = 1$ s and $t_{off} = 4$ s, for 250 cycles), the pauses between pulses providing enough time to avoid the overheating of the surface, enabling better control of the intensity of the micro-discharges and delivering smoother coatings, as the gas evolution is also limited [22]. Generally, the PEO treatment regime and duration directly influence the thickness and the roughness of the PEO produced coatings. In fact, the applied potential is the main factor controlling these characteristics. In the DC regime, in the range of low potentials, up to 150 V, the thickness increases slightly at the beginning of the oxidation, and after the breakdown potential is reached, the coating growth is accelerated. In this case, the Ti PEO DC coatings reach an average value of 35 μm. In the PC treatment, where it is possible to control the duration and intensity of the discharges by adjusting the t_{on}, t_{off}, and *d.c.*, the thickness of the TiO_2 coatings reached only 12–16 μm, taking into account that the total oxidation time is also reduced compared to DC regime.

Based on the EDS spectra (Figure 4a,b), it was found that both PEO-produced coatings consist of titanium, oxygen, and silicon, the last element arriving from the chemical composition of the electrolyte. For the HAP electrophoretic deposition, the Ti PEO PC-modified samples were selected, as they provide a surface with fewer defects compared to the DC-treated ones. The morphology of the HAP-modified sample (Figure 2c) indicates the presence of fine HAP uniformly distributed onto the surface of the oxide coating in brighter areas, filling the majority of gaps and pores. In this way, the surface roughness is reduced to approx. 0.2 μm. The EDS spectrum (Figure 4c) confirms the presence of HAP particles since the characteristic peaks for calcium and phosphorous are identified in addition to the ones for Ti and O, increasing in this way the biocompatibility degree of the modified surfaces [23].

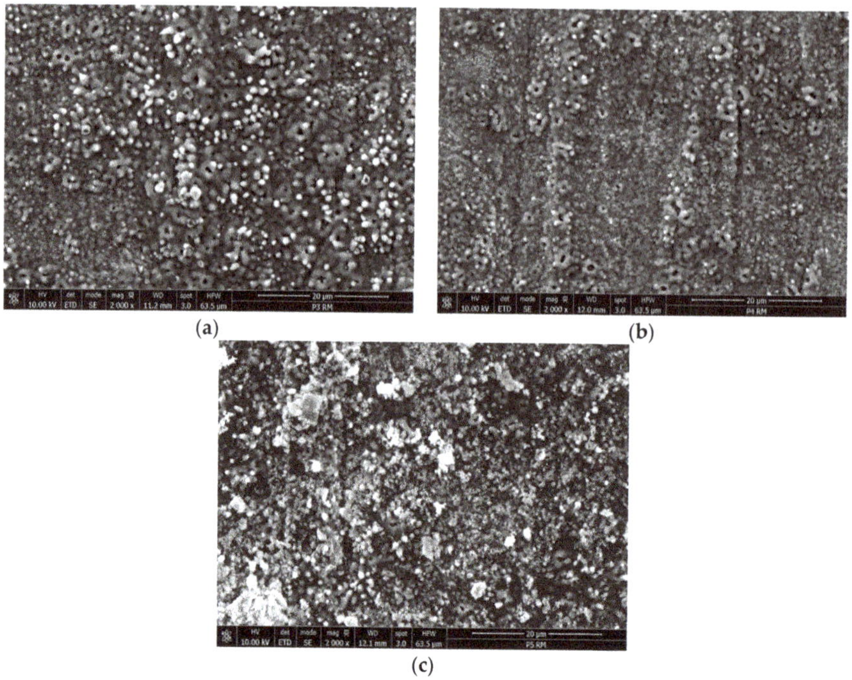

Figure 2. SEM micrographs of PEO-modified Ti-based samples: (**a**) Ti PEO DC; (**b**) Ti PEO PC; (**c**) Ti PEO PC + HAP.

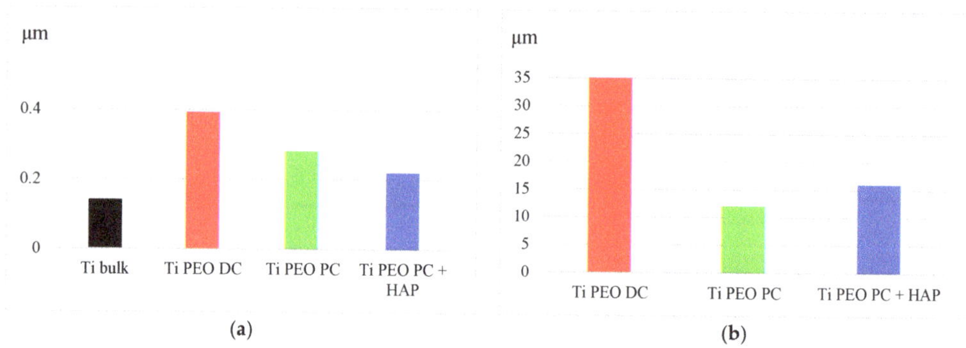

Figure 3. (**a**) Average surface roughness (*Ra*) of Ti substrate, Ti PEO DC; Ti PEO PC; Ti PEO PC + HAP; (**b**) Coating thickness of Ti PEO DC; Ti PEO PC; Ti PEO PC + HAP.

Figure 5 displays the representative diffraction patterns of the PEO-modified samples, including the one obtained for pure titanium substrate, as a reference. The untreated Ti is entirely formed of hexagonal α-phase (denoted Titanium in Figure 5), with the preferred orientation of the crystallites in (002), (101), and (013) directions. Since the XRD analysis detects the full-thickness coating, also reaching the substrate, indicating that the thickness of the coatings is not very high, the main characteristic peaks for titanium are also visible in the PEO-modified samples' patterns. A change in the intensity of the Ti (002) reflection was observed in the case of PEO-treated samples, and in addition to the titanium characteristic peaks, the TiO_2 phase, in the form of anatase, is detected for all treated samples. The dominant reflection for anatase is identified at 2θ angles of 25° (101). This phase typically forms during the PEO treatment in alkaline solutions with high electrical conductivity as a reaction between Ti^{4+} and OH^- due to the high pressure and temperature reached in the discharge channels. It has been previously reported that the applied voltage value directly influences the crystallinity degree of the titanium surfaces [24]. Consequently, there are no significant differences between the PEO-treated samples regarding the XRD patterns since all of the samples were prepared using similar voltage values in the same electrolyte. The peaks corresponding to HAP were not identified in the XRD patterns, as the weight percent of hydroxyapatite on the PEO-treated samples is under the detection limit of this method. Additional phases derived from the electrolyte chemical composition (such as Si-containing phases) may be found in the coating, either in an amorphous or crystalline structure, but the content is below the detection limit of XRD [25].

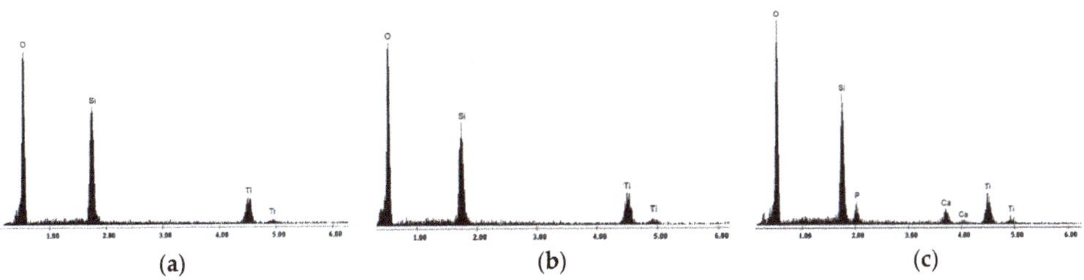

Figure 4. EDS spectra of Ti-based samples: (**a**) Ti PEO DC; (**b**) Ti PEO PC; (**c**) Ti PEO PC + HAP.

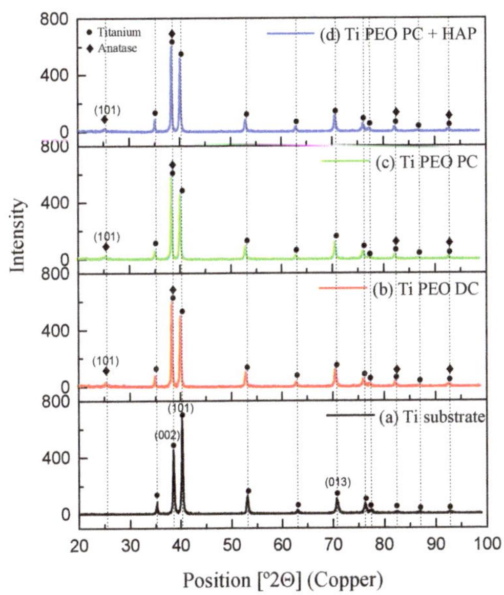

Figure 5. XRD diffraction patterns of Ti-based samples: (**a**) Ti substrate, (**b**) Ti PEO DC; (**c**) Ti PEO PC; (**d**) Ti PEO PC + HAP.

3.2. Corrosion Behavior

The corrosion resistance of the titanium-modified samples and titanium substrate used as reference was assessed with OCP measurements, potentiodynamic polarization, and EIS tests. The OCP variation during 600 s in Ringer's solution at 37 °C can be found in Figure 6. The titanium substrate stabilizes at less noble potential compared to the PEO-treated samples. The most promising behavior in terms of corrosion performance can be observed for the Ti PEO DC, where higher OCP values are identified. The potentiodynamic polarization curves, in the semilogarithmic form (Figure 7), deliver important information regarding the corrosion behavior of the titanium-modified surfaces. Significant differences in both corrosion current density and potential can be observed between the PEO-modified samples and titanium substrate. The anodic branch of the polarization curves was shifted to lower current densities after the PEO treatment, compared to the Ti substrate, denoting higher stability. In fact, analyzing the values presented in Table 2, an increase in the corrosion potential of about 0.23 V is observed for the Ti PEO DC compared to the Ti substrate, while the corrosion current density decreases up to two orders of magnitude due to the presence of a compact TiO_2 layer formed during the PEO process. Similar behavior was also observed for the PC-treated samples. In the case of HAP-modified titanium, even if the corrosion potential presents a more negative value, the addition of HAP does not alter the corrosion performance, and the current density remains low, comparable with the PEO-treated samples. Furthermore, the corrosion rates decreased significantly for the samples treated with PEO. Although the Ti PEO DC sample presents larger micro-pores and roughness compared to the other modified samples, which should increase the probability for the adsorption of Cl^- anions on the surface, the better corrosion behavior may be attributed to the thicker oxide layer previously formed in the PEO process. Based on the potentiodynamic polarization curves, the corrosion resistance increases in the order of Ti substrate < Ti PEO PC + HAP < Ti PEO PC $\approx$ Ti PEO DC. To investigate more deeply the corrosion mechanism of the titanium-based samples, EIS measurements were performed in the same electrolyte and similar testing conditions. Figure 8 gives the results of impedance measurements represented as Nyquist plots (Figure 8a,b), absolute impedance Bode plots (Figure 8c), and phase angle Bode plots (Figure 8d).

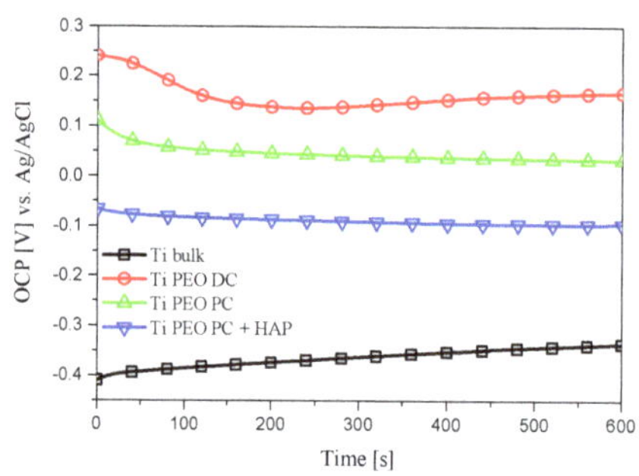

Figure 6. OCP measurements of the Ti-based samples for 600 s in Ringer's solution at 37 °C.

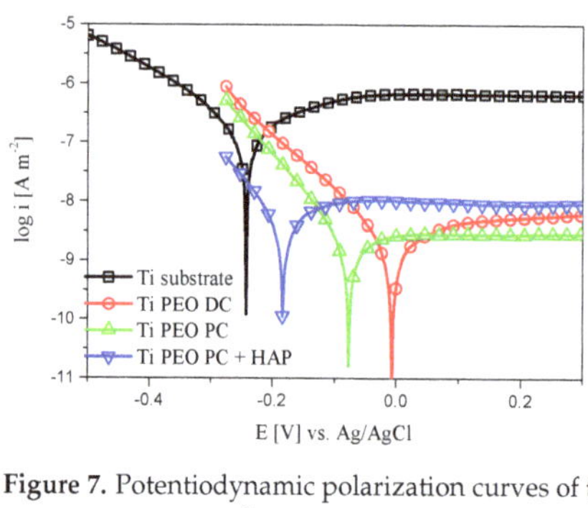

Figure 7. Potentiodynamic polarization curves of the Ti-based samples in Ringer's solution at 37 °C, scan rate 0.16 mV s^{-1}.

Table 2. Corrosion parameters were estimated from the Tafel extrapolation method.

Sample	i_{corr}, [A cm^{-2}]	E_{corr}, [V] vs. Ag/AgCl	Corr. Rate [mm Year^{-1}]
Ti substrate	687×10^{-10}	−0.242	598×10^{-6}
Ti PEO DC	3.30×10^{-10}	−0.007	2.87×10^{-6}
Ti PEO PC	3.64×10^{-10}	−0.077	3.17×10^{-6}
Ti PEO PC + HAP	10.1×10^{-10}	−0.187	8.77×10^{-6}

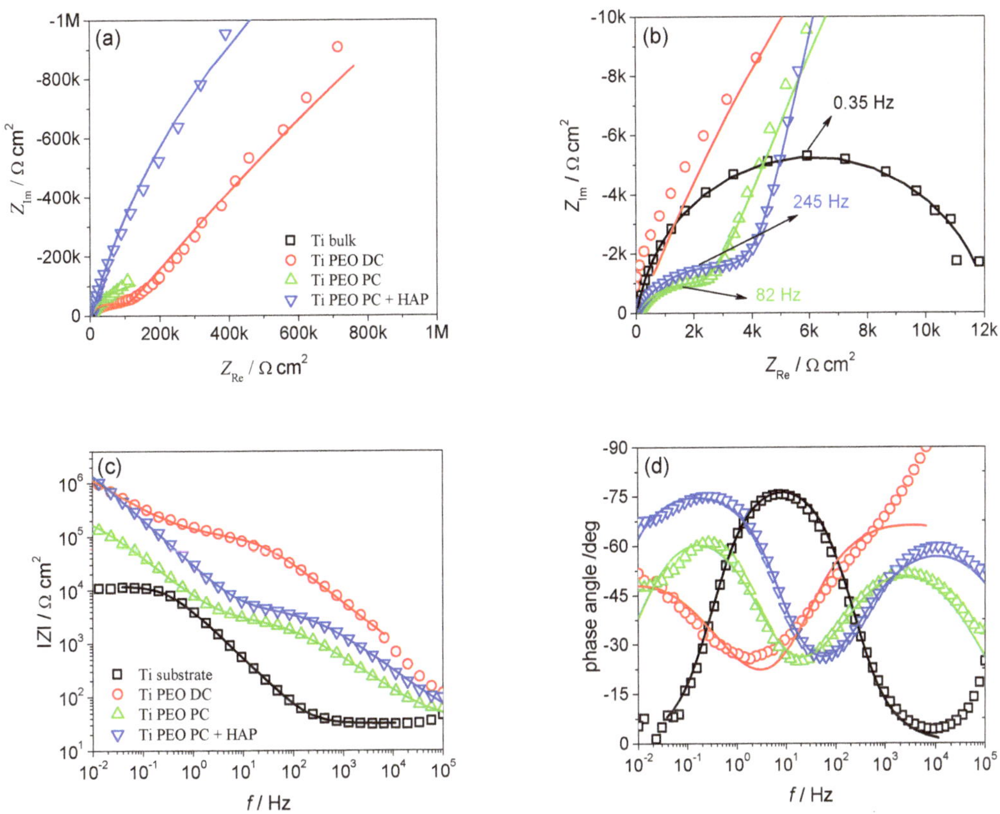

Figure 8. Results of experimental EIS measurements (symbols) in Ringer's solution at 37 °C for untreated Ti and PEO-treated Ti samples: (**a**) Nyquist plots; (**b**) detail of the high-frequency part of Nyquist plots; (**c**) absolute impedance plots and (**d**) phase angle plots. Continuous lines represent the results of modeling to the EEC given in Figure 8.

The impedance data of untreated Ti substrate indicate the presence of one semicircle in the whole frequency range (Figure 8b), corresponding to a single time constant, as indicated by a single maximum in the phase angle plot (Figure 8d). In contrast, the PEO-treated Ti samples show the appearance of a smaller radius semicircle at high frequencies, followed by a larger radius semicircle at low frequencies (Figure 8a,b), indicating the presence of two-time constants. Two maxima are also observed in the phase angle plots (Figure 8d) at low and, respectively, high frequencies. This behavior is consistent with a two-layer structure often observed in the PEO coatings [5,9,26], where the high-frequency semicircle corresponds to the outer porous layer and the low-frequency semicircle to the compact inner layer. The highest absolute impedances were observed for the Ti PEO DC and HAP-modified samples, two orders of magnitude higher than for untreated Ti and one order of magnitude higher than in the absence of HAP. This indicates similar corrosion resistance for Ti PEO DC and the samples treated using pulsed current and modified with HAP.

Figure 9 gives the two electrical equivalent circuits used to model the experimental EIS data of untreated and PEO-treated Ti samples. Both models use a constant phase element (CPE) instead of a capacitor for a better approximation of the non-ideal capacitive response of real systems. The impedance of a CPE is given by $Z_{CPE} = T^{-1}(j\omega)^{-n}$, where T is a parameter expressed in F s^{n-1} cm^{-2}, related to the capacitance, j is the imaginary unit, ω is the angular frequency in radians and n is an exponent between 0 and 1, describing the constant phase angle of CPE, equal to $-(n \times 90)$. Depending on the n values, a CPE behaves between the limiting cases of a resistor (if $n = 0$) or a capacitor (if $n = 1$). The model used for the untreated Ti sample (Figure 9a) has one time constant and contains the solution resistance *Rs* and a parallel connection of *R* and *CPE* corresponding to the resistance and constant phase element of the passive film on the Ti surface. The model chosen for PEO-treated samples (Figure 9b) has two time constants and includes, in addition to *Rs*, two parallel connections *R1-CPE1* and *R2-CPE2* corresponding to the resistance and capacitance of the outer porous layer and the compact inner layer, respectively. Since untreated Ti shows a single passive layer on its surface, the impedance characteristics of this layer could be assimilated to that of the inner layer in the case of PEO-treated Ti samples.

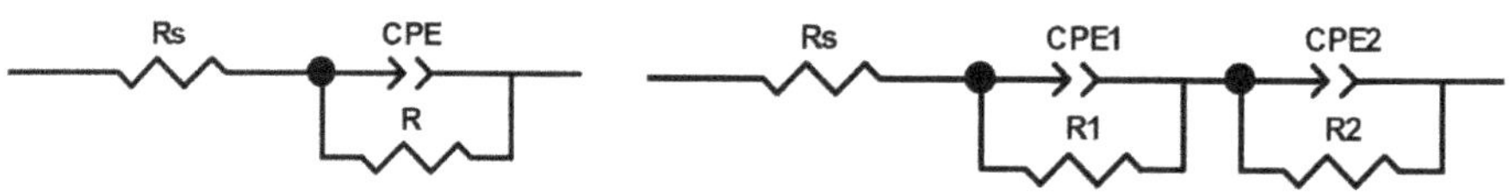

(a) (b)

Figure 9. The electrical equivalent circuit was used to model the impedance data of (**a**) untreated titanium substrate and (**b**) PEO-treated titanium.

Table 3 gives the calculated EIS parameters for untreated and PEO-modified samples, together with the experimental errors and the goodness-of-fit expressed by the Chi-squared (χ^2) value. The low values of both errors and χ^2 indicate good quality of fit of the experimental EIS data to the proposed electrical equivalent circuits.

Table 3. EIS parameters were obtained by fitting the experimental data of untreated Ti and PEO-treated Ti substrate.

Parameter	Ti Substrate	Ti PEO DC	Ti PEO PC	Ti PEO PC + HAP
R_S/Ω	31.8 (0.7%)	20.6 (0.7%)	32.6 (1.5%)	30.6 (2.7%)
CPE-T_1/F cm^{-2} s^{n-1}	-	4.18×10^{-7} (3.7%)	5.20×10^{-6} (7.03%)	1.41×10^{-6} (7.3%)
n_1	-	0.69 (1.4%)	0.67 (1.1%)	0.70 (1.0%)
R_1/Ω cm^2	-	11.6×10^4 (6.9%)	0.25×10^4 (3.1%)	0.40×10^4 (2.4%)
CPE-T_2/F cm^{-2} s^{n-1}	4.35×10^{-5} (1.0%)	5.42×10^{-6} (4.9%)	3.54×10^{-5} (1.2%)	7.30×10^{-6} (1.0%)
n_2	0.91 (0.2%)	0.66 (5.6%)	0.78 (1.0%)	0.86 (0.5%)
R_2/Ω cm^2	1.2×10^4 (0.9%)	8.14×10^6 (88%)	0.29×10^6 (5.7%)	5.02×10^6 (11.5%)
χ^2	2.5×10^{-3}	1.9×10^{-2}	4.5×10^{-3}	5.1×10^{-3}

Data from Table 3 clearly show that for all PEO-treated samples, the resistance of the inner layer (R_2) is higher than that of the outer layer (R_1), indicating that the compact barrier layer is primarily responsible for the corrosion resistance of PEO coatings. For Ti PEO DC and Ti PEO PC samples, the resistance of the inner film is about two orders of magnitude higher than that of the outer layer, while it increases three orders of magnitude after HAP deposition. The highest resistance of the outer layer is observed for the Ti PEO DC sample (11.6×10^4 Ω cm^2) and the lowest for the Ti PEO PC sample (0.25×10^4 Ω cm^2) with a slight improvement after HAP deposition (0.40×10^4 Ω cm^2). These results are correlated with the layer thickness, as the Ti PEO DC coating exhibits the highest thickness. Based on the impedance data, it has been concluded that the corrosion resistance increases in the order of Ti substrate < Ti PEO PC < Ti PEO PC + HAP ≈ Ti PEO DC.

3.3. Sliding Wear Tests

Figure 10 displays the relationship between the coefficient of friction (COF) and the sliding distance obtained during the tribological tests of the titanium substrate and PEO-modified samples versus the counter body 100Cr6 ball. It can be observed that the most important variation of COF takes place in the first 250 m of the testing distance for all the samples. After this interval, the COF values remain almost stable for the rest of the measurement. Comparing the COF variation during the tests, it can be concluded that the smallest values are achieved for the Ti substrate, as it provides the smoothest initial surface (Ra = 0.15 μm), leading to lower friction forces in contact. In this case, the difference between the starting COF (0.2) and the maximum reached value (0.65) denotes the poor mechanical properties, low hardness, and high wear rate of the titanium in this specific testing system due to the increase in the contact area between the sample's surface and the counter body. In comparison, the DC and PC PEO-treated samples exhibit similar behavior, with a minimum COF value, at the beginning of the test, of around 0.5 ± 0.05 and an average COF value of around 0.8. This performance may be attributed to the initial rougher surface of the samples due to the presence of pores and grooves generated during the PEO treatment and possible detachment of particles (friability of the outer layer), but once the measurement enters the steady state, the COF remains stable. The plots presented in Figure 8 indicate that the PEO treatment leads to an increase in the COF values due to the abrasive action of the rough treated surface on the counter facing 100Cr6 ball. Even if the COF value is higher for the PEO-treated samples, the histogram presented in Figure 11 indicates lower wear rates. In both cases, the estimated wear rate is 50% lower compared to the titanium substrate. In the case of the HAP-modified sample, the COF values are significantly reduced, starting with a minimum value of 0.33 and reaching an average COF value of 0.65, similar to the Ti substrate. The deposition of HAP onto the PC PEO-treated samples resulted in a decrease in surface roughness due to the incorporation of HAP in the pores and grooves, leading to a reduced contact area between the sample and the counter body. Moreover, the wear rate is lower, compared to the titanium substrate values, as the presence of HAP leads to a more compact surface and may form, during testing, a tribofilm that reduces the contact friction between the two surfaces [27].

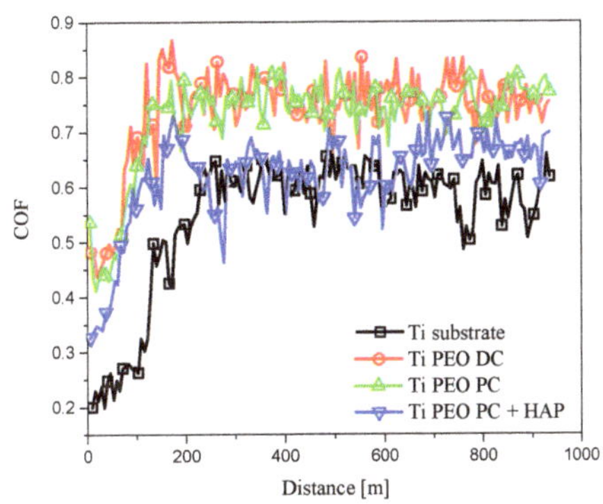

Figure 10. COF evolution for Ti-based samples vs. 100Cr6, F = 5 N, 300 rot/min.

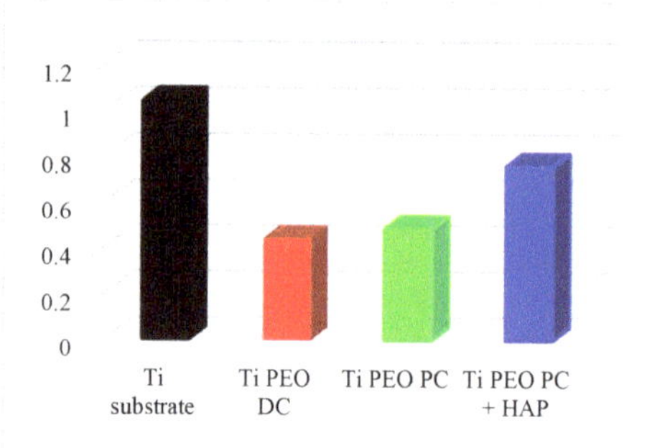

Figure 11. Wear rate [10^4 mm^3 N^{-1} m^{-1}] of Ti substrate, Ti PEO DC, Ti PEO PC, and Ti PEO PC + HAP under the load of 5 N and the sliding distance of around 1000 m.

4. Conclusions

PEO coatings on high-purity titanium substrates were successfully produced from an alkaline electrolyte using two different polarization conditions. A two-step method is proposed to fabricate HAP-modified coatings. The deposition of HAP was performed by electrophoresis from an additive-free suspension. SEM micrographs revealed the formation of a TiO_2 layer with uneven microstructure during the PEO treatment and the homogeneous deposition of HAP particles by electrophoresis. The TiO_2 phase, in the form of anatase, is detected for all the PEO-treated samples using the XRD analysis. The corrosion resistance of the PEO coatings is significantly higher compared to untreated titanium, in the Ringer's solution, at 37 °C. In the case of HAP-modified titanium samples, even if the corrosion potential presents a more negative value, the addition of HAP does not alter the corrosion performance, and corrosion current densities remain low, comparable to the ones obtained for PEO-treated samples. The highest absolute impedances were observed for the Ti PEO DC and HAP-modified samples, two orders of magnitude higher than for untreated Ti and one order of magnitude higher than in the absence of HAP. The coefficient of friction in the dry sliding wear tests of the coatings against a steel ball (100Cr6), using a load of 5 N, is the average value of ~0.8 compared with ~0.6 for the untreated titanium. Even if the COF values are higher for the PEO-treated samples, the estimated wear rates are 50% lower compared to the titanium substrate. The developed PEO-modified coatings may be suitable for biomedical applications.

Author Contributions: Conceptualization, R.M.; methodology, R.M. and M.B.; validation, R.M. and A.K.; formal analysis, R.M., M.B., D.B. and N.S.N.; investigation, R.M., D.B., N.S.N. and I.D.U.; data curation, R.M. and A.K.; writing—original draft preparation, R.M.; writing—review and editing, R.M., M.B., D.B., N.S.N., A.K. and I.D.U. All authors have read and agreed to the published version of the manuscript.

Funding: This research received no external funding.

Institutional Review Board Statement: Not applicable.

Informed Consent Statement: Not applicable.

Data Availability Statement: Not applicable.

Conflicts of Interest: The authors declare no conflict of interest.

References

1. Mortazavi, G.; Jiang, J.; Meletis, E.I. Investigation of the plasma electrolytic oxidation mechanism of titanium. *Appl. Surf. Sci.* **2019**, *488*, 370–382. [CrossRef]
2. Hartjen, P.; Hoffmann, A.; Henningsen, A.; Barbeck, M.; Kopp, A.; Kluwe, L.; Precht, C.; Quatela, O.; Gaudin, R.; Heiland, M.; et al. Plasma Electrolytic Oxidation of Titanium Implant Surfaces: Microgroove-Structures Improve Cellular Adhesion and Viability. *In Vivo* **2018**, *32*, 241–247. [CrossRef] [PubMed]
3. Aliofkhazraei, M.; Macdonald, D.D.; Matykina, E.; Parfenov, E.; Egorkin, V.; Curran, J.; Troughton, S.; Sinebryukhov, S.; Gnedenkov, S.; Lampke, T.; et al. Review of plasma electrolytic oxidation of titanium substrate: Mechanism, properties, applications, and limitation. *Appl. Surf. Sci. Adv.* **2021**, *5*, 100121. [CrossRef]
4. Aliasghari, S.; Skeldon, P.; Thompson, G.E. Plasma electrolytic oxidation of titanium in a phosphate/silicate electrolyte and tribological performance of the coatings. *Appl. Surf. Sci.* **2014**, *316*, 463–476. [CrossRef]
5. Hou, F.; Gorthy, R.; Mardon, I.; Tang, D.; Goode, C. Low voltage environmentally friendly plasma electrolytic oxidation process for titanium alloys. *Sci. Rep.* **2022**, *12*, 6037. [CrossRef]
6. Alves, S.A.; Bayon, R.; Igartua, A.; de Viteri, V.S.; Rocha, L.A. Tribocorrosion behaviour of anodic titanium oxide films produced by plasma electrolytic oxidation for dental implants. *Lubr. Sci.* **2013**, *26*, 500–513. [CrossRef]
7. Molaei, M.; Nouri, M.; Babaei, K.; Fattah-Alhosseini, A. Improving surface features of PEO coatings on titanium and titanium alloys with zirconia particles: A review. *Surf. Interfaces* **2021**, *22*, 100888. [CrossRef]
8. Garcia-Cabezon, C.; Rodriguez-Mendez, M.L.; Borras, V.A.; Raquel, B.; Cabello, J.C.R.; Fonseca, A.I.; Martin-Pedrosa, F. Application of Plasma Electrolytic Oxidation Coating on Powder Metallurgy Ti-6Al-4V for Dental Implants. *Metals* **2020**, *10*, 1167. [CrossRef]
9. Molaei, M.; Fattah-Alhosseini, A.; Kashavarz, M.K. Influence of different sodium-based additives on corrosion resistance of PEO coatings on pure Ti. *J. Asian Ceram. Soc.* **2019**, *7*, 247–255. [CrossRef]
10. Rokosz, K.; Hryniewicz, T.; Malorny, W. Characterization of coatings created on selected titanium alloys by plasma electrolytic oxidation. *Adv. Mater. Sci.* **2016**, *16*, 6–16. [CrossRef]
11. Rokosz, K.; Hryniewicz, T.; Gaiaschi, S.; Chapon, P.; Raaen, S.; Dudek, Ł.; Pietrzak, K.; Malorny, W.; Ciuperca, R. Characterisation of porous coatings formed on titanium under AC plasma electrolytic oxidation. *MATEC Web Conf.* **2018**, *178*, 03008. [CrossRef]
12. Sobolev, A.; Zinigrad, M.; Borodianskiy, K. Ceramic coating on Ti-6Al-4V by plasma electrolytic oxidation in molten salt: Development and characterization. *Surf. Coat. Technol.* **2021**, *408*, 126847. [CrossRef]
13. Shokouhfar, M.; Dehghanian, C.; Montazeri, M.; Baradaran, A. Preparation of ceramic coating on Ti substrate by plasma electrolytic oxidation in different electrolytes and evaluation of its corrosion resistance: Part II. *Appl. Surf. Sci.* **2012**, *258*, 2416–2423. [CrossRef]
14. Santos-Coquillat, A.; Martinez-Campos, E.; Mohedano, M.; Martínez-Corriá, R.; Ramos, V.; Arrabal, R.; Matykina, E. In vitro and in vivo evaluation of PEO-modified titanium for bone implant applications. *Surf. Coat. Technol.* **2018**, *347*, 358–368. [CrossRef]
15. Cheng, Y.L.; Wu, X.C.; Xue, Z.G.; Matykina, E.; Skeldon, P.; Thompson, G. Microstructure, corrosion and wear performance of plasma electrolytic oxidation coatings formed on Ti–6Al–4V alloy in silicate-hexametaphosphate electrolyte. *Surf. Coat. Technol.* **2013**, *217*, 129–139. [CrossRef]
16. Ahmadnia, S.; Aliasghari, S.; Ghorbani, M. Improved Electrochemical Performance of Plasma Electrolytic Oxidation coating on Titanium in Simulated Body Fluid. *J. Mater. Eng. Perform.* **2019**, *28*, 4120–4127. [CrossRef]
17. Ulasevich, S.A.; Kulak, A.I.; Poznyak, S.K.; Karpushenkov, S.A.; Lisenkov, A.D.; Skorb, E.V. Deposition of hydroxyapatite-incorporated TiO_2 coating on titanium using plasma electrolytic oxidation coupled with electrophoretic deposition. *RSC Adv.* **2016**, *6*, 62540. [CrossRef]
18. Parfenov, E.; Parfenova, L.; Mukaeva, V.; Farrakhov, R.; Stotskiy, A.; Raab, A.; Danilko, K.; Rameshbabu, N.; Valiev, R. Biofunctionalization of PEO coatings on titanium implants with inorganic and organic substances. *Surf. Coat. Technol.* **2020**, *404*, 126486. [CrossRef]
19. Zhu, L.; Petrova, R.S.; Gashinski, J.P.; Yang, Z. The effect of surface roughness on PEO-treated Ti-6Al-4V alloy and corrosion resistance. *Surf. Coat. Technol.* **2017**, *325*, 22–29. [CrossRef]
20. He, T.; Cao, C.; Xu, Z.; Li, G.; Cao, H.; Liu, X.; Zhang, C.; Dong, Y. A comparison of micro-CT and histomorphometry of PEO-coated titanium implants in a rat model. *Sci. Rep.* **2017**, *7*, 16270. [CrossRef]
21. Frutuoso, F.G.d.O.; Vitoriano, J.D.O.; Alves, C. Controlling plasma electrolytic oxidation of titanium using current pulses compatible with the duration of microdischarges. *Results Mater.* **2022**, *15*, 100310. [CrossRef]
22. Bertuccioli, C.; Garzone, A.; Martini, C.; Morri, A.; Rondelli, G. Plasma Electrolytic Oxidation (PEO) Layers from Silicate/Phosphate Baths on Ti-6Al-4V for Biomedical Components: Influence of Deposition Conditions and Surface Finishing on Dry Sliding Behaviour. *Materials* **2019**, *9*, 614. [CrossRef]
23. Mashtalyar, D.V.; Nadaraia, K.V.; Gnedenkov, A.S.; Imshinetskiy, I.M.; Piatkova, M.A.; Pleshkova, A.I.; Belov, E.A.; Filonina, V.S.; Suchkov, S.N.; Sinebryukhov, S.L.; et al. Bioactive Coatings Formed on Titanium by Plasma Electrolytic Oxidation: Composition and Properties. *Materials* **2020**, *13*, 4121. [CrossRef]
24. Friedemann, A.E.R.; Gesing, T.M.; Plagemann, P. Electrochemical rutile and anatase formation on PEO surfaces. *Surf. Coat. Technol.* **2017**, *315*, 139–149. [CrossRef]

25. Sobolev, A.; Kossenko, A.; Borodianskiy, K. Study of the Effect of Current Pulse Frequency on Ti-6Al-4V Alloy Coating Formation by Micro Arc Oxidation. *Materials* **2019**, *12*, 3983. [CrossRef] [PubMed]
26. Kostelac, L.; Pezzato, L.; Settimi, A.G.; Franceschi, M.; Gennari, C.; Brunelli, K.; Rampazzo, C.; Dabalà, M. Investigation of hydroxyapatite (HAP) containing coating on grade 2 titanium alloy prepared by plasma electrolytic oxidation (PEO) at low voltage. *Surf. Interfaces* **2022**, *30*, 101888. [CrossRef]
27. Avila, J.D.; Stenberg, K.; Bose, S.; Bandyopadhyay, A. Hydroxyapatite reinforced Ti6A4V composites for load-bearing implants. *Acta Biomater.* **2021**, *123*, 379–392. [CrossRef]

MDPI AG
Grosspeteranlage 5
4052 Basel
Switzerland
Tel.: +41 61 683 77 34

Materials Editorial Office
E-mail: materials@mdpi.com
www.mdpi.com/journal/materials

www.ingramcontent.com/pod-product-compliance
Lightning Source LLC
Chambersburg PA
CBHW061507210926
51850CB00008B/1839
9783725820665